TORQUE CONTROL

Edited by **Moulay Tahar Lamchich**

Torque Control
Edited by Moulay Tahar Lamchich

Published by InTech
Janeza Trdine 9, 51000 Rijeka, Croatia

Publishing Process Manager Katarina Lovrecic

Technical Editor Teodora Smiljanic

Cover Designer InTech Design Team

Additional hard copies can be obtained from orders@intechopen.com

Torque Control, Edited by Moulay Tahar Lamchich
p. cm.
ISBN 978-953-307-428-3

For EU product safety concerns:
IN TECH d.o.o., Prolaz Marije Krucifikse Kozulić 3, 51000 Rijeka, Croatia,
info@intechopen.com or visit our website at intechopen.com.

Meet the editor

Moulay Tahar Lamchich was born in Marrakech, Morocco on December 4, 1966. He has presented his thesis in September 1991 in the field of electrotechnics and received his third cycle degree. After several years of research on international projects he received his PhD from the University Cadi Ayyad in Marrakech, Morroco. He is presently a professor at the same Faculty, at the department of physics, and the person in charge of research group on electrotechnics and industrial electronics and robotics. He is a promotor of several doctoral and professional trainings.

His main activities and research interests are based on short-circuit mechanical effects in substation structures, active power filters, machine drivers, static converters, artificial intelligence techniques, decentralized energy production based on renewable and novel energy and hybrid systems. His major research can be summarized in three categories: transmission and distribution of electric energy, power electronic structures and applications, control of electrical machines.

He has mentored six theses and has published more than forty technical papers in international journals and conferences.

Contents

Preface

Modern electrical drive systems are composed principally of; motors, power electronics components, transformers, analog/digital controllers and sensors or observers. The improvements and the enormous advances in the field of power electronic (semiconductor components), converters technology and software/implementation technology have enabled advanced and complex control techniques. With these advances and robust control algorithms improved, considerable research effort is devoted for developing optimal techniques of speed and torque control for induction machines.

Also, torque control has been, for a long time, a remarkable field for industrial and academic research. Different speed and torque control techniques were developed for many types of induction machines and for various applications. As it can be confirmed from the increasing number of conferences and journals on speed and torque control of induction motors, it is certain that the optimal and robust torque control are a significant guidance for technology development especially for applications based speed variators.

This book is the result of inspirations and contributions from many researchers (a collection of 9 works) which are, in majority, focalised around the Direct Torque Control and may be comprised of three sections: different techniques for the control of asynchronous motors and double feed or double star induction machines, oriented approach of recent developments relating to the control of the Permanent Magnet Synchronous Motors, and special controller design and torque control of switched reluctance machine.

In the first section, composed on chapters 1 to 5, the recent developments of the induction machines control are widely developed: after presentation of the machine model (asynchronous, doubly fed or double star), and control method adopted (specially Field Oriented Control, Direct Torque Control or their association) some ideas for improved performance are provided. In this object, the impact study, of different structures of static power converters (Current Source Inverter, Multi-level inverter), on the dynamic and system performance is exploited. In addition, the contributions made by the use of artificial intelligence techniques either for the design of the controllers or in the development of switching tables, in the case of the Direct Torque Control, are amply explained. Furthermore the study of special phenomena (flux weakening phase) or other types of machines for specific applications (Double feed or Double star induction machines) is presented.

The second section, composed of two chapters, is on behaviour of Permanent Magnet machine controlled by direct torque control method. Different techniques are used to improve torque ripple reduction and harmonic noises in PMSM. Also, as an improvement approach, different filters types are discussed and a RLC low pass is tested in order to eliminate the harmonics.

The last section includes two research articles on development of reluctance motors. The investigations are focused on the controller design and the implementation of sensorless reluctance drive with direct torque control and also on the torque ripple which can be minimized through magnetic circuit design or switched reluctance motors control.

Finally, in my capacity, as the Editor of this book, I would like to thank and appreciate the chapter authors, who ensured the quality of their best works submitted. Most of the results presented in the book have already been presented on many international conferences and will make this book useful for students and researchers who will contribute to further development of the existing technology.

I hope all will enjoy the book.

February 10, 2011

Pr. M.T. Lamchich
Department of Physic
Laboratory of Electronic and Instrumentation
Faculty of Sciences Semlalia - University Cadi Ayyad
Marrakech - Morocco

Part 1

Different Techniques for the Control of Asynchronous Motors and Double Feed or Double Star Induction Machines

1

Torque Control of CSI Fed Induction Motor Drives

Aleksandar Nikolic
Electrical Engineering Institute "Nikola Tesla", Belgrade
Serbia

1. Introduction

An electric drive is an industrial system which performs the conversion of electrical energy to mechanical energy (in motoring) or vice versa (in generator braking) for running various processes such as: production plants, transportation of people or goods, home appliances, pumps, air compressors, computer disc drives, robots, music or image players etc. About 50% of electrical energy produced is used in electric drives today.

Electric drives may run at constant speed or at variable speed. Nowadays, most important are variable speed drives, especially in Europe where according to the Ecodesign for Energy-Using Products Directive (2005/32/EC) (the "EuP Directive") and its regulation regarding electric motors (Regulation 640/2009/EC) on 1 January 2015 - motors with a rated output of 7.5-375kW must meet higher energy efficiency standards, or meet the 2011 levels and be equipped with a variable speed drive.

The first motor used in variable speed applications was DC motor drive, since it is easily controllable due to the fact that commutator and stator windings are mechanically decoupled.

The cage rotor induction motor became of particular interest as it is robust, reliable and maintenance free. It has lower cost, weight and inertia compared to commutator DC motor of the same power rating. Furthermore, induction motors can work in dirty and explosive environments. However, the relative simplicity of the induction motor mechanical design is contrasted by a complex dynamic structure (multivariable, nonlinear, important quantities not observable).

In the last two decades of the 20th century, the technological improvements in power semiconductor and microprocessor technology have made possible rapid application of advanced control techniques for induction motor drive systems. Nowadays, torque control of induction motor is possible and has many advantages over DC motor control, including the same system response and even faster response in case of the latest control algorithms. Two most spread industrial control schemes employs vector or field-oriented control (FOC) and direct torque control (DTC).

Current source inverters (CSI) are still viable converter topology in high voltage high power electrical drives. Further advances in power electronics and usage of new components like SGCT (Symmetric Gate Commutated Thyristor), gives the new possibilities for this type of converter in medium voltage applications. Power regeneration during braking, what is a one of main built-in feature of CSI drives, is also merit for high power drives. Despite the above advantages, the configuration based on a thyristor front-end rectifier presents a poor and variable overall input power factor (PF) since the current is not sinusoidal, but trapezoidal waveform. Also, the implementation of CSI drive systems with on-line control capabilities is more complex than for voltage source inverters (VSI), due to the CSI gating requirements. Regarding mentioned disadvantages, CSI drives are of interest for research in the field of torque control algorithms, such as vector control (or FOC) and direct torque control (DTC). This chapter will present basic FOC and DTC algorithms for CSI drives and show all features and disadvantages of those control schemes (sluggish response, phase error, large torque ripples, need for adaptive control, etc.). Using recent analysis tools like powerful computer simulation software and experiments on developed laboratory prototype two new FOC and DTC solutions will be presented in the chapter. The proposed FOC enables CSI drive to overcome mentioned inconveniences with better dynamic performances. This enhancement relies on fast changes of the motor current, without phase error, similar to the control of current regulated voltage source PWM inverter. The realized CSI drive has more precise control, accomplished by the implemented correction of the reference current. This correction reduces the problem of the incorrect motor current components produced by the non-sinusoidal CSI current waveform. On the other side, proposed DTC algorithm is completely new in the literature and the only such a control scheme intended for CSI induction motor drives. Presented DTC is based on the constant switching frequency, absence of coordinate transformation and speed sensor on the motor shaft. Furthermore, since flux estimator is based only on DC link measurements, there is not necessity for any sensor on the motor side which is one of main drive advantages. In this case, by combination of vector control and basic DTC, a robust algorithm is developed that has a faster torque response and it is simpler for implementation.

2. Characteristics of current source inverters

The most prevailing industrial drive configuration in low voltage range is based on IGBT transistors as power switches and voltage-source inverter (VSI) topology. On the other side, the induction motor drives with thyristor type current-source inverter (CSI, also known as auto sequentially commutated inverter, Fig. 1) possess some advantages over voltage-source inverter drive, but it has a larger torque ripples since the current wave-form is not sinusoidal. Furthermore, due to the nature of the CSI operation, the dynamic performance that exists in VSI PWM drives could not be achieved. But, CSI permits easy power regeneration to the supply network under the breaking conditions, what is favorable in large-power induction motor drives. At low voltage range (up to 1kV) this type of inverter is very rare and abandoned, but this configuration is still usable at high power high voltage range up to 10kV and several MW. In traction applications bipolar thyristor structure is replaced with gate turn-off thyristor (GTO). Nowadays, current source inverters are very popular in medium-voltage applications, where symmetric gate-commutated thyristor (SGCT) is utilized as a new switching device with advantages in PWM-CSI drives (Wu, 2006). New developments in the field of microprocessor control and application in electrical drives gives possibility for employment of very complex and powerful control algorithms. Torque control of CSI fed induction motor drives becomes also viable and promising solution, since some of CSI control disadvantages could be overcome using improved mathematical models and calculations.

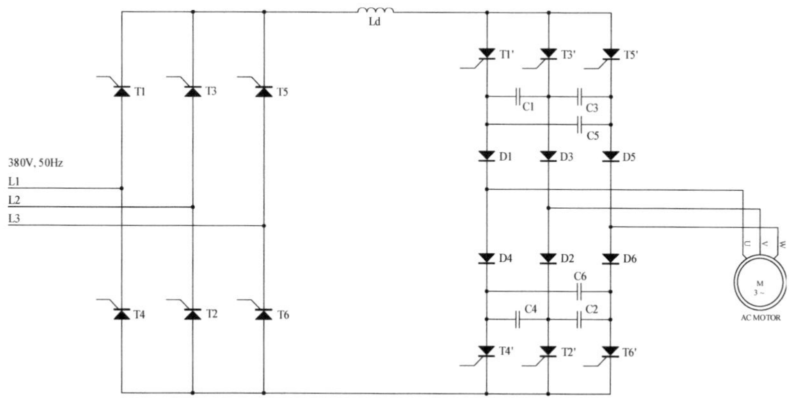

Fig. 1. Basic CSI topology using thyristors as power switches

Two most important torque control schemes are presented, namely FOC and DTC. Both control schemes will be shown with all variations known from literature, including those proposed by previous research work of author. All presented torque control algorithms, basic and proposed by author, are analyzed and verified by simulations and experiments.

3. Vector control

In the past, DC motors were used extensively in areas where variable-speed operation was required, since their flux and torque could be controlled easily by the field and armature current. However, DC motors have certain disadvantages, which are due to the existence of the commutator and brushes. On the other side, induction motors have less size for the same power level, has no brushes and commutator so they are almost maintenance free, but still has disadvantages. The control structure of an induction motor is complicated since the stator field is revolving. Further complications arise due to the fact that the rotor currents or rotor flux of a squirrel-cage induction motor cannot be directly monitored.

The mechanism of torque production in an AC and DC machine is similar. Unfortunately, that similarity was not emphasized before 1971, when the first paper on field-oriented control (FOC) for induction motors was presented (Blaschke, 1971). Since that time, the technique was completely developed and today is mature from the industrial point of view. Today field oriented controlled drives are an industrial reality and are available on the market by several producers and with different solutions and performance.

3.1 Basic vector control of CSI drives

Many strategies have been proposed for controlling the motion of CSI fed induction motor drives (Bose, 1986; Novotny & Lipo, 1988; Wu et al., 1988; Deng & Lipo, 1990; Vas, 1990). The vector control has emerged as one of the most effective techniques in designing high-performance CSI fed induction motor drives. Compared to the PWM VSI drives CSI has advantage in the reversible drives, but it has a larger torque ripples since the current wave-form is not sinusoidal. Furthermore, due to the nature of the CSI operation, the dynamic performance that exists in PWM drives are not achieved with the existing vector control algorithms.

The well-known ("basic") structure of a CSI fed induction motor drive with indirect vector control is shown in Fig. 2 (Bose, 1986; Vas, 1990).

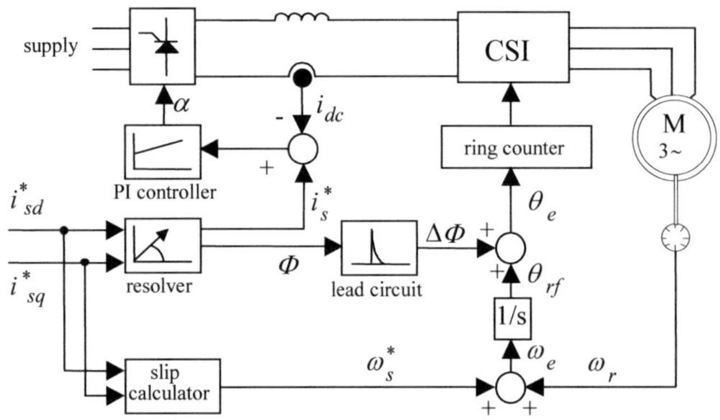

Fig. 2. Indirect vector control of a CSI fed induction machine

This method makes use of the fact that satisfying the slip relation is a necessary and sufficient condition to produce field orientation, i.e. if the slip relation is satisfied (denoted as slip calculator in Fig. 2), current *d* component will be aligned with the rotor flux. Current commands are converted to amplitude and phase commands using resolver (rectangular to polar coordinate transformation). The current amplitude command is directly employed as the reference for the current PI controller intended for controlling the input converter (three phase full wave bridge rectifier). The phase command is passed through a lead circuit such that phase changes are added into the inverter frequency channel, since these instantaneous phase changes are not contained in the slip frequency command signal coming from the slip calculator.

3.2 Proposed vector control of CSI drives

In the vector controlled CSI drives found in (Bose, 1986; Wu et al., 1988; Deng & Lipo, 1990; Vas, 1990; Novotny & Lipo, 1996) and shown in previous chapter, the problems of the speed response are reported. This is influenced by the instantaneous phase error and, as a result, these configurations have slower torque response compared to the current regulated PWM drives. In addition to the phase error, the commutation delay and the non-sinusoidal supply that is inhered in CSI operation must be generally compensated for, to achieve acceptable vector control. To overcome these disadvantages the phase error elimination and the reference current correction should be performed.

In this chapter, the vector control algorithm that eliminates the two drawbacks is shown (Nikolic & Jeftenic, 2006). The suggested algorithm produces the performance of the CSI drive that exists in the PWM vector controlled drives. That enables this simple and robust configuration to be used in applications where reversible operation is a merit.

The necessity for the phase error elimination can be explained with the help of the following phasor diagram:

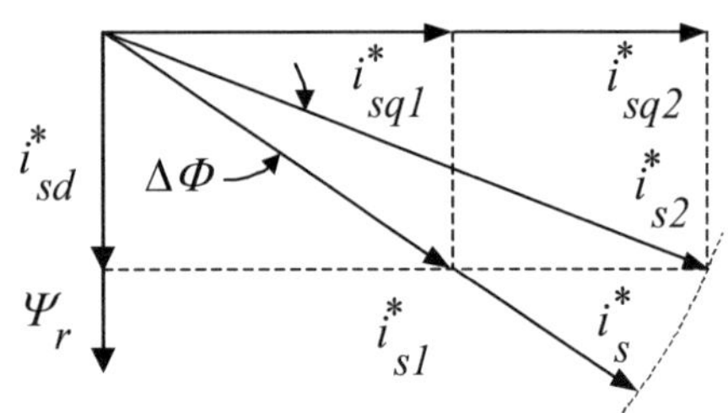

Fig. 3. Phasor diagram with shown phase error

When the torque command is stepped from i_{sq1}* to i_{sq2}* (with a constant i_{sd}*), the current vector should instantaneously change from i_{s1}* to i_{s2}*. The slip frequency should also change immediately. The resolver does give the correct amplitude and the new slip frequency will be obtained by the slip calculator. However, although the phase change $\Delta\Phi$ is added by a lead circuit as shown in Fig. 2, since the instant phase changes are not contained in the slip frequency command signal coming from the slip calculator (Bose, 1986; Deng & Lipo, 1990; Novotny & Lipo, 1996), the stator current command will correspond to the vector i_s* in Fig. 3, and there will be a phase error in the vector control system. This would result in an instantaneous loss of the field-orientation that produces a very sluggish response of both flux and torque. This problem could be overcome by the proposed algorithm, which unifies features of both PWM and CSI converter. The resolver is still used to calculate the rectifier reference current, but for the inverter thyristors control, a method used in the current controlled PWM inverter is implemented. Instead of a lead circuit (shown in Fig. 2), the new algorithm includes a synchronous to stator transformation (T^{-1}) to transfer the d-q commands to the three-phase system. This is essential for achieving a fast torque response, since the torque value is determined by the fundamental harmonic of the stator current.
For correct firing of the thyristors in the inverter, the switching times should be properly determined to ensure that the phase angle of the motor current matches the phase angle of the reference currents in a-b-c system. The reference sinusoidal currents obtained as a result of transformation T^{-1} are divided by the value obtained on the resolver output to produce currents of unity amplitudes. Introduction of these currents into the comparator with trigger level equal to 0.5 gives the proper thyristor conduction time of 120 degrees. This is illustrated in Fig. 4, where i_a* is unity sinusoidal current, i_a is scaled CSI output current and i_{a1} is fundamental stator current.

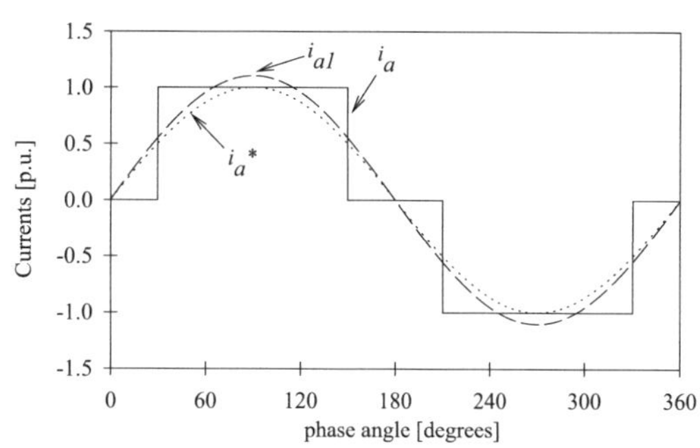

Fig. 4. Waveforms of the reference current and fundamental stator current

The algorithm possesses the additional advantage regarding the practical realization. In digital control system the lead circuit divides the difference of the two succeeding samples with the sampling time. Since the sampling time is small, this operation produces the computational error. The phasor diagram from Fig. 3 with removed phase error is presented in Fig. 5.
Without the phase error ($\Delta\Phi$ = 0), the step of the torque command produces the new stator current command (i_s* = i_{s2}*). Due to non-sinusoidal currents of the CSI, the average values of the motor d-q currents i_{sd_av} and i_{sq_av} and the resulting stator current vector is greater than the corresponding references shown in Fig. 5. To improve proposed algorithm and avoid improper resultant d-q motor currents, the rectifier reference current correction is performed.

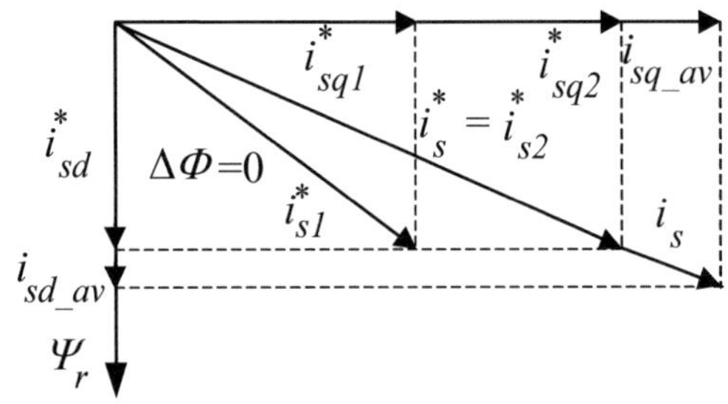

Fig. 5. Phasor diagram without phase error

In the vector controlled induction motor drive fed by a CSI a problem of incorrect copying of the d-q references to the motor exist. As stated earlier, the reason is non-sinusoidal current waveform produced by a CSI. The ideal CSI current is a quasi-square waveform (shown in Fig. 4). The Fourier analysis of this waveform gives the expression:

$$i_a = \frac{2\sqrt{3}}{\pi} \cdot I_d \cdot \left(\sin(\omega t) - \frac{1}{5}\sin(5\omega t) - \frac{1}{7}\sin(7\omega t) + \cdots \right) \tag{1}$$

The previous relation shows that the fundamental component of AC output current has the amplitude 10 percent greater than the value of DC link current. For correct reproduction of the d-q references and satisfactory vector control it is not sufficient to adjust only the phase of the fundamental motor current and the phase angle of the generated commands. The fine-tuning of the motor currents in d-q frame is required.

To avoid supplementary hardware and software, a procedure that relies only on the values calculated off-line is proposed. The corresponding relation between the mean values of the motor currents in d-q frame and the commanded d-q currents is calculated. For proposed correction it is not sufficient to use the difference between currents of 10% from (1), because the correction depends on the phase angle of the d-q components and the inverter commutation process. At lower speed, the commutation process could be neglected since it is much shorter than the motor current cycle. Taking all this in consideration, the rectifier reference current is corrected concerning the reference amplitude, the phase angle and the commutation duration. The rectifier reference current formed in that manner is now introduced to the current controller to obtain suitable motor d-q currents and achieve desired vector control.

The calculation starts from the fundamental reference current from the resolver:

$$i_s^* = \sqrt{(i_{sd}^*)^2 + (i_{sq}^*)^2} \tag{2}$$

and the phase angle (also obtained from the resolver):

$$\Phi = \arctan\left(i_{sd}^* / i_{sq}^*\right) \tag{3}$$

Since the inverter commutation process is not neglected, the waveform of the inverter output current is represented by a trapezoidal approximation analyzed in (Cavalini et al., 1994) with adequate precision. Trapezoidal waveform is very near to the real current cosine waveform due to the short commutation period, as explained in (Bose, 1986). This approximation assumes that during the commutation period the inverter current rises with a constant rate of change. For rated rectifier current I_d the current rate of change during the commutation is equal to I_d/t_c, where t_c is corresponding commutation time calculated from the values of the commutation circuit components. The adequate commutation angle μ could be obtained as a product of the inverter frequency ω_e and particular commutation time. This time interval is determined from the current rate of change I_d/t_c and the reference value of the DC link current i_s*, therefore the commutation angle is:

$$\mu = \omega_e \cdot \frac{t_c}{I_d} \cdot i_s^* \tag{4}$$

Since the inverter current is periodical, the trapezoidal waveform in all three phases could be represented on a shorter angle interval with the following equations:

$$i_a(\theta,\mu,\Phi) = \begin{cases} 0 & \theta < \Phi - \frac{\pi}{3} \\ \frac{I_d}{\mu} \cdot \left(\theta - \Phi + \frac{\pi}{3}\right) & \theta < \Phi - \frac{\pi}{3} + \mu \\ I_d & \theta < \Phi + \frac{\pi}{3} \\ I_d - \frac{I_d}{\mu} \cdot \left(\theta - \Phi - \frac{\pi}{3}\right) & \theta < \Phi + \frac{\pi}{3} + \mu \\ 0 & \theta < \pi \end{cases} \tag{5}$$

$$i_b(\theta,\mu,\Phi) = i_a(\theta - \frac{2\pi}{3}, \mu, \Phi) \tag{6}$$

$$i_c(\theta,\mu,\Phi) = i_a(\theta - \frac{4\pi}{3}, \mu, \Phi) \tag{7}$$

where θ ranges from 0 to π, μ is the commutation angle and Φ is the phase angle obtained from (3). The instantaneous values of d-q currents are solved by a three phase to d-q frame transformation T:

$$\begin{bmatrix} i_{sd}(\theta,\mu,\Phi) \\ i_{sq}(\theta,\mu,\Phi) \end{bmatrix} = \frac{2}{3} \cdot \mathbf{T} \cdot \begin{bmatrix} i_a(\theta,\mu,\Phi) \\ i_b(\theta,\mu,\Phi) \\ i_c(\theta,\mu,\Phi) \end{bmatrix} = \tag{8}$$

$$= \frac{2}{3} \cdot \begin{bmatrix} \sin(\theta) & \sin(\theta - 2\pi/3) & \sin(\theta - 4\pi/3) \\ \cos(\theta) & \cos(\theta - 2\pi/3) & \cos(\theta - 4\pi/3) \end{bmatrix} \cdot \begin{bmatrix} i_a(\theta,\mu,\Phi) \\ i_b(\theta,\mu,\Phi) \\ i_c(\theta,\mu,\Phi) \end{bmatrix}$$

The average values of the currents in d and q axis obtained from (8) on the range from 0 to π be:

$$i_{sd\_av}(\mu,\Phi) = \frac{1}{\pi} \cdot \int_0^{\pi} i_{sd}(\theta,\mu,\Phi)\, d\theta \tag{9}$$

$$i_{sq\_av}(\mu,\Phi) = \frac{1}{\pi} \cdot \int_0^{\pi} i_{sq}(\theta,\mu,\Phi)\,\mathrm{d}\theta \tag{10}$$

The amplitude of the motor current vector in polar coordinates could be determined using the average values obtained from (9) and (10):

$$i_s(\mu,\Phi) = \sqrt{i_{sd\_av}^2(\mu,\Phi) + i_{sq\_av}^2(\mu,\Phi)} \tag{11}$$

The difference between reference amplitude calculated from (2) and the resulting stator amplitude obtained from (11) is shown in Fig. 4. To avoid this difference, the corresponding correction factor f_{cor} is introduced as a ratio of the reference (2) and the actual motor current (11):

$$f_{cor}(\mu,\Phi) = \frac{i_s^*}{i_s(\mu,\Phi)} \tag{12}$$

For simulation and practical realization purposes, the correction factor f_{cor} is computed from (2) - (12), and placed in a look-up table with the following restrictions:

- i_{sd}* is constant,
- i_{sq}* is changed only to its rated value with i_s* limited to 1 p.u.
- for given references, all possible values of Φ and μ are calculated using (3) and (4), respectively.

The rectifier reference current that provides the correct values of motor current d-q components is now:

$$i_{ref} = i_s^* \cdot f_{cor}(\mu,\Phi) \tag{13}$$

The interdependence between correction factor f_{cor}, commutation angle μ and phase angle Φ is presented in Fig. 6 as a 3-D graph.

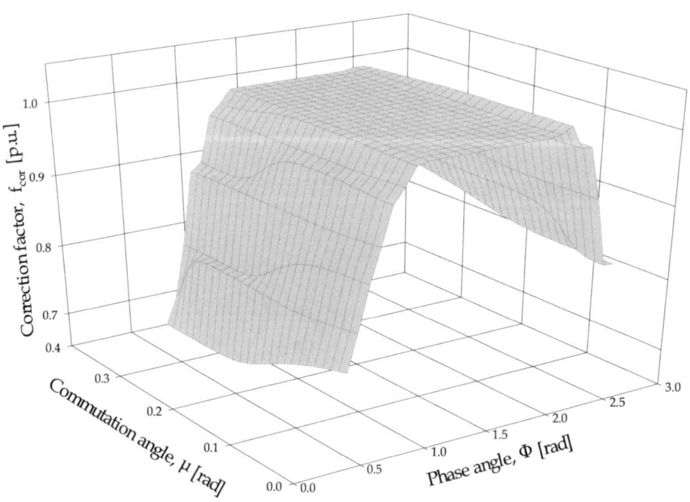

Fig. 6. Correction factor, commutation angle and phase angle interdependence

The calculated results of the current correction in d-axis and q-axis are presented in Fig. 7a and Fig. 7b respectively. The corrected currents are given along with references and motor average d-q currents (values without correction). The flux command is held constant (0.7 p.u.), while torque command is changed from -0.7 p.u. to 0.7 p.u.

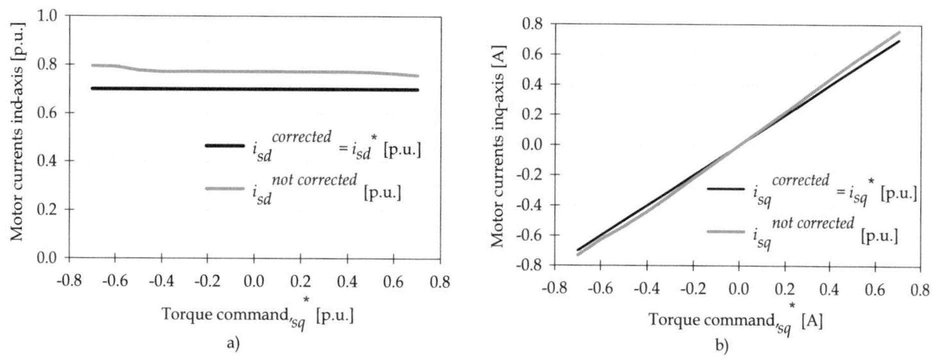

Fig. 7. Calculated motor current corrected in d-axis and q-axis (a,b respectively)

From the previous analysis the new resolver with current correction is formed as shown in Fig. 8. This structure is used both in the simulations and the experiments. The new resolver is consisted of the block "Cartesian to polar" (the coordinate transformation) and the block "Correction" that designates the interdependence given in Fig. 6. As stated before, this interdependence is placed in a 3-D look-up table using (2)-(12).

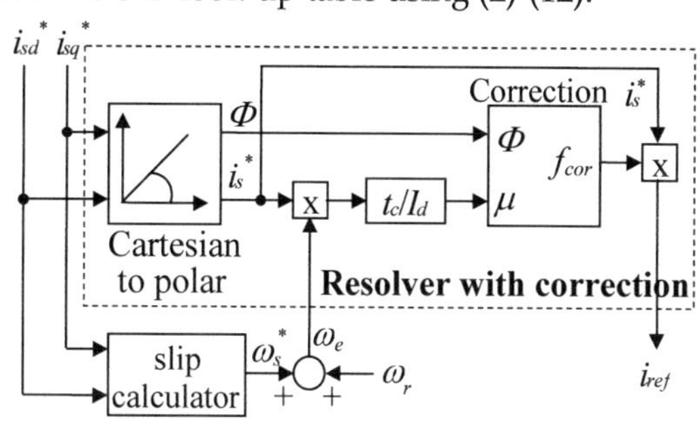

Fig. 8. New resolver with current correction

To analyze dynamic performances of the proposed CSI drive, the torque response of the "basic" structure shown in Fig. 2 is compared to the response of the new vector control algorithm. This is done by simulations of these two configurations' mathematical models in Matlab/Simulink. The first model represents the drive with basic arrangement and the second is the drive with new control algorithm. The simulation of both models is done with several initial conditions. Magnetizing (d-axis) current for rated flux has been determined from the motor parameters and its value (0.7p.u.) is constant during simulations. The rated q-axis current has been determined from the magnetizing current and the rated full-load current using (2). At first, simulations of both models are started with d-axis command set to 0.7p.u, no-load and all initial conditions equal to zero. When the rotor flux in d-axis approaches to the steady state, the machine is excited. This value of d-axis flux is now initial for the subsequent simulations. For the second simulation the pulse is given as a torque command, with the amplitude of 0.2p.u. and duration of 0.5s. With no-load, the motor will be accelerated from zero speed to the new steady-state speed (0.2p.u.), which is the initial condition for the next simulation. Finally, the square wave torque command is applied to both models with equal positive and negative amplitudes (±0.2p.u) and the observed dynamic torque response is extracted from the slope of the speed (Lorenz, 1986). The square wave duty cycle (0.9s) is considerably greater than the rotor time constant (T_r = 0.1s), hence the rotor flux could be considered constant when the torque command is changed. Fig. 9 shows torque, speed and rotor flux responses of both models. It could be noticed that the torque response of the basic structure is slightly slower (Fig. 9a), while the proposed algorithm gives almost instantaneous torque response (Fig. 9b). This statement could be verified clearly from the speed response analysis. In both cases the torque command is the same. In the new model this square wave torque command produces speed variations from 0.2p.u. to 0.6p.u. with identical slope of the speed. But, in the basic model at the end of the first cycle the speed could not reach 0.6p.u. for the same torque command due to the fact

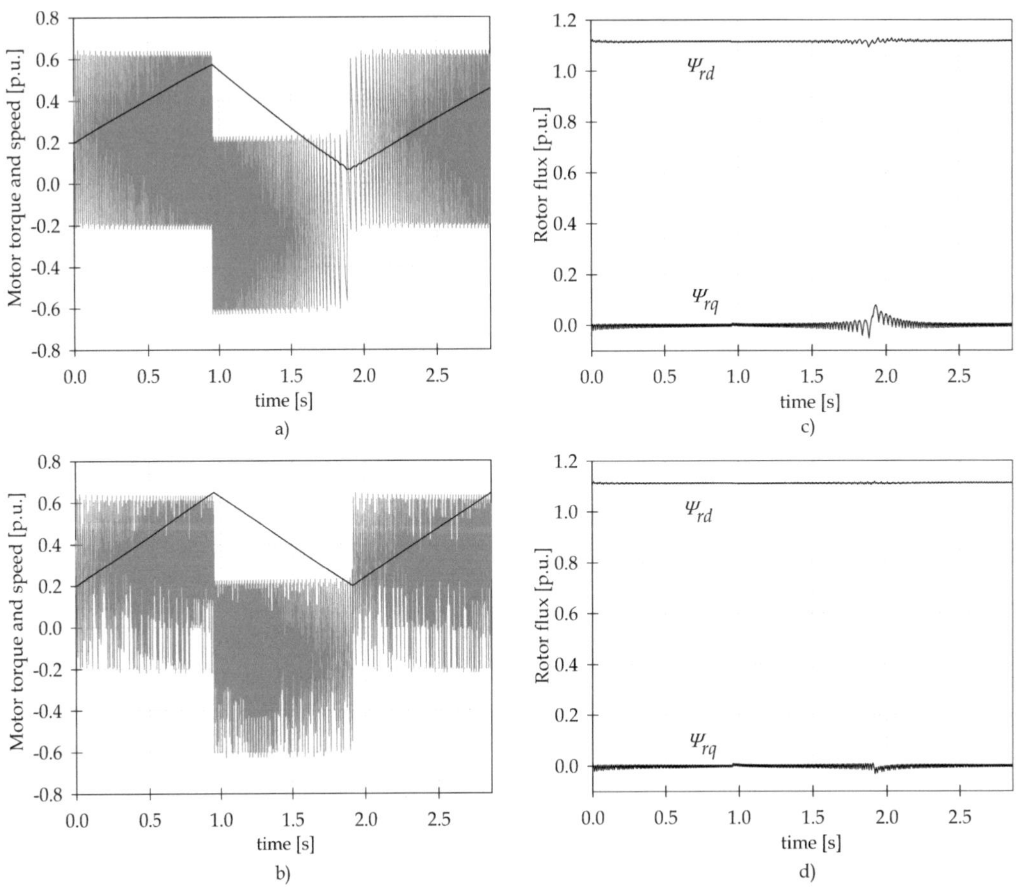

Fig. 9. Torque, speed and rotor flux of the basic structure (a), (c) and of the proposed algorithm (b), (d)

that torque response is slower. Also, in the next cycle (negative torque command) the speed does not return to 0.2p.u. for the same reason. From different slopes of the speed in these two models it could be concluded that proposed algorithm produces quicker torque response. The rotor q-axis flux disturbance in transient regime that exists in the basic model (Fig. 9c) is greatly reduced by the proposed algorithm in the new model (Fig. 9d). It could be seen that some disturbances also exist in the case of d-axis flux, but they are almost disappeared in the new model.

To illustrate the significance and facilitate the understanding of theoretical results obtained in the previous section, a prototype of the drive is constructed. The prototype has a standard thyristor type frequency converter digitally controled via Intel's 16-bit 80C196KC20 microcontroller. Induction motor used in laboratory is 4kW, 380V, 50Hz machine. The speed control of the drive and a prototype photo are shown in Fig. 10. Simplicity of this block diagram confirms that the realized control algorithm is easier for a practical actualization.

The proposed circuit for the phase error elimination is at first tested on the simulation model. The simulation is performed in such a manner that C code for a microcontroller could be directly written from the model. The values that are read from look-up tables in a real system (cosine function, square root) are also presented in the model as tables to properly emulate calculation in the microcontroller.

Fig. 11a shows waveforms of the unity sinusoidal references (i_a* and i_b*) while Fig. 11b indicates inverter thyristors switching times with changed switching sequence when the phase is changed (0.18s, marked with an arrow). On these diagrams it could be observed that thyristors T_1 and T_2 are switched to ON state when unity references i_a* and i_b* reach 0.5 p.u., respectively. Fig. 11c,d represents the instant phase variation of the currents in a and b phases after the reference current is altered. The corresponding currents without command changes are displayed with a thin line for a clear observation of the instant when the phase is changed.

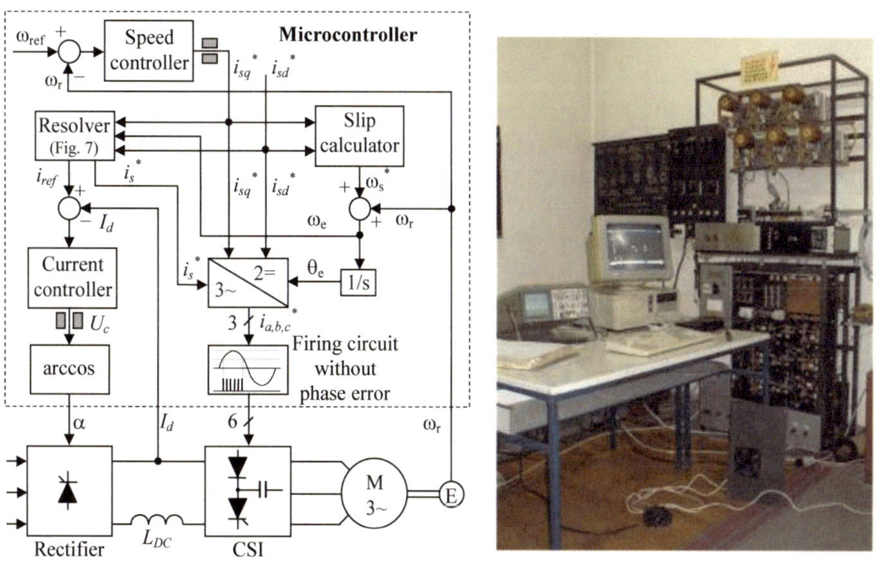

Fig. 10. CSI fed induction motor drive with improved vector control algorithm: control block diagram (left), laboratory prototype (right)

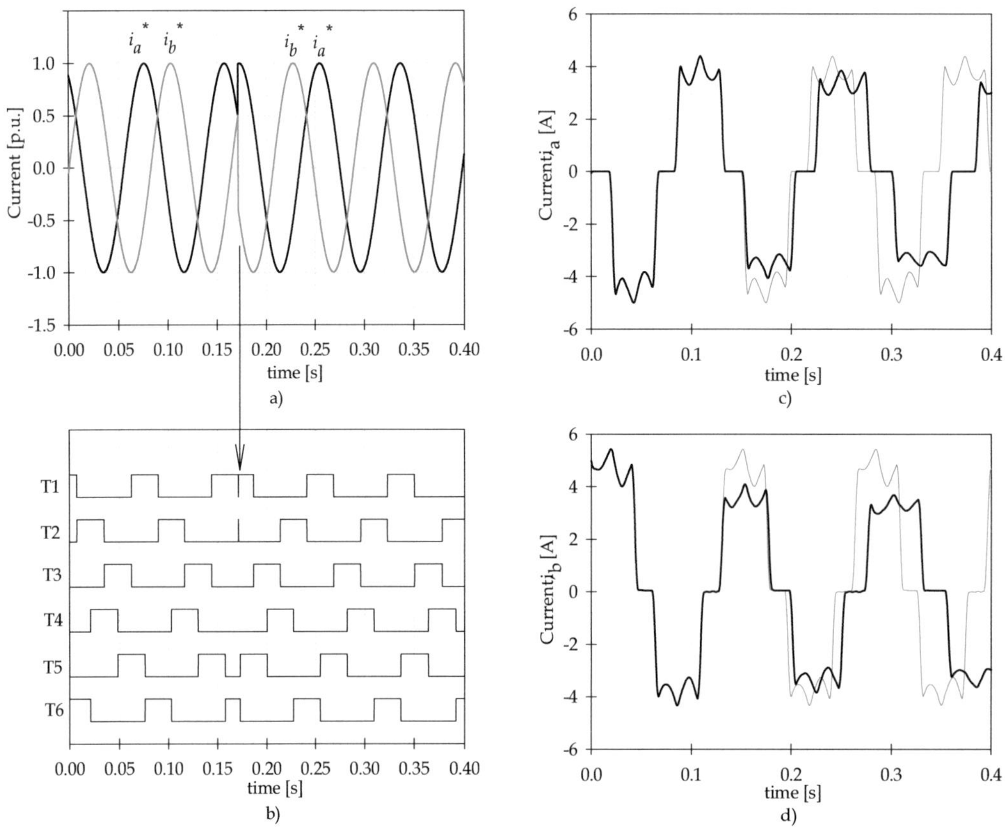

Fig. 11. Results of the phase error elimination (a,b - simulation, c,d - experimental)

The effects of the reference current correction are given by the specific experiment. To estimate d and q components, the motor currents in *a* and *b* phases and the angle θ_e between *a*-axis and *d*-axis are measured. This angle is obtained in the control algorithm (Fig. 10) as a result of a digital integration:

$$\theta_e(n) = \theta_e(n-1) + \omega_e \cdot T_s \tag{14}$$

where *n* is a sample, T_s is the sample time and ω_e is excitation frequency. The integrator is reset every time when θ_e reaches 0 or 360 degrees. The easiest way for acquiring the value of this angle is to change the state of the one microcontroller's digital output at the instants when the integrator is reset. On the time range between two succeeding pulses the angle is changed linearly from 0 to 360 degrees (for one rotating direction). Since only this time range is needed for determine the currents in d and q axis, the reset signal from the digital output is processed to the external synchronization input of the oscilloscope. In that way the motor phase currents are measured only on the particular time (angle) range. The corresponding currents in d-q axes are calculated from (8) using for θ_e, i_a and i_b experimentally determined values.

The experimental results are given in Fig. 12 with disabled speed controller.

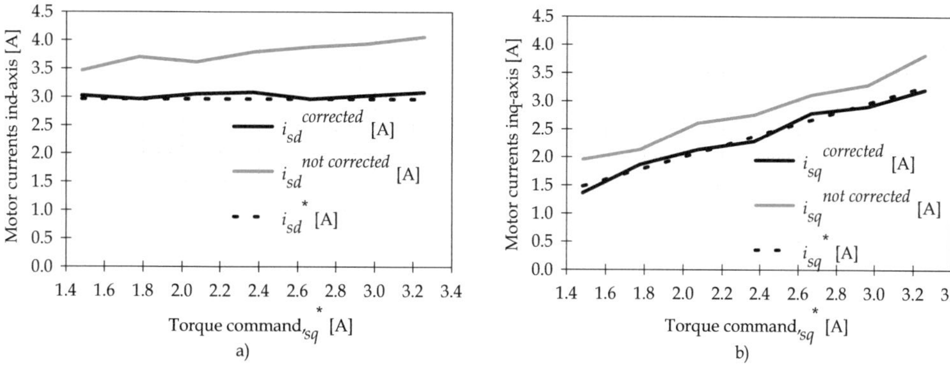

Fig. 12. Experimental results of the motor current correction in d-axis and q-axis (a,b respectively)

The flux reference was maintained constant at 2.96A (0.7p.u.) and torque command was changed from 1.5A (0.35p.u.) to 3.3A (0.78 p.u.). The inverter output frequency is retained the same during experiment (≈20Hz) by varying the DC motor armature current. From Fig. 12 it could be seen that for the proposed algorithm average values of d-q components in the p.u. system are almost equal to corresponding references. On the other side, in the system without correction there is a difference up to 15%, which confirmed the results obtained from calculations shown in Fig. 7. This difference produces steady state error, what makes such a system unacceptable for vector control in high performance applications.
On the Fig. 13 the motor speed and rotating direction changes are shown with enabled speed controller. The reference speed is swapped from -200min^{-1} to +200min^{-1}.

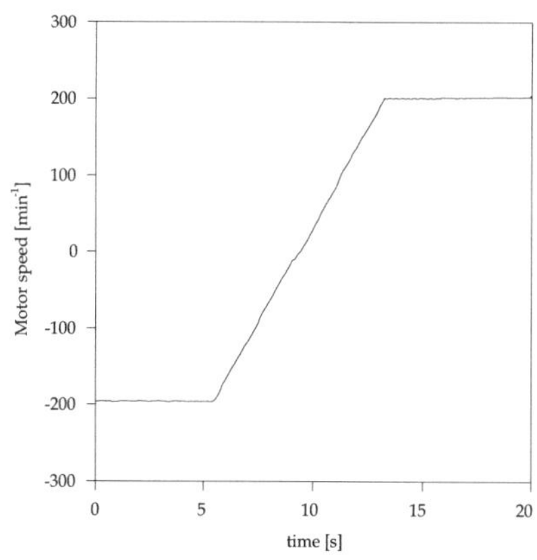

Fig. 13. The motor speed reversal

In Fig. 14 the influence of the load changes to the speed controller is presented. As a load, DC machine (6kW, 230VDC, controlled by a direct change of the armature current via 3-phase rectifier) is used. At first, the induction motor works unloaded in a motor region (M) with the reference speed of -200min^{-1} that produces the torque command current i_{sq1}* = -1.57A. After that, the DC machine is started with its torque in the same direction with rotating direction of the induction motor. That starts the breaking of the induction motor and it goes to the generator region (G). In this operating region the power from DC link returns to the supply network. The reference torque command current changes its value and sign (i_{sq2}* = 1.72A). When DC machine is switched off, the induction motor goes to the motor region (M) and the reference torque command current is now i_{sq3}* = -1.48A.

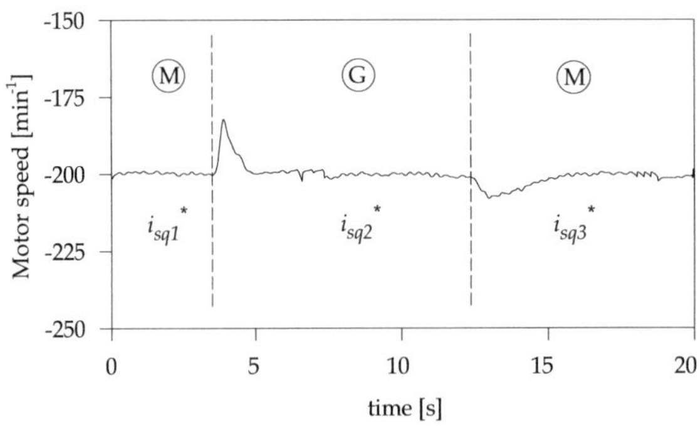

Fig. 14. The load changes at motor speed of -200min^{-1}

4. Direct torque control

The direct torque control (DTC) is one of the actively researched control schemes of induction machines, which is based on the decoupled control of flux and torque. DTC provides a very quick and precise torque response without the complex field-orientation block and the inner current regulation loop (Takahashi & Noguchi, 1986; Depenbrok, 1988). DTC is the latest AC motor control method (Tiitinen et al., 1995), developed with the goal of combining the implementation of the V/f-based induction motor drives with the performance of those based on vector control. It is not intended to vary amplitude and frequency of voltage supply or to emulate a DC motor, but to exploit the flux and torque producing capabilities of an induction motor when fed by an inverter (Buja et al., 1998).

4.1 Direct torque control concepts

In its early stage of development, direct torque control is developed mainly for voltage source inverters (Takahashi & Noguchi, 1986; Tiitinen et al., 1995; Buja, 1998). Voltage space vector that should be applied to the motor is chosen according to the output of hysteresis controllers that uses difference between flux and torque references and their estimates. Depending on the way of selecting voltage vector, the flux trajectory could be a circle (Takahashi & Noguchi, 1986) or a hexagon (Depenbrok, 1988) and that strategy, known as Direct Self Control (DSC), is mostly used in high-power drives where switching frequency is need to be reduced.

Controllers based on direct torque control do not require a complex coordinate transform. The decoupling of the nonlinear AC motor structure is obtained by the use of on/off control, which can be related to the on/off operation of the inverter power switches. Similarly to direct vector control, the flux and the torque are either measured or mostly estimated and used as feedback signals for the controller. However, as opposed to vector control, the states of the power switches are determined directly by the estimated and the reference torque and flux signals. This is achieved by means of a switching table, the inputs of which are the torque error, the stator flux error and the stator flux angle quantized into six sections of 60°. The outputs of the switching table are the settings for the switching devices of the inverter. The error signal of the stator flux is quantized into two levels by means of a hysteresis comparator. The error signal of the torque is quantized into three levels by means of a three stage hysteresis comparator (Fig. 15).

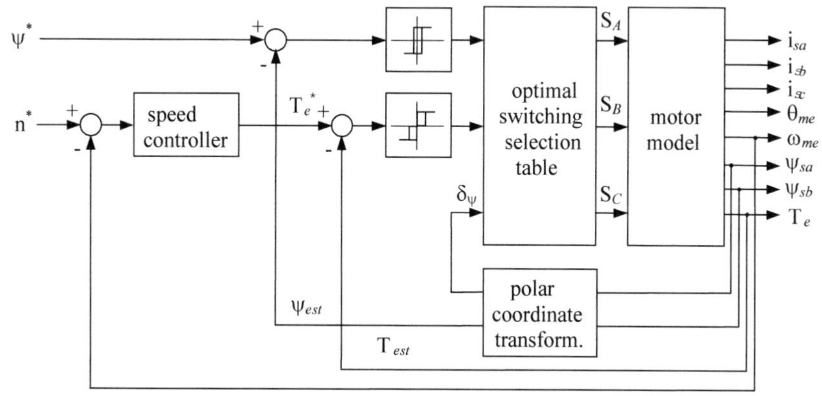

Fig. 15. Basic concept of direct torque control

The equation for the developed torque may be expressed in terms of rotor and stator flux:

$$T_e = \frac{M}{L_s \cdot L_r - M^2} \cdot |\vec{\psi}_s| \cdot |\vec{\psi}_r| \cdot \sin(\delta_\psi) \tag{15}$$

where δ_ψ is the angle between the stator and the rotor flux linkage space phasors. For constant stator and rotor flux, the angle δ_ψ may be used to control the torque of the motor. For a stator fixed reference frame ($\omega_e = 0$) and $R_s = 0$ it may be obtained that:

$$\psi_s = \frac{1}{T_n} \int_0^t u_s \cdot dt \tag{16}$$

The stator voltage space phasor may assume only six different non zero states and two zero states, as shown in Fig. 16. The change of the stator flux vector per switching instant is therefore determined by equation (16) and Fig. 16. The zero vectors V_0 and V_7 halt the rotation of the stator flux vector and slightly decrease its magnitude. The rotor flux vector, however, continues to rotate with almost synchronous frequency, and thus the angle δ_ψ changes and the torque changes accordingly as per (15). The complex stator flux plane may be divided into six sections and a suitable set of switching vectors identified as shown in Table 1, where $d\Psi$ and dT_e are stator flux and torque errors, respectively, while $S_{1,\dots,6}$ are sectors of 60° where stator flux resides.

Further researches in the field of DTC are mostly based on reducing torque ripples and improvement of estimation process. This yields to development of sophisticated control algorithms, constant switching schemes based on space-vector modulation (Casadei et al., 2003), hysteresis controllers with adaptive bandwidth, PI or fuzzy controllers instead of hysteresis comparators, just to name a few.

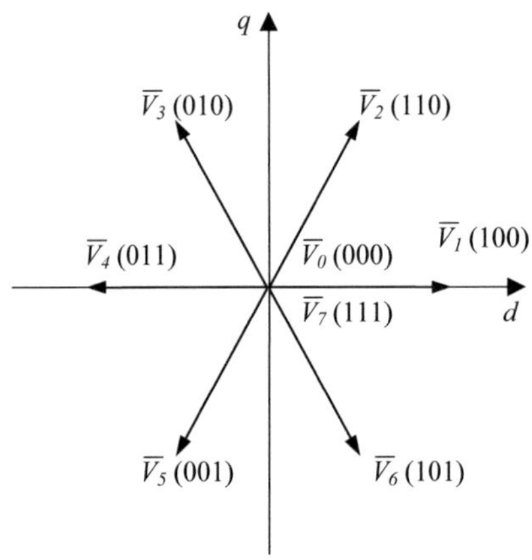

Fig. 16. Voltage vectors of three phase VSI inverter

dΨ	dT_e	S_1 -π/6, π/6	S_2 π/6, π/2	S_3 π/2, 2π/3	S_4 2π/3, -2π/3	S_5 -2π/3, -π/2	S_6 -π/2, -π/6
1	1	V_2	V_3	V_4	V_5	V_6	V_1
	0	V_0	V_7	V_0	V_7	V_0	V_7
	-1	V_6	V_1	V_2	V_3	V_4	V_5
0	1	V_3	V_4	V_5	V_6	V_1	V_2
	0	V_7	V_0	V_7	V_0	V_7	V_0
	-1	V_5	V_6	V_1	V_2	V_3	V_4

Table 1. Optimal switching vectors in VSI DTC drive

4.2 Standard DTC of CSI drives

Although the traditional DTC is developed for VSI, for synchronous motor drives the CSI is proposed (Vas, 1998; Boldea, 2000). This type of converter can be also applied to DTC induction motor drive (Vas, 1998), and in the chapter such an arrangement is presented. The induction motor drives with thyristor type CSI (also known as auto sequentially commutated inverter) possess some advantages over voltage-source inverter drive. CSI permits easy power regeneration to the supply network under the breaking conditions, what is favorable in large-power induction motor drives. In traction applications bipolar thyristor structure is replaced with gate turn-off thyristor (GTO). Nowadays, current source inverters are popular in medium-voltage applications (Wu, 2006), where symmetric gate-commutated thyristor (SGCT) is utilized as a new switching device (Zargari et al., 2001) with advantages in PWM-CSI drives.

DTC of a CSI-fed induction motor involves the direct control of the rotor flux linkage and the electromagnetic torque by applying the optimum current switching vectors. Furthermore, it is possible to control directly the modulus of the rotor flux linkage space vector through the rectifier voltage and the electromagnetic torque by the supply frequency of CSI. Basic CSI DTC strategy (Vas, 1998) is shown in Fig. 17.

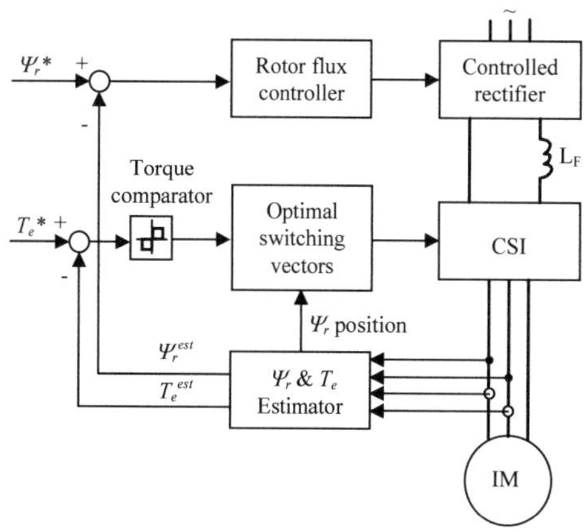

Fig. 17. DTC of CSI drive based on hysteresis control

The stator flux value, needed for DTC control loop, is not convenient to measure directly. Instead of that, the motor flux estimation is performed. In the voltage-based estimation method, the motor flux can be obtained by integrating its back electromotive force (EMF). The EMF is calculated from the motor voltage and current (17) and the only motor parameter required is the stator winding resistance. In practice, this simple integration is replaced by more sophisticated closed-loop estimators using filtering techniques, adaptive integration or even observers and Extended Kalman filters (Holtz, 2003).

$$\overline{\psi}_S^S = \int_0^t \left(\overline{u}_S^S - R_S \cdot \overline{i}_S^S \right) dt + \overline{\psi}_{S0}^S \tag{17}$$

For DTC of CSI fed induction motor drive, the appropriate optimal inverter current-switching vectors (Fig. 18) are produced by using an optimal current-switching table similarly to the table given for VSI drive (Table 2). The main difference is that in CSI exist only one hysteresis comparator for torque and only one zero switching current vector.

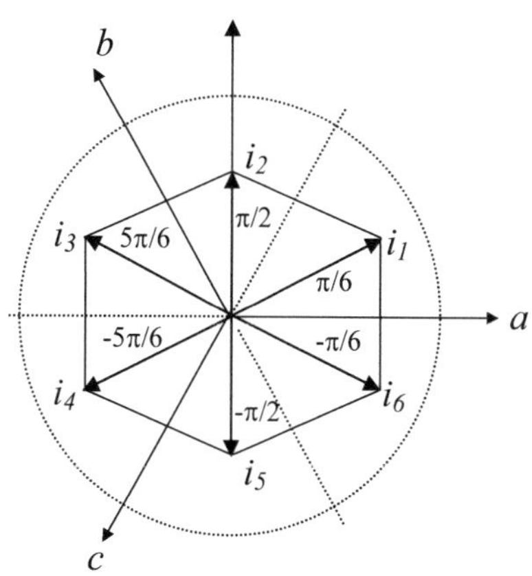

Fig. 18. Current vectors in CSI

dT_e	S_1 0, $\pi/3$	S_2 $\pi/3$, $2\pi/3$	S_3 $2\pi/3$, π	S_4 $-\pi$, $-2\pi/3$	S_5 $-2\pi/3$, $-\pi/3$	S_6 $-\pi/3$, 0
1	i_2	i_3	i_4	i_5	i_6	i_1
0	i_0	i_0	i_0	i_0	i_0	i_0
-1	i_6	i_1	i_2	i_3	i_4	i_5

Table 2. Optimal switching vectors in CSI DTC drive

4.3 Proposed DTC of CSI drives

In DTC schemes, the presence of hysteresis controllers for flux and torque determines variable-switching-frequency operation for the inverter. Furthermore, using DTC schemes a fast torque response over a wide speed range can be achieved only using different switching tables at low and high speed. The problem of variable switching frequency can be overcome by different methods (Vas, 1998; Casadei et al., 2003). In (Casadei et al., 2003), a solution based on a stator flux vector control (SFVC) scheme has been proposed. This scheme may be considered as a development of the basic DTC scheme with the aim of improving the drive performance. The input commands are the torque and the rotor flux, whereas the control variables are the stator flux components. The principle of operation is based on driving the stator flux vector toward the corresponding reference vector defined by the input commands. This action is carried out by the space-vector modulation (SVM) technique, which applies a suitable voltage vector to the machine in order to compensate the stator flux vector error. In this way it is possible to operate the induction motor drive with a constant switching frequency.

In proposed DTC CSI drive shown in Fig. 19 the inputs are rotor flux and torque as in VSI presented in (Casadei et al., 2003), but now as a control variable the stator flux angle α_s is used (Nikolic & Jeftenic, 2008).

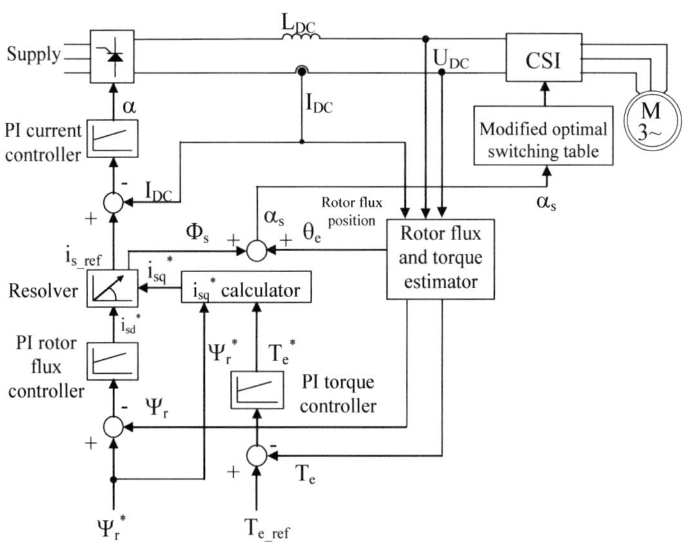

Fig. 19. Proposed constant-switching DTC strategy in CSI fed induction motor drive

Although this configuration could remind on field-oriented control, the main difference is absence of coordinate transformation since it is not necessary to use coordinate transformation to achieve correct firing angle as in vector control of the same drive (Nikolic & Jeftenic, 2006). Identical result would be obtained when phase angle Φ_s between d-q current references and rotor flux vector angle $\theta_e = \arctan(\Psi_{r\beta}/\Psi_{r\alpha})$ are summed and resulting angle α_s is than used to determine sector of 60 degrees where resides rotor flux vector. In that way, phase angle Φ_s acts as a torque control command. When reference torque is changed, i_{sq}^* is momentary changed. Phase angle Φ_s "moves" stator current vector i_s in direction determined by the sign of torque reference and its value accelerate or decelerate flux vector movement according to the value of the reference torque (Fig. 20).

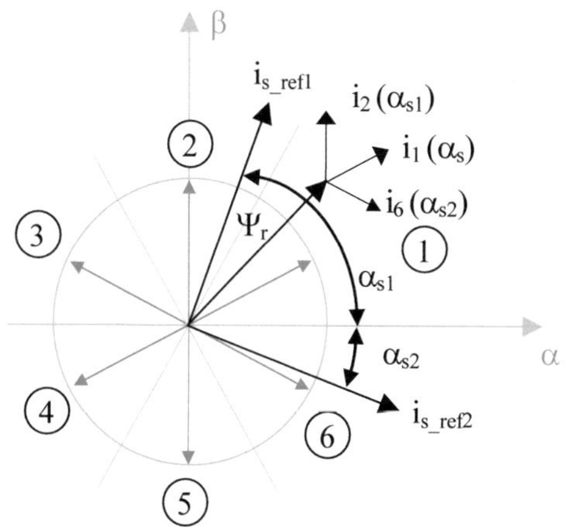

Fig. 20. Selecting proper current vector in proposed DTC algorithm

This modification implies somewhere different switching table for activating inverter switches from that shown in Table 2. Now α_s (angle between referent α-axis and reference current vector i_s) determines which current vector should be chosen: i_2 for torque increase, i_6 for torque decrease or i_1 for keeping torque at the current value.

Current vector	Angle range (degrees)
i_1	$\alpha_s > 0°$ and $\alpha_s \leq 60°$
i_2	$\alpha_s > 60°$ and $\alpha_s \leq 120°$
i_3	$\alpha_s > 120°$ and $\alpha_s \leq 180°$
i_4	$\alpha_s > 180°$ or $\alpha_s \leq -120°$
i_5	$\alpha_s > -120°$ and $\alpha_s \leq -60°$
i_6	$\alpha_s > -60°$ and $\alpha_s \leq 0°$

Table 3. Optimal switching table in proposed DTC

It is necessary to emphasize the importance of zero space vectors. In VSI there are two zero voltage vectors: V_0 denotes case when all three switches from the one half of inverter are switched ON while V_7 represent state when switches are OFF. Contrary, in CSI (using analogy to the VSI) zero current vector i_0 represent case when all thyristors are OFF. That could lead to both torque and motor speed decrease. Due to the nature of commutation in CSI, it is convenient to keep the selected current vector at instants when zero current vector is chosen.

The voltage and the current of CSI fed induction motor, necessary for stator flux calculation, can be reconstructed from the DC link quantities knowing the states of the conducting inverter switches. In one duty cycle of the output current CSI has six commutations. In that case six intervals of 60 degrees can be defined in which the current and the voltage changes its values. In every interval the current from DC link flows through two inverter legs and two motor phase windings. The motor line voltage is equal to the DC voltage on the inverter input reduced for the voltage drop on the active semiconductors, i.e. serial connection of the thyristor and diode in each inverter leg (Fig. 1). This voltage drop is forward voltage and for diodes it is about 0.7V-0.8V and for thyristors it is about 1V-1.5V. In this algorithm the average value of the overall forward voltage is used (2V), but for the practical realization it is chosen from the semiconductors datasheets or determined experimentally. It can be generally concluded that the voltage drop on the corresponding thyristor-diode par could have the following values in dependence of the conducting thyristor T_x, where x = 1,…,6:

- $V_{TDx} = V_F$, when T_x is conducting,
- $V_{TDx} = 0.5 \cdot U_{DC}$, when conducts thyristor from the same half-bridge where T_x is,
- $V_{TDx} = U_{DC} - V_F$, when conducts thyristor from the same inverter leg where T_x is.

These results are used for the voltage calculation in all conducting intervals, and they are summarized in Table 4. Prior to the flux estimation, the currents and voltages given in the Table 4 should be converted to α-β stationary frame.

The resistance of the stator windings, needed for stator flux calculation, can be easily determined from the simple experiment when the motor is in the standstill. When only thyristors T_1 and T_6 conducts, the DC current will flow through motor phases a and b. Since the motor is in the standstill, the only voltage drop is on the stator resistance R_s:

$$U_{ab} = 2 \cdot R_s \cdot i_a \tag{18}$$

when the windings are Y-connected. Generally, for the motor voltage value calculated from Table 4 and any type of the motor winding connection, the stator resistance is:

$$R_s = \frac{U_{ab}}{k_s \cdot i_a} = \frac{U_{DC} - 2 \cdot V_F}{k_s \cdot I_{DC}} \tag{19}$$

where k_s = 1 for Delta connection and k_s =2 for Y connection. Relation (19) can be easily implemented in the control software if the thyristors T_1 and T_6 are switched ON prior the motor start and the stator resistance is determined from the measured DC link current and voltage and the knowing voltage drop on the thyristor-diode par using Table 4.

	Active Thyristors	i_a	i_b	U_{ab}	U_{bc}
1	T1,T6	I_{DC}	0	$U_{DC} - 2 \cdot V_F$	$-0.5 \cdot U_{DC} + V_F$
2	T1,T2	I_{DC}	$-I_{DC}$	$0.5 \cdot U_{DC} - V_F$	$0.5 \cdot U_{DC} - V_F$
3	T3,T2	0	$-I_{DC}$	$-0.5 \cdot U_{DC} + V_F$	$U_{DC} - 2 \cdot V_F$
4	T3,T4	$-I_{DC}$	0	$-U_{DC} + 2 \cdot V_F$	$0.5 \cdot U_{DC} - V_F$
5	T5,T4	$-I_{DC}$	I_{DC}	$-0.5 \cdot U_{DC} + V_F$	$-0.5 \cdot U_{DC} + V_F$
6	T5,T6	0	I_{DC}	$0.5 \cdot U_{DC} - V_F$	$-U_{DC} + 2 \cdot V_F$

Table 4. Motor current and voltage determined only by DC link measurements

The main feedback signals in DTC algorithm are the estimated flux and torque. They are obtained as outputs of the estimator operating in stator reference frame. This estimator at first performs electro-motive force (EMF) integration (17) to determine the stator flux vector and than calculates the flux amplitude and find the sector of 60 degrees in α-β plane where flux vector resides, according to the partition shown in Fig. 20. After the stator current and voltage are determined by previously explained reconstruction of stator voltages and currents, pure integrator in (17) yields flux vector, which components are subsequently limited in amplitude to the magnitude values of the stator flux references. The trajectory of flux vector is not circular in the presence of DC offset. Since its undisturbed radius equals Ψ_s^*, the offset components tend to drive the entire trajectory toward one of the $\pm\Psi_s^*$ boundaries. A contribution to the EMF offset vector can be estimated from the displacement of the flux trajectory (Holtz, 2003), as:

$$EMF_{\alpha\beta}^{off} = \frac{1}{\Delta t} \cdot \left(\Psi_{\alpha\beta_{\max}} + \Psi_{\alpha\beta_{\min}} \right) \tag{20}$$

where the maximum and minimum values in (20) are those of the respective components $\Psi_{s\alpha}$ and $\Psi_{s\beta}$, and Δt is the time difference that defines a fundamental period. The signal EMF_{off} is fed back to the input of the integrator so as to cancel the offset component in EMF. The input of the integrator then tends toward zero in a quasi-steady state, which makes the estimated offset voltage vector equal the existing offset Ψ_{s0} in (17). The trajectory of Ψ_s is now exactly circular, which ensures a precise tracking of the EMF offset vector. Since offset drift is mainly a thermal effect that changes the DC offset very slowly, the response time of the offset estimator is not at all critical. It is important to note that the dynamics of stator flux estimation do not depend on the response of the offset estimator (Holtz, 2003).
The estimated rotor flux is calculated from the stator flux estimate using motor parameters and reconstructed stator current:

$$\hat{\Psi}_{r\alpha\beta} = \frac{L_r}{L_m} \cdot \hat{\Psi}_{s\alpha\beta} - \frac{L_s \cdot L_r - L_m^2}{L_m} \cdot i_{s\alpha\beta} \tag{21}$$

and its position in α-β reference frame is determined by:

$$\theta_e = \arctan\left(\frac{\Psi_{r\beta}}{\Psi_{r\alpha}} \right) \tag{22}$$

Finally, from the estimated stator flux and reconstructed current vector the motor torque is:

$$T_e = \frac{3}{2} p \cdot \left(i_{s\beta} \Psi_{s\alpha} - i_{s\alpha} \Psi_{s\beta} \right) \tag{23}$$

where the stator flux and current vectors are given in stationary α-β frame, and p denotes the number of poles.
The simulation model is developed in Matlab/SIMULINK, using SimPowerSystems block library that allows a very real representation of the power section (rectifier, DC link, inverter and induction motor). All electrical parameters (inductance of DC link, motor parameters) are the same as in real laboratory prototype, also used for testing previosly explained FOC algorithm. Rated flux is 0.8Wb and rated torque is 14Nm. Rectifier reference current is limited to 12A and reference torque is limited to 150% of rated torque (20Nm).

Comparison between the basic and proposed DTC of CSI induction motor drive are shown in Fig. 21, using the same mathematical model of CSI drive as used for FOC algorithm. Proposed DTC shows much better torque response from motor standstill.

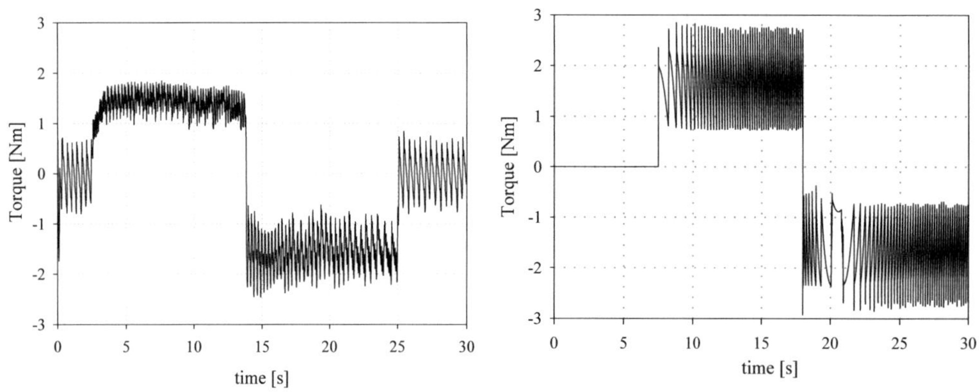

Fig. 21. Torque response for basic (left) and proposed (right) DTC algorithm

Dynamical performances of DTC algorithm are analyzed at first with rated flux and zero torque reference, than drive is accelerated up to 1000rpm. The speed is controlled in closed-loop via digital PI controller (proportional gain: $K_P = 5$, integral gain: $K_I = 0.5$, torque limit = 20Nm, controller sampling time: Ts = 4ms). The speed reference is set using succeeding scheme: +1000rpm at t = 0.25s, than –1000rpm at t = 0.6s. As could be seen from Fig. 22, a fast torque response is achieved with correct torque reference tracking and slow rotor flux ripple around the reference value (<1.5%).
For experimental purposes, the same laboratory prototype of CSI drive is used as explained before. The CSI feeds a 4kW induction motor and, as a mechanical load, the 6kW DC machine with controlled armature current is used (Nikolic & Jeftenic, 2008). The presented algorithm is not dependent on the motor power or the type of switching devices and it could be applied to any current source converter topology. The low-power induction motor and standard type thyristors are used just for the simplicity of the laboratory tests.
The torque response is analyzed both with direct torque demand and under the closed-loop speed control. Speed controller is implemented with soft start on its input and sample time of 20ms. Torque limit on the controller output is ±5Nm and is determined in such a way that under the maximum torque value slip is equal to maximal slip for current control:

$$s_{max} = \frac{1}{\omega_e \cdot T_r} = 0{,}0405 \qquad (24)$$

where ω_e is synchronous frequency (314rad/s) and T_r is rotor time constant (78.7ms).
Since rotor flux is not measured but determined by estimation, its value is checked with that obtained from simulation. The comparison between simulated and estimated rotor flux whith zero speed (torque) reference and rated flux reference are given in Fig. 23 (a). Good performance of the flux estimator, necessary for proper direct torque control, could be observed from Fig. 23 (b), where flux trajectory is shown starting from zero to its rated value. Almost circular flux trajectory with equal amplitudes in both α and β axes assures correct offset compensation.

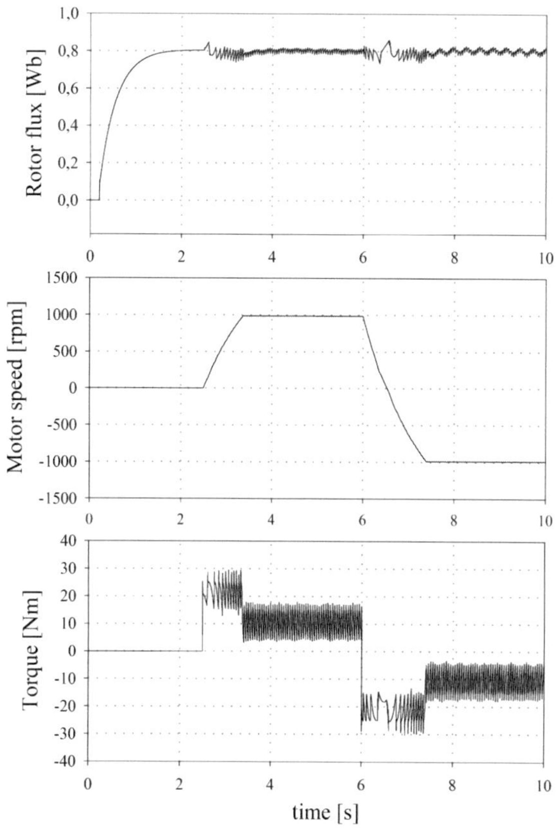

Fig. 22. Simulation results for the proposed DTC method

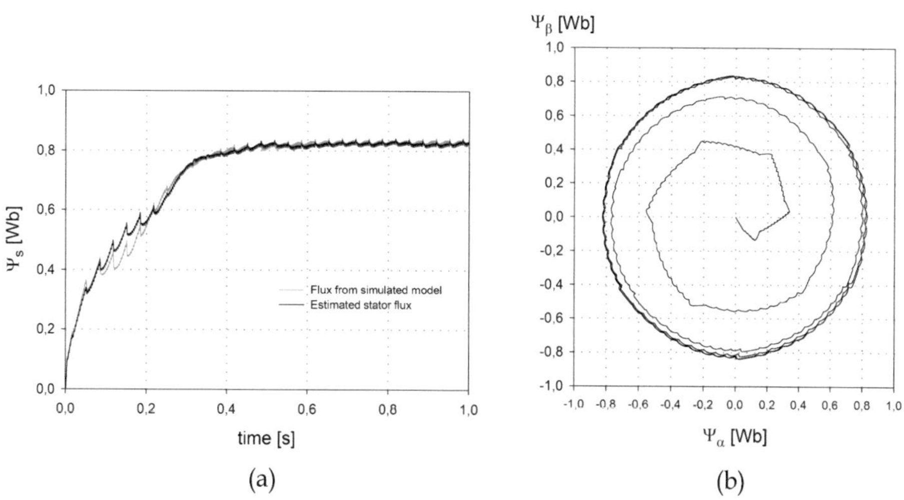

Fig. 23. Rotor flux response (a) and its trajectory (b) during motor start-up

Motor speed and torque response when the speed control loop is closed is shown in Fig. 24. Response tests are performed during motor accelerating from 0rpm to 300rpm, than from 300rpm to 500rpm and back to 300rpm and 0rpm.

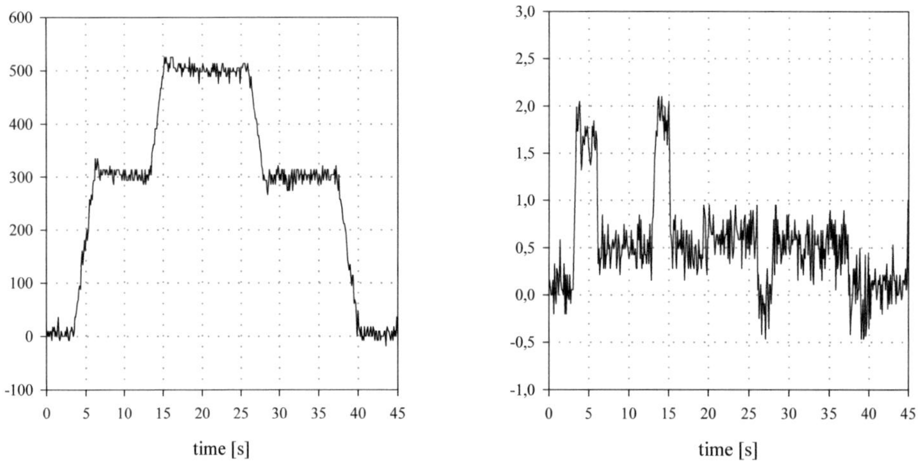

(a) (b)

Fig. 24. Motor speed (a) and torque (b) response under different speed references

5. Conclusions

In this chapter two main torque control algorithms used in CSI fed induction motor drives are presented, namely FOC and DTC. The first one is precise vector control (FOC) algorithm. The explained inconveniences of the vector controlled induction motor drives fed by a CSI could be overcome with the new vector control algorithm. The main advantage of the suggested algorithm compared to that known from the literature is better dynamic performances of the proposed CSI drive. This enhancement relies on the fast changes of the motor current, without phase error, similar to the control of current regulated voltage source PWM inverter. The realized CSI drive has more precise control, accomplished by the implemented correction of the reference current. This correction reduces the problem of the incorrect motor d-q currents values produced by the non-sinusoidal CSI current waveform.
Next, the two different methods of direct torque control in CSI fed induction motor drive are presented. Contrary to the well-known hysteresis control derived from VSI drive, new DTC algorithm based on the constant switching frequency is proposed. Merit of such a solution in comparison to the vector control of the same drive is absence of coordinate transformation and speed sensor on the motor shaft. Furthermore, since flux estimator is based only on DC link measurements, there is not necessity for any sensor on the motor side which is one of main drive advantages. In this case, by combination of vector control and basic DTC, a robust algorithm is developed that has a faster torque response and it is simpler for implementation. Moreover, algorithm is less sensitive to the parameter variation than standard FOC on the same drive. Contrary to the slip calculation using rotor time constant, proposed algorithm uses stator resistance for flux calculation and its value could be checked every time when motor is stopped using explained method for reconstruction

based only on DC link measurements. Other motor parameters (windings and mutual inductances) are used only when flux reference is changed and their values have no influence on the performance of the flux estimator due to the offset compensation. The validity of all presented torque control algorithms was proven by simulations and experimental results on developed laboratory prototype of CSI drive.

6. References

Blaschke, F. (1971). A new method for the structural decoupling of A.C. induction machines, *Proceedings of IFAC Symposium on Multivariable Technical Control Systems*, pp. 1-15, ISBN 0720420555, Duesseldorf, Germany, October 1971, American Elsevier, New York

Bose, BK. (1986). *Power Electronics and AC Drives*, Prentice-Hall, ISBN 0-13-686882-7, New Jersey, USA

Novotny, DW & Lipo, TA. (1996). *Vector Control and Dynamics of AC Drives*, Oxford University Press, ISBN 978-0-19-856439-3, New York, USA

B, Wu; SB, Dewan & Sen PC. (1988). A Modified Current-Source Inverter (MCSI) for a Multiple Induction Motor Drive System. *IEEE Transactions on Power Electronics*, Vol. 3, No. 1, January 1988, pp. 10-16, ISSN 0885-8993

Lorenz, RD. (1986). Tuning of Field-Oriented Induction Motor Controllers for High-Performance Applications, *IEEE Transactions on Industry Applications*, Vol. IA-22, No. 2, March 1986, pp. 293-297, ISSN 0093-9994

Deng, D & Lipo, TA. (1990). A Modified Control Method for Fast Response Current Source Inverter Drives, *IEEE Transactions on Industry Applications*, Vol. IA-22, No. 4, July 1986, pp. 653-665, ISSN 0093-9994

Vas, P. (1990). *Vector Control of AC Machines*, Clarendon Press, ISBN-10: 0198593708, ISBN-13: 978-0198593706, Oxford, New York, USA

Cavallini, A.; Loggini, M. & Montanari, GC. (1994). Comparison of Approximate Methods for Estimate Harmonic Currents Injected by AC/DC Converters, *IEEE Transactions on Industial Electronics*, Vol. 41, No. 2, April 1994, pp. 256-262, ISSN 0278-0046

Nikolic, A. & Jeftenic, B. (2006). Precise Vector Control of CSI Fed Induction Motor Drive, *European Transactions on Electrical Power*, Vol.16, March 2006, pp. 175-188, ISSN 0170-1703

Zargari, N. R.; Rizzo, S. C.; Xiao, Y.; Iwamoto, H.; Satoh, K. & Donlon, J. F. (2001). A new current-source converter using a symmetric gate-commutated thyristor (SGCT), *IEEE Transactions on Industry Applications*, Vol.37, 2001, pp. 896-903, ISSN 0093-9994

Takahashi, I. & Noguchi, T. (1986). A New Quick-Response and High-Efficiency Control Strategy of an Induction Motor, *IEEE Transactions on Industry Applications*, Vol. 22, No. 5, September/October 1986, pp. 820-827, ISSN 0093-9994

Depenbrok, M. (1988)., Direct Self-Control (DSC) of Inverter-Fed Induction Machine, *IEEE Transactions on Power Electronics*, Vol. PE-3, No. 4, October 1988, pp. 420-429, ISSN 0885-8993

Tiitinen, P.; Pohjalainen, P.& Lalu, J. (1995). The next generation motor control method: Direct Torque Control (DTC), *European Power Electronics Journal*, Vol.5, March 1995, pp. 14-18, ISSN 09398368

Buja, G.; Casadei, D. & Serra, G. (1998). Direct Stator Flux and Torque Control of an Induction Motor: Theoretical Analysis and Experimental Results, in *Proceedings of the IEEE International Conference on Industrial Electronics IECON '98*, pp. T50-T64, Vol. 1, ISBN 0-7803-4503-7, Aachen, Germany, August/September 1998, IEEE, New Jersey

Vas, P. (1998). *Sensorless Vector and Direct Torque Control*, Oxford University Press, ISBN 0-19-856465-1, New York, USA

Boldea, I. (2000). Direct Torque and Flux (DTFC) of A.C. Drives: A Review. *Proceedings of the 9th Conference EPE-PEMC 2000*, pp. 88-97, ISBN 80-88922-18-6, Kosice, Slovakia, September 2000, EPE-PEMC, Budapest

Casadei, D.; Serra, G.; Tani, A.; Zarri, L. & Profumo, F. (2003). Performance Analysis of a Speed-Sensorless Induction Motor Drive Based on a Constant-Switching-Frequency DTC Scheme, *IEEE Transactions on Industry Applications*, Vol.39, 2003, pp. 476-484, ISSN 0093-9994

Wu, B. (2006). *High-power converters and AC drives*, John Wiley & Sons, Inc., Hoboken, ISBN-13 978-0-471-73171-9, ISBN-10 0-471-73171-4, New Jersey, USA

Nikolic, A. & Jeftenic, B. (2008). Different Methods for Direct Torque Control of Induction Motor Fed From Current Source Inverter, *WSEAS Transactions on Circuits and Systems*, Volume 7, Issue 7, January 2008, pp. 738-748, ISSN 1109-2734

Holtz, J. (2003). Drift- and Parameter-Compensated Flux Estimator for Persistent Zero-Stator-Frequency Operation of Sensorless-Controlled Induction Motors, *IEEE Transactions on Industrial Applications*, Vol.39, 2003, pp. 1052-1060, ISSN 0093-9994

2

Direct Torque Control Based Multi-level Inverter and Artificial Intelligence Techniques of Induction Motor

Lamchich Moulay Tahar and Lachguer Nora
University Cadi Ayyad/ Faculty of Sciences Semlalia/Department of Physic/Laboratory of Electronic and Instrumentation
Morocco

1. Introduction

With the enormous advances in converters technology and the development of complex and robust control algorithms, considerable research effort is devoted for developing optimal techniques of speed control for induction machines (IM). Also, induction motor control has traditionally been achieved using field oriented control (FOC). This method involves the transformation of the stator currents into a synchronously rotating dq reference frame that is aligned with one of the stator fluxes, typically the rotor flux. In this reference frame, the torque and flux producing components of the stator currents are decoupled, such that the d-axis component of the stator current controls the rotor flux magnitude and the q-axis component controls the output torque.
The implementation of this system however is complicated and furthermore FOC, in particularly indirect method, which is widely used, is well known to be highly sensitive to parameters variations due to the feed-forward structure of its control system.
Another induction motor control technique known as a Direct Torque Control (DTC) was introduced in the mid 1980s, by Takahachi and Noguchi, for low and medium power applications; also Direct Self Control (DSC) was proposed by Depenbrock for high power applications.
DTC has a relatively simple control structure yet performs at least as good as the FOC technique. It is also known that DTC drive is less sensitive to parameters de-tuning (only stator resistor is used to estimate the stator flux) and provides a high dynamic performances than the classical vector control (fastest response of torque and flux).
This method allows a decoupled control of flux and torque without using speed or position sensors, coordinate transformation, Pulse Width Modulation (PWM) technique and current regulators. This type of command involves nonlinear controller type hysteresis, for both stator flux magnitude and electromagnetic torque, which introduces limitations such as a high and uncontrollable switching frequency. This controller produces a variable switching frequency and consequently large torque and flux ripples and high currents distortion.
The DTC is mostly used in the objective to improve the reduction of the undulations or the flux's distortion, and to have good dynamic performances. It's essentially based on a localization table which allows selecting the vector tension to apply to the inverter according

to the position of the stator flux vector and of the direct control of the stator flux and the electromagnetic torque.
The general structure of the asynchronous motor with DTC and speed regulation and using multilevel inverter is represented by the following figure.

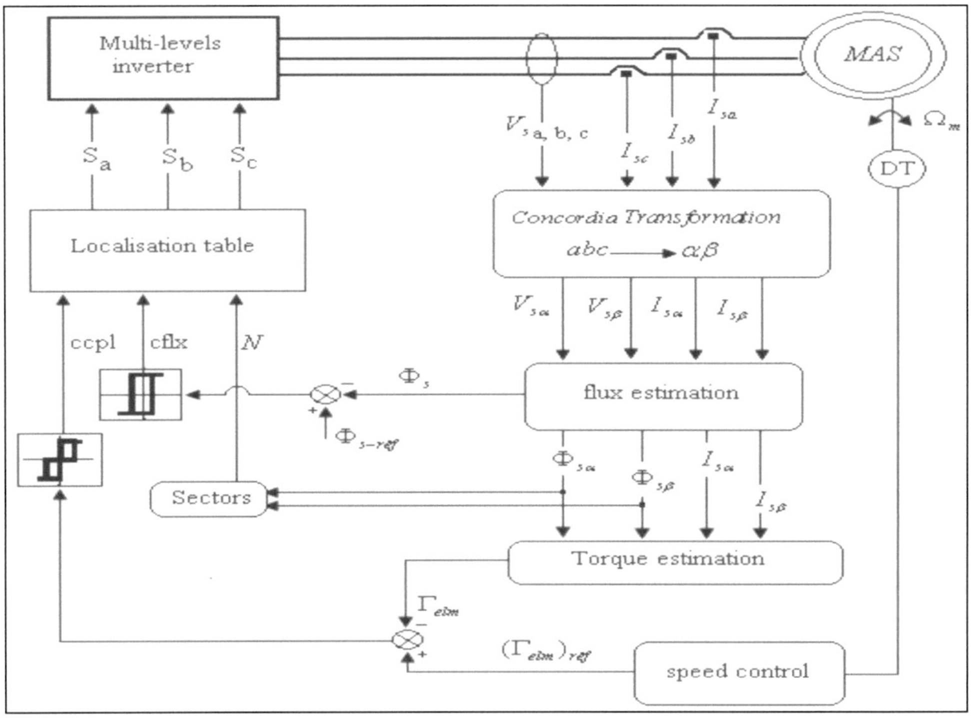

Fig. 1. General structure of the asynchronous motor with DTC and speed regulation

Also, the use of multi-level inverters and artificial techniques contribute to the performances amelioration of the induction machine control. In fact, the use of three level inverter (or multi-level inverter) associated with DTC control can contribute to more reducing harmonics and the ripple torque and to have a high level of output voltage.
Also, in last years, much interest has focused on the use of artificial intelligence techniques (neural networks, fuzzy logic, genetic algorithms,…) in identification and non linear control systems. This is mainly due to their ability learning and generalisation.
It become a number of papers appeared in literature interest to improving the performance of DTC applied to induction motor drive.
Among the different control strategies that were applied to achieve improved performance include:

- The switching frequency is maintained constant by associating the DTC to the space vector modulation;
- The space voltage is divided into twelve sectors instead of six with the classic DTC, and used some changes of the switching table.

Many researches have been performed using the multi-level inverter and, for example, some articles described a novel DTC algorithm suited for a three level inverter, and proposed a very simple voltage balancing algorithm for the DTC scheme.

Also, different other strategies using the artificial intelligence techniques were introduced, in order to achieve the objective that improving the performance of DTC:

- The direct torque control using a fuzzy logic controller to replace the torque and stator flux linkage hysteresis loop controller, space vector modulation, and fuzzy stator resistance estimator is more developed;
- The artificial neural network replacing the convectional switching table in the DTC of induction motor is also widely detailed.

In this chapter, all these points will be deeply developed and some simulation results, using Matlab/Simulink environment and showing the advantages of these approaches, will be presented. In the 1st section, we present the description of DTC method applied to the induction motor, as well as the simulation results will be illustrate the effectiveness of this method. In 2nd section, in the objective to improve the performance of DTC, the technique of multi-level inverter fed induction motor has been analyzed and simulation results show the performance of this approach. In 3rd section, we present the fuzzy logic direct torque control with two approaches: pulse width modulation and space vector modulation, also a model of artificial neural network is applied in DTC.

In the latest sections, the association of three-level inverter with fuzzy/Neural speed corrector for direct torque control of induction motor is developed.

2. Direct flux-torque control fundamentals

The direct torque control is principally a non-linear control in which the inverter switching states are imposed through a separate control of stator flux and electromagnetic torque of the motor. The inverter command is instantaneous and it replaces then the decoupling through the vectorial transformation. One of the most important characteristics of the DTC is the non-linear regulation of stator flux and electromagnetic torque with variables structures or by hysteresis.

The flux regulation is imperative for an efficient control of the induction machine torque and in the DTC, the stator flux regulation is chosen because it's easier to estimate, and partly it has a faster dynamics than the rotor flux. By adjusting the stator flux, we also adjust the rotor flux.

As in the other control methods, which use a direct regulation of the flux, the flux nominal value is imposed as a constant reference, for speeds lower than the nominal value. For higher speeds, a flux reference value, decreasing proportionally with speed; is imposed. On the other hand, the quality of rotation speed, and/or position, control of the modern actuators depends directly on the toque control.

2.1 Stator flux control

The IM equations, in a stator reference frame, are defined by:

$$\begin{cases} \bar{V}_s = R_s \bar{I}_s + \dfrac{d\bar{\phi}_s}{dt} \\ \bar{V}_r = 0 = R_r \bar{I}_r + \dfrac{d\bar{\phi}_r}{dt} - j\omega \bar{\phi}_r \\ \bar{\phi}_s = L_s \bar{I}_s + M_{sr} \bar{I}_r \\ \bar{\phi}_r = L_r \bar{I}_r + M_{sr} \bar{I}_s \end{cases} \tag{1}$$

where R_s and R_r are the stator and rotor resistances.
L_s and L_r are the mutual stator and rotor inductances.
The stator flux is estimated from the measure of stator current and voltage and their transformation in the $\alpha\beta$ subspace. So:

$$\Phi_{s\alpha} = \int_0^t (V_{s\alpha} - R_s I_{s\alpha})dt \qquad \Phi_{s\beta} = \int_0^t (V_{s\beta} - R_s I_{s\beta})dt \tag{2}$$

The stator flux module and the linkage phase are given by:

$$\Phi s = \sqrt{\Phi_{s\alpha}^2 + \Phi_{s\beta}^2} \qquad \alpha_s = arctg(\frac{\phi_{s\beta}}{\phi_{s\alpha}}) \tag{3}$$

On a sampling period T_e, and by neglecting the term $(R_s I_s)$ in equation of stator flux, valid hypothesis for high speeds, the evolution of this last one is given by the vector Vs during Te:

$$\Delta\Phi_s = \Phi_s - \Phi_{s0} = V_s T_e \tag{4}$$

Φ_{s0} is the initial stator flux at the instant t_0.
So, the variation of the stator flux is directly proportional to the stator voltage, thus the control is carried out by varying the stator flux vector by selecting a suitable voltage vector with the inverter.
A two level hysteresis comparator could be used for the control of the stator flux. So, we can easily control and maintain the flux vector Φ_s in hysteresis bound as shown in Figure.2.
The output of this corrector is represented by a Boolean variable *cflx* which indicates directly if the amplitude of flux must be increased $(cflx = 1)$ or decreased $(cflx = 0)$ so as to maintain: $\left|(\Phi_s)_{réf} - \Phi_s\right| \leq \Delta\Phi_s$, with $(\Phi_s)_{réf}$ the flux reference value and $\Delta\Phi_s$ the width of the hysteresis corrector.

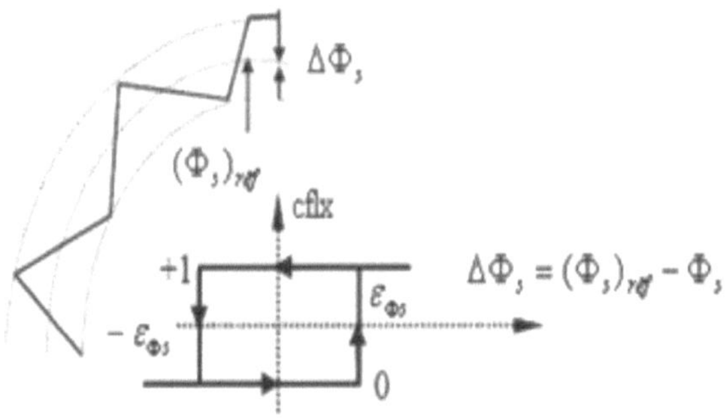

Fig. 2. Flux hysteresis corrector

2.2 Torque control

The electromagnetic torque expression is defined as follws, where γ represents the angle between the rotor and stator flux vectors:

$$\Gamma_{elm} = p\frac{L_m}{\sigma L_s L_r}\|\Phi_s\|\|\Phi_r\|\sin(\gamma) \quad (5)$$

where p is the number of pole pair
L_m: mutual inductance
σ: leakage coefficient (Blondel coefficient)
We deduct that the torque depends on the amplitude and the position of stator and rotor flux vectors.
On the other hand, the differential equation linking the stator flux and the rotor flux of motor is given by:

$$\frac{d\Phi_r}{dt} + (\frac{1}{\sigma\tau_r} - j\omega)\Phi_r = \frac{L_m}{\sigma\tau_r L_s}\Phi_s \quad (6)$$

From this equation, the flux Φ_r tracks the variations of the flux Φ_s with a time constant $\sigma\tau_r$.
In controlling perfectly the stator flux vector, from the vector V_s, in module and in position, we can control the amplitude and the relative position of the rotor flux vector and consequently the electromagnetic torque. This is possible only if the command period T_e of the voltage V_s is very lower to time constant $\sigma\tau_r$.
The expression of the electromagnetic torque is only obtained from the stator flux components $\Phi_{s\alpha}$, $\Phi_{s\beta}$ and currents $I_{s\alpha}$, $I_{s\beta}$:

$$\Gamma_{elm} = p(\phi_{s\alpha}\, i_{s\beta} - \phi_{s\beta}\, i_{s\alpha}) \quad (7)$$

For the control of the electromagnetic torque, we can use a three level hysteresis comparator which permits to have the two senses of motor rotation. The output of this corrector is represented by a Boolean variable *Ccpl* indicating directly if the amplitude of the torque must be increased, decreased or maintained constant $(ccpl = 1, -1, 0)$.

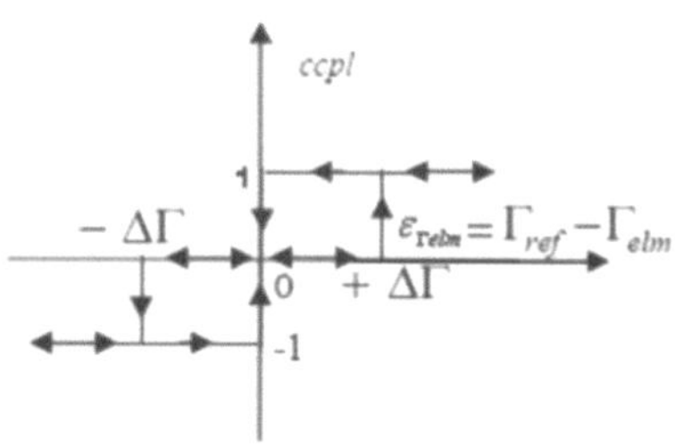

Fig. 3. Three level hysteresis comparator

2.3 Control strategy of DTC based two-level voltage inverter

Direct Torque Control of IM is directly established through the selection of the appropriate stator vector to be applied by the inverter. To do that, in first state, the estimated values of stator flux and torque are compared to the respective references, and the errors are used through hysteresis controller.
The phase plane is divided, when the IM is fed by two-level voltage inverter with eight sequences of the output voltage vector, into six sectors.

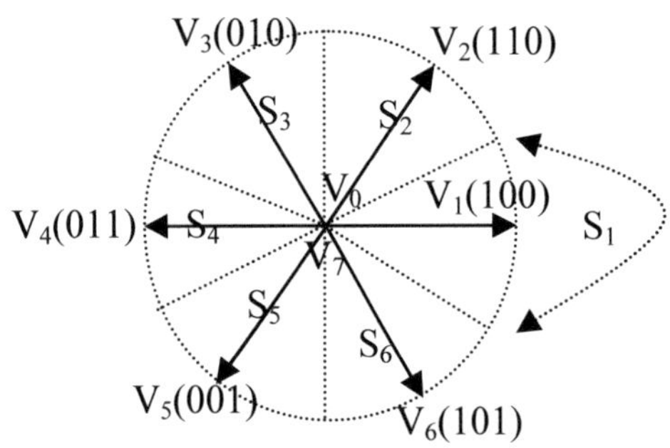

Fig. 4. Stator vectors of tensions delivered by a two level voltage inverter

When the flux is in a sector (i), the control of flux and torque can be ensured by the appropriate vector tension, which depends on the flux position in the reference frame, the variation desired for the module of flux and torque and the direction of flux rotation:

	Φs increase, Γ_{elm} increase	Φs increase, Γ_{elm} decrease	Φs decrease, Γ_{elm} increase	Φs decrease, Γ_{elm} decrease
Vector tension selected	V_{i+1}	V_{i-1}	V_{i+2}	V_{i-2}

Table 1. Selection of vector tension

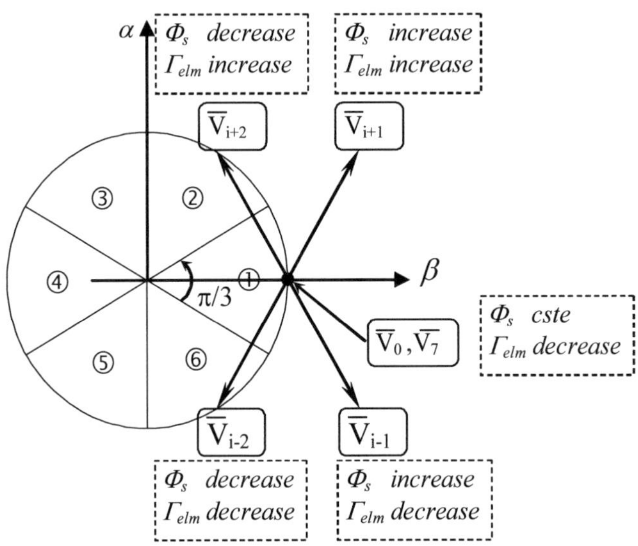

Fig. 5. Selection of vector tension

The null vectors (V_0, V_7) could be selected to maintain unchanged the stator flux.

According to the table 2, the appropriate control voltage vector (imposed by the choice of the switching state) is generated:

Cflx	ccpl	S_1	S_2	S_3	S_4	S_5	S_6
1	1	V_2	V_3	V_4	V_5	V_6	V_1
	0	V_7	V_0	V_7	V_0	V_7	V_0
	-1	V_6	V_1	V_2	V_3	V_4	V_5
0	1	V_3	V_4	V_5	V_6	V_1	V_2
	0	V_0	V_7	V_0	V_7	V_0	V_7
	-1	V_5	V_6	V_1	V_2	V_3	V_4

Table 2. Voltage vector selected (for each sector S_i)

The following figure shows the selected voltage vector for each sector to maintain the stator flux in the hysteresis bound.

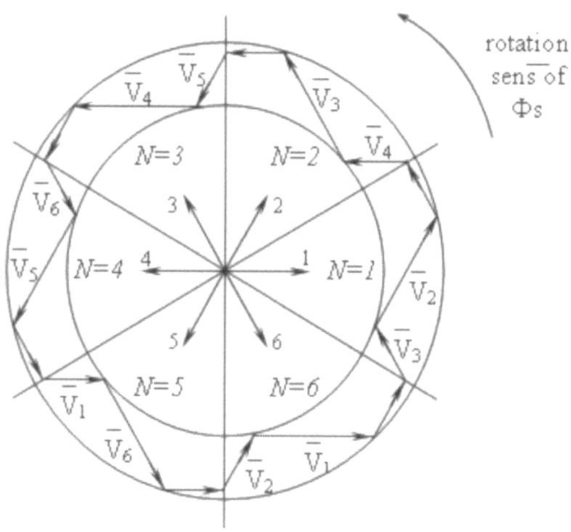

Fig. 6. Selection of vector tension

2.4 Simulation results

Simulations were performed to show the behavior of the asynchronous motor fed by two-level inverter and controlled by Direct Torque Control.

The torque reference value is deduced from the regulation of the IM speed using a PI corrector. We have chosen to present the results corresponding to the rotation speed evolution, the electromagnetic torque, the flux evolution in the αβ subspace and the stator currents.

The obtained simulation results show that:

- trajectory of the stator flux, represented by its two components in the αβ phase plane, is in a circular reference (Figure 7)
- phase current obtained by this strategy is quasi-sinusoidal (Figure 7)
- speed track its reference with good performance (Figure 8)
- overshoot on torque is limited by saturation on the reference value (Figure 8)

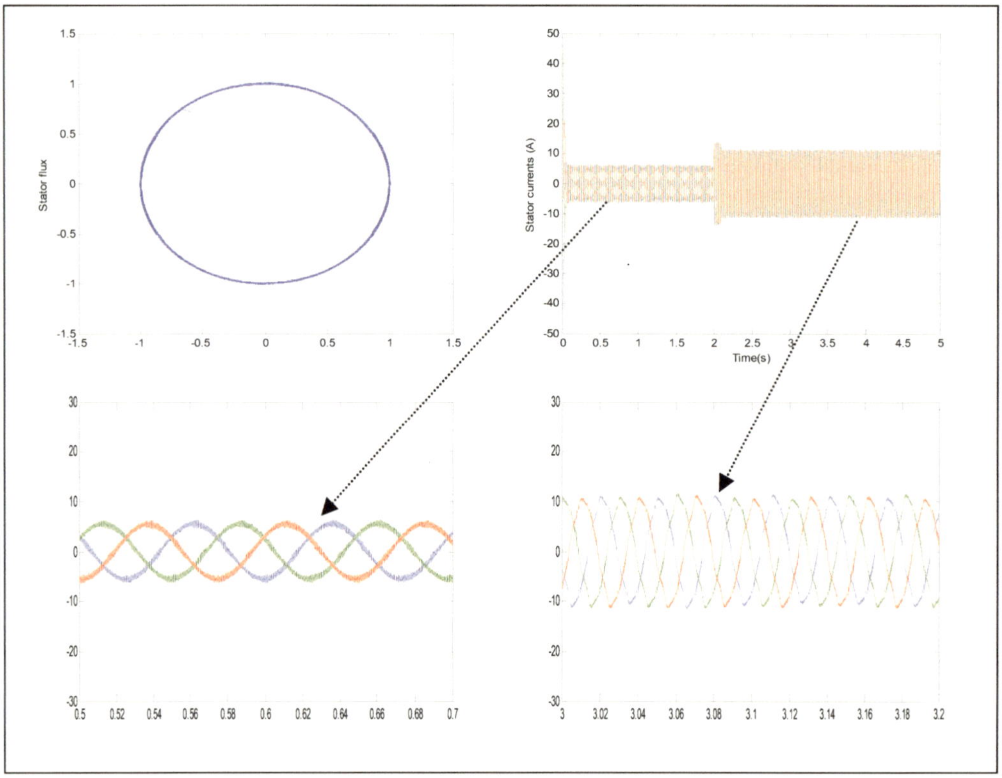

Fig. 7. Stator flux in the αβ phase plane and stator current time evolution

Fig. 8. Time evolution of speed and electromagnetic torque

3. DTC of Induction motor fed by multilevel inverter

Multilevel inverter present a big interest in the field of the high voltages and the high powers of the fact that they introduce less distortion and weak losses with relatively low switching frequency.

Three level inverter (or multilevel) can be used in the command DTC, what allows to reduce advantage the harmonics, to have a high level of output voltage and can contribute to more reducing harmonics and the ripple torque. In that case, the space of voltages is subdivided into twelve sectors (instead of six with the classic DTC) and by considering the method of the virtual vectors, three sections with small, medium and large vectors can be exploited.

We can also subdivide the space of voltages into only six sectors by adopting a technique which employs only twelve active voltage space vectors, corresponding to the small and large vectors and consequently without using the null or the medium space vectors.

3.1 Vectors tensions and phase level sequences of a three level inverter

The structure of the so called diode clamped three level inverter associated with the asynchronous motor is shown by figure 9.

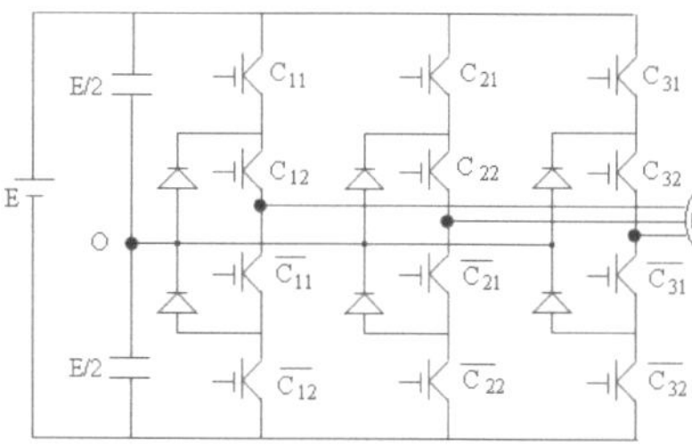

Fig. 9. Three level inverter structure

To analyze the potential generated by this three states inverter, every arm is schematized by three switches which permit to independently connect the stator inputs to the source potentials (represented by E/2, 0 and -E/2).

The interrupters (IGBTs) are switched in pairs consisting of (C_{11}, C_{12}), (C_{12}, $\overline{C}_{11}$) and ($\overline{C}_{11}$, $\overline{C}_{12}$). When, as example, the upper pair (C_{11}, C_{12}) is turned, the output is connected to the positive rail of the DC bus.

By making a transformation into $\alpha\beta$ (or dq) subspace, a resulting voltage vector is defined and associated to the spatial position of the stator flux. Then, the different states number of this vector is 19, since some of the 27 possible combinations produce the same voltage vector. There are three different inverter states that will produce the zero voltage vector and two states for each of the six inner voltage vectors (called small vector).

The figure 10 shows the various discreet positions, in the $\alpha\beta$ subspace, of the tension vector generated by a three level inverter.

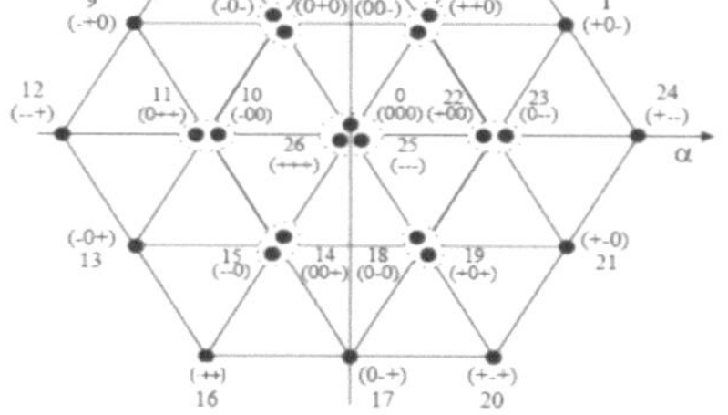

Fig. 10. Tension vectors generated by a three level inverter

3.2 Selection of voltages vectors for the control of the stator flux amplitude

As noted previously, the space evolution of the stator flux vector could be divided into twelve sectors i (Figure 11), instead of six with the classical DTC, with i= [1, 12] of 30° each, or into six sectors without using the medium vectors.

When the stator flux vector is in a sector i, the control of the flux and the torque can be assured by selecting one of 27 possible voltages vectors.

The difference between each of the inverter states that generate the same voltage vectors is in the way the load is connected to the DC bus. The analysis of the inverter states show that:

- the large vectors, such as V_{24} (+--), correspond to only the positive and negative rails of the DC bus are used and consequently have no effect on the neutral point potential;
- in the case of the medium vectors, the load is connected to the positive rail, neutral point and negative rail. The affect on the neutral point depends on the load current;
- there are two possible states of each of the small voltage vectors which can be used to control the neutral point voltage. As an example, small vector V_{22} (+00) causes capacitor C_1 to discharge and C_2 to charge and as a result the voltage of the neutral point starts to rise.

Depending on the stator flux position (sector) and the values of the outputs of torque and flux controllers, $\varepsilon_{\Phi s}$ and $\varepsilon_{\Gamma elm}$ respectively, the optimal vector is selected, from all available vectors. The first sector could be chosen between -15° and 15° or 0° and 30°. Figure 11 present the space plane for the second case.

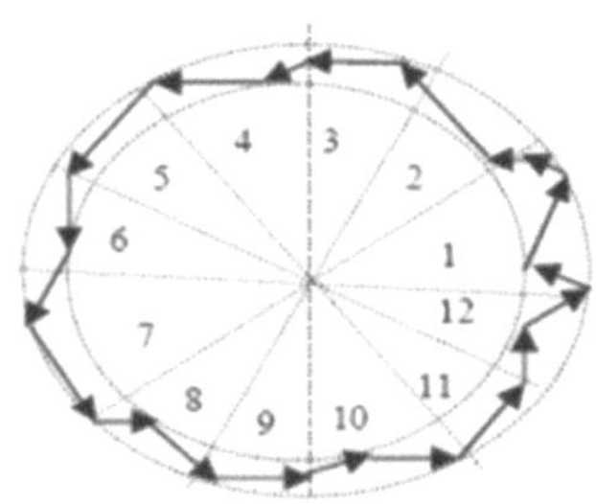

Fig. 11. Selection of vectors tensions Vs corresponding to the control of the magnitude Φ_s for a three level inverter.

3.3 Elaboration of the control switching table

The elaboration of the command structure is based on the hysteresis controller output relating to the variable flux (Cflx) and the variable torque (Ccpl) and the sector N corresponding to the stator flux vector position.

The exploitation of the first degree of freedom of the inverter, is made by the choice of vectors apply to the machine among 19 possibilities, during a sampling period. For the rebalancing of the capacitive middle point, the phase level sequence is chosen among all the possibilities associated with every voltage vector adopted. This establishes the second degree of freedom which must be necessarily used.

The switching table is elaborated depending on the technique adopted for the switching states choice.

3.3.1 Switching table based on a natural extension of classical DTC

This control scheme, which uses only twelve active voltage space vectors corresponding to the sections with small and large vectors and without using the null and medium space vectors, is a natural extension of classical DTC for a two level inverter.
We can consider the case where stator flux is achieved by using two-level hysteresis comparator and electromagnetic torque by using 4-level hysteresis. The inverter state is considered as high if the output of torque comparator is high or equal to two and otherwise, the state is low.
We can note that the choice of one of the two same states corresponding to the level low is relating to the capacitor voltage balancing.
Table 3 represents, in this case, the switching table.

Φ_s	Γ_{elm}	S_1	S_2	S_3	S_4	S_5	S_6
↑	↑↑	V_2^H	V_3^H	V_4^H	V_5^H	V_6^H	V_1^H
	↑	V_2^L	V_3^L	V_4^L	V_5^L	V_6^L	V_1^L
	↓	V_6^L	V_1^L	V_2^L	V_3^L	V_4^L	V_5^L
	↓↓	V_6^H	V_1^H	V_2^H	V_3^H	V_4^H	V_5^H
↓	↑↑ ↑	V_3^H	V_4^H	V_5^H	V_6^H	V_1^H	V_2^H
	↓	V_3^L	V_4^L	V_5^L	V_6^L	V_1^L	V_2^L
	↓↓	V_5^L	V_6^L	V_1^L	V_2^L	V_3^L	V_4^L
		V_5^H	V_6^H	V_1^H	V_2^H	V_3^H	V_4^H

Table 3. Switching table with twelve active voltage space vectors

As shown by figure 10, the high vectors V_1^H, V_2^H, V_3^H, V_4^H, V_5^H and V_6^H are represented respectively by the configuration states of the inverter (+--),(++-),(-+-),(--+),(-++) and (+-+).

3.3.2 Switching table with twelve sectors

The space voltage vector diagram, for the three-level inverter, is divided into twelve sectors by using the diagonal between the adjacent medium and long vector.
According to the errors of torque and the stator flux linkage, the optimal vector is selected, from all 19 different available vectors (figure 12). The first sector is then chosen between -15° and 15°.

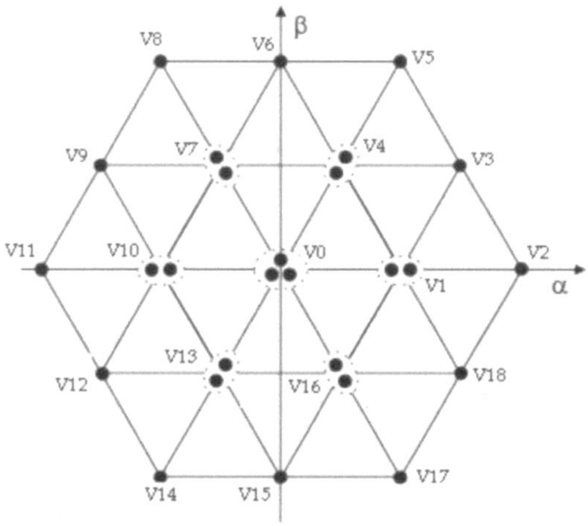

Fig. 12. Space voltage vector diagram (case of twelve sectors).

In analysing the effect of each available voltage vector, it can be seen that the vector affects the torque and flux linkage with the variation of the module and direction of the selected vector. For example, to increase the torque and flux V_3, V_4 and V_5 can be selected, but the action on the increasing torque and flux respectively of V_5 and of V_3 is the biggest.
Table 4 represents one of the solutions adapted to choice the optimal selected voltage vector for each sector. In this case, stator flux and torque are achieved by using respectively three levels and four levels hysteresis comparator. This technique doesn't use the null voltage vector for dynamics raisons.

Φ_s	Γ_{elm}	S_1	S_2	S_3	S_4	S_5	S_6	S_7	S_8	S_9	S_{10}	S_{11}	S_{12}
1	2	V_5	V_6	V_8	V_9	V_{11}	V_{12}	V_{14}	V_{15}	V_{17}	V_{18}	V_2	V_3
	1	V_3	V_5	V_6	V_8	V_9	V_{11}	V_{12}	V_{14}	V_{15}	V_{17}	V_{18}	V_2
	-1	V_{18}	V_2	V_3	V_5	V_6	V_8	V_9	V_{11}	V_{12}	V_{14}	V_{15}	V_{17}
	-2	V_{17}	V_{18}	V_2	V_3	V_5	V_6	V_8	V_9	V_{11}	V_{12}	V_{14}	V_{15}
0	2	V_7	V_7	V_{10}	V_{10}	V_{13}	V_{13}	V_{16}	V_{16}	V_1	V_1	V_4	V_4
	1	V_4	V_4	V_7	V_7	V_{10}	V_{10}	V_{13}	V_{13}	V_{16}	V_{16}	V_1	V_1
	-1	V_{16}	V_1	V_1	V_4	V_4	V_7	V_7	V_{10}	V_{10}	V_{13}	V_{13}	V_{16}
	-2	V_{13}	V_{16}	V_{16}	V_1	V_1	V_4	V_4	V_7	V_7	V_{10}	V_{10}	V_{13}
-1	2	V_8	V_9	V_{11}	V_{12}	V_{14}	V_{15}	V_{17}	V_{18}	V_2	V_3	V_5	V_6
	1	V_9	V_{11}	V_{12}	V_{14}	V_{15}	V_{17}	V_{18}	V_2	V_3	V_5	V_6	V_8
	-1	V_{12}	V_{14}	V_{15}	V_{17}	V_{18}	V_2	V_3	V_5	V_6	V_8	V_9	V_{11}
	-2	V_{14}	V_{15}	V_{17}	V_{18}	V_2	V_3	V_5	V_6	V_8	V_9	V_{11}	V_{12}

Table 4. Switching table with twelve sectors

This approach and others, for establishing of the optimal switching table and taking into account all the factors such as the capacitors balance, system dynamic and system reliability, must be deeply analysed and tested. Also, it's difficult, in this case, to select the optimal voltage vectors; however, the use of artificial intelligence techniques will add their superiority to some extent.

4. Direct torque control based fuzzy / neural network

By analyzing the structure of the switching table, we can notice that it can be translated in the form of fuzzy rules. Consequently, we can replace the switching table and the hysteresis comparator by a fuzzy system. The fuzzy character of this system allows flexibility in the choice of the fuzzy sets of the input and the capacity to introduce knowledge of the human expert there.
Also, as the DTC uses algorithms to select a large number of statements inverter switches, neural networks can accomplish this task after a learning phase. The neural network selector inputs will be proposed as the position of the flux stator vector, the error between its estimated value and the reference one, and the difference between the estimated and reference values of electromagnetic torque. The next figure shows an example of this structure.

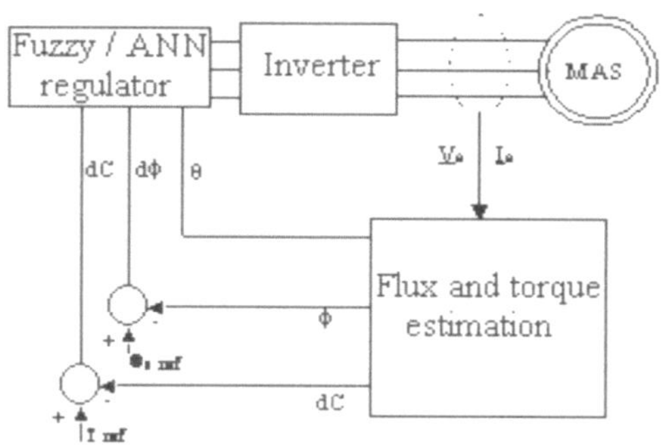

Fig. 13. Switching table based Fuzzy / ANN technique

4.1 Direct torque control based fuzzy logic

The principle of fuzzy direct torque control (FDTC) consists to replace, in conventional DTC, the torque and stator flux hysteresis controllers and the switching table by a fuzzy system.
In this case, two approaches can be presented to illustrate the strategy of FDTC of the induction motor fed by two-level inverter.
We can consider three variables input fuzzy logic controllers; the stator flux error, electromagnetic torque error and angle of stator flux, however, the choice of the output deferred according to the approach utilized. The output could be the voltage space vector, for FDTC based PWM, or the magnitude and argument of voltage vector for space vector modulation.

4.1.1 FDTC based Pulse Width Modulation (PWM)

The fuzzy logic controller blocks using PWM inverter is shown in the following figure. These blocks are composed of two main parts: fuzzification and fuzzy rules base, since no defuzzification is needed because, in this case, the output of fuzzy controller is the actual PWM voltage vector sequence and these states are directly the results of fuzzy rules.

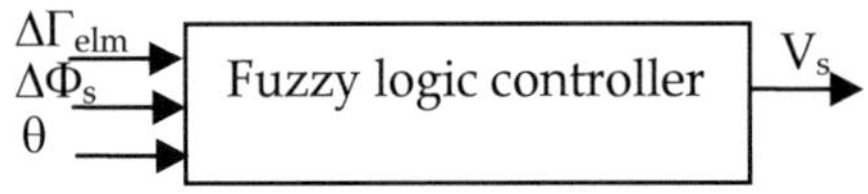

Fig. 14. Fuzzy logic controller based PWM

4.1.1.1 Fuzification

Based on the switching table of the conventional DTC, the universe of discourse for each three inputs of the fuzzy logic controller has been divided into: two fuzzy sets (NP), for stator flux error, three fuzzy sets (NZP), for electromagnetic torque error, and seven fuzzy sets ($\theta_0, \theta_1, ..., \theta_7$) for angle of flux stator.
These fuzzy sets are defined by the delta and trapezoidal membership functions and are presented by the following figure.

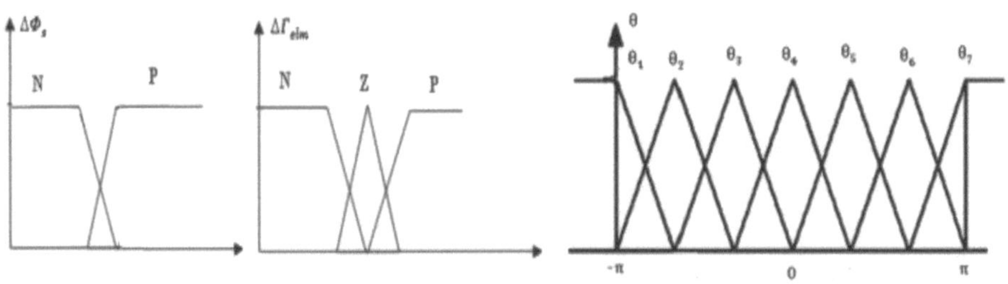

Fig. 15. Membership functions

4.1.1.2 Fuzzy rules base

The table can be expressed by fuzzy rules given by:
The ith rule R_i : if $\Delta\Phi$ is A_i and $\Delta\Gamma$ is B_i and θ is C_i then n is N_i.
A_i, B_i and C_i are the fuzzy sets of the variables $\Delta\Phi_s$, $\Delta\Gamma_{elm}$ and θ
n is the inverter switching state.
The inference method used is Mamdani's procedure based on min-max decision.
These rules are resumed by the following table.

	θ_1		θ_2		θ_3		θ_4		θ_5		θ_6		θ_7	
$\Delta\Phi_s$ / $\Delta\Gamma_{elm}$	P	N	P	N	P	N	P	N	P	N	P	N	P	N
P	V_5	V_6	V_6	V_1	V_1	V_2	V_2	V_3	V_3	V_4	V_4	V_5	V_5	V_6
Z	V_0	V_7	V_7	V_0	V_0	V_7	V_7	V_0	V_0	V_7	V_7	V_0	V_0	V_7
N	V_3	V_2	V_4	V_3	V_5	V_4	V_6	V_5	V_1	V_6	V_2	V_1	V_3	V_2

Table 5. Fuzzy rules

4.1.2 FDTC based Space Vector Modulation (SVM)

Using space vector modulation permit, in addition to the advantages obtained by the fuzzy logic controller (reduction of the torque, stator flux and current ripples and to get a fast torque response), to maintain constant the switching frequency. With this strategy two fuzzy controller of Mamdani could be used to control the magnitude and argument of voltage vector reference. For this technique, two controllers (next figure) are used concerning the variables magnitude and argument of vector tension.

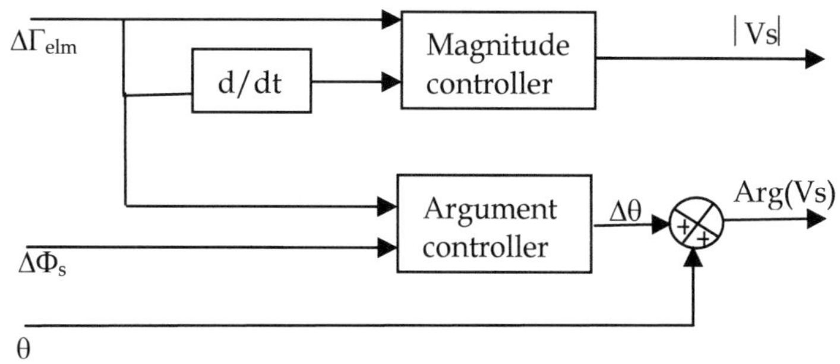

Fig. 16. Fuzzy logic controller based SVM

In the following figure, the membership functions of the variables $\Delta\Phi s$ and $\Delta\Gamma_{elm}$ are presented.

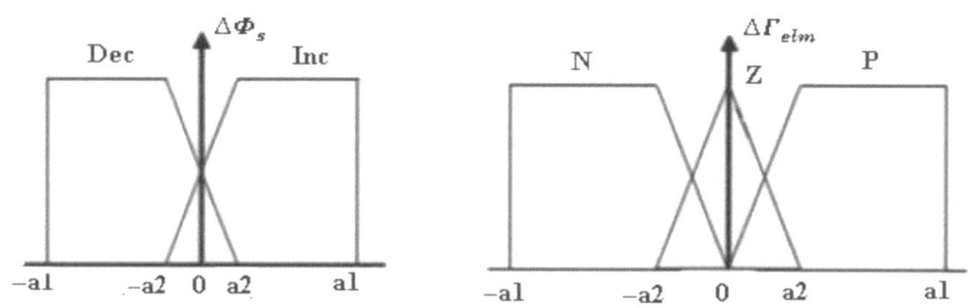

Fig. 17. Membership functions for $\Delta\Phi s$ and $\Delta\Gamma_{elm}$

We consider, in this case, two fuzzy sets functions (D: Decrease, I: Increase) for the stator flux and electromagnetic torque errors and three membership functions (N: Negative, Z: Zero, P: Positive) for the variation of the electromagnetic torque error.
The fuzzy rules of the argument fuzzy controller are presented in the following table.

$\Delta\theta$		$\Delta\Phi_s$	
		Dec	Inc
$\Delta\Gamma_{elm}$	Dec	$\mu(-2\pi/3)$	$\mu(-\pi/3)$
	Inc	$\mu(2\pi/3)$	$\mu(\pi/3)$

Table 6. Fuzzy rules of argument controller

$\mu(\theta)$ is the membership function for the output variable of argument fuzzy controller defined as represented by the following figure.

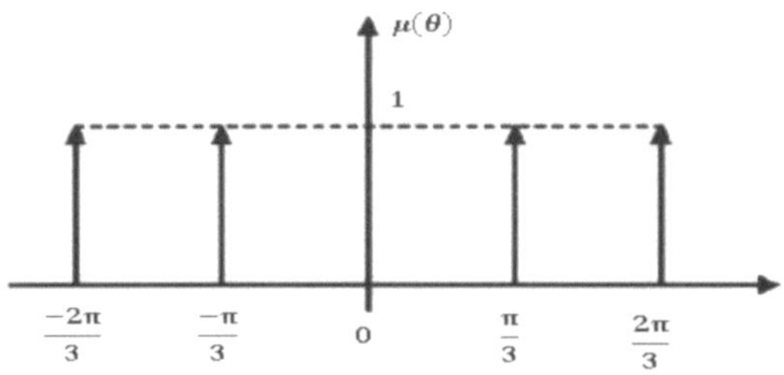

Fig. 18. Membership function for output argument controller

The voltage vectors in conventional DTC have constant amplitude in opposite with FDTC based space vector modulation where this amplitude is modified versus the torque and its derivative. Then, the fuzzy rules of the amplitude fuzzy controlled take form:
If $\Delta\Gamma_{elm}$ decrease and $d(\Delta\Gamma_{elm})/dt$ is negative then magnitude vector is small.
Consequently, these different rules are resumed in the following table, where the fuzzy sets used are, N: Negative, M: Medium, Z: Zero, P: Positive, S: Small and L: Large.

V		$\frac{d(\Delta\Gamma_{elm})}{dt}$		
		N	Z	P
$\Delta\Gamma_{elm}$	Dec	S	S	M
	Inc	M	L	L

Table 7. Fuzzy rules of amplitude fuzzy controller

Finally, the fuzzy sets of output magnitude fuzzy controller are defined by delta and trapezoidal membership functions as shown by this figure.

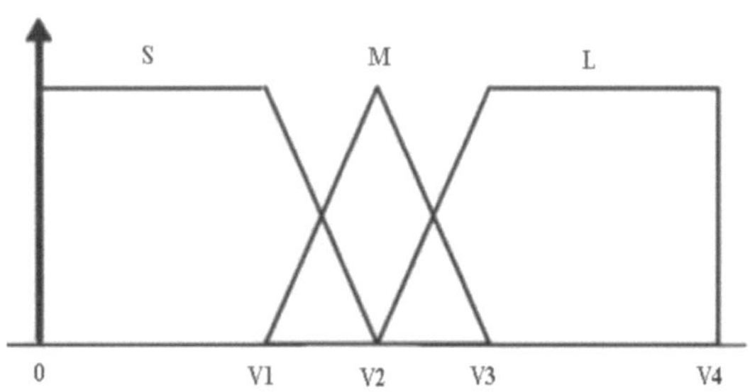

Fig. 19. Membership function for output magnitude controller

4.2 Direct torque control based artificial neural networks

Among the other intelligence techniques can improving the performance of system control and are recently showing good promise for applications in power electronics and motion control system, the use of Artificial Neural Network (ANN).

Different techniques based ANN are exploited for the control of IM; particularly, in the field of the IM Direct Torque Control, many types of these techniques are adopted. The most popular of ANN, used in DTC, is the multilayer feed forward network, trained by the back propagation algorithm, which is composed on the input layer, output layer, and several hidden layers.

Also, as the switching table has an important role in the DTC, for increasing the execution speed of the system, ANN is applied to emulate the classical switching table of the DTC obtaining the optimal switching patterns.

The switching table has as inputs the electromagnetic torque error, the stator flux error and the angle of the flux, and as output the voltage space vector to be generated by the inverter. Since this switching lookup table only depends on these inputs and not on the parameters of the IM, it can be trained off-line. Therefore, the inputs of switching table will be converted to digital signals, for reducing the training patterns and increasing the execution speed of the training process. Thus, one bit (1 or 0) represents the flux error, two bits (11 for state 1 , 00 for state 0 or 01 for state -1) the torque error and three bits the region of stator flux.

The structure of the ANN as a part of DTC is presented by figure 20, which has six inputs nodes corresponding to the digital variables (three for angle flux, one for flux error and two for torque error), six neurons in the first hidden layer, five neurons in the second hidden layer and three neurons in the output layer.

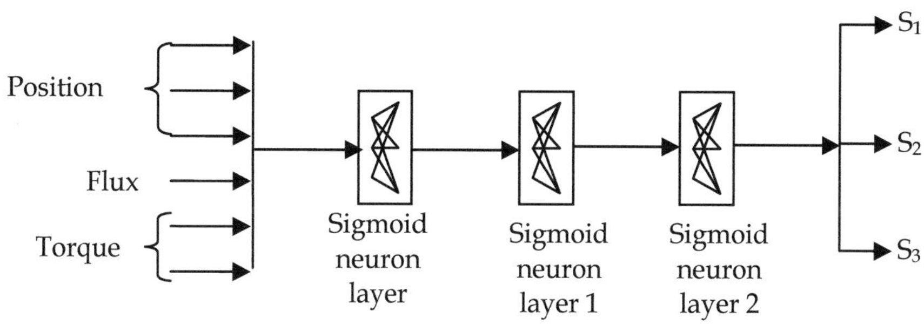

Fig. 20. Structure of the ANN

After completion of the training procedure, the network performance off-line with an arbitrary input pattern will be tested to ensure successful training. After that, the weights and biases are down-loaded to the prototype network substituting the traditional switching lookup table as a part of DTC.

An example of the ANN combined with Fuzzy inference system for the control of the IM speed will be presented in the next section.

5. Control of asynchronous motor speed based on a fuzzy / neural corrector

These last years, a most interest concerned the use of the artificial intelligence techniques (neural networks, fuzzy logic, genetic algorithms) which have the potential to provide an improved method of deriving non-linear models, have self adapting capabilities which make them well suitable to handle non-linearities, uncertainties and parameter variations.

The simplest of these methods are based on the learning of an already existing conventional controller; others methods operate a learning off-line of the process inverse model or of a reference model either completely on-line.

5.1 Description of the technique adopted for IM Speed control

As example, we have chosen to develop the case where a conventional neural controller (CNC) associated with a reference model (MRAS) for the learning phase is used to control the IM speed.

The parameters of the CNC are adjusted by minimising the error (e=u'-u) between the outputs of the MRAS and CNC as shown in the following figure.

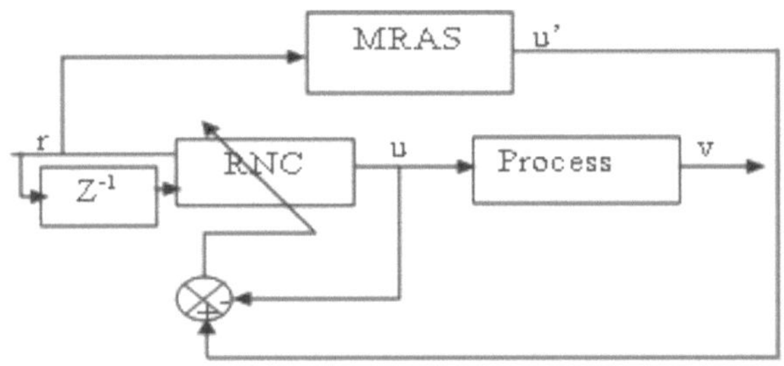

Fig. 21. Neural corrector for the IM speed control

Once the learning phase is carried out, the weights obtained are used for the neural controller, in the second phase, without the reference model.

Neural network, coupled with the fuzzy logic, speed control will be so efficient and robust. In this case, the reference model is represented by a fuzzy logic corrector (FLC) with two inputs: the error and the derivative of the error (next figure).

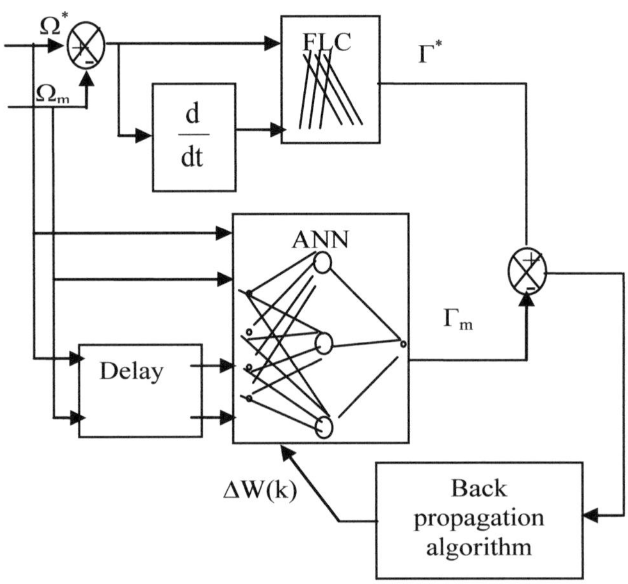

Fig. 22. IM Speed control based Fuzzy / neural corrector

The neural corrector architecture, shown by figure 23, presents 4 inputs, 3 neurons for the hidden coat with activation function type sigmoid and one output with linear activation function.

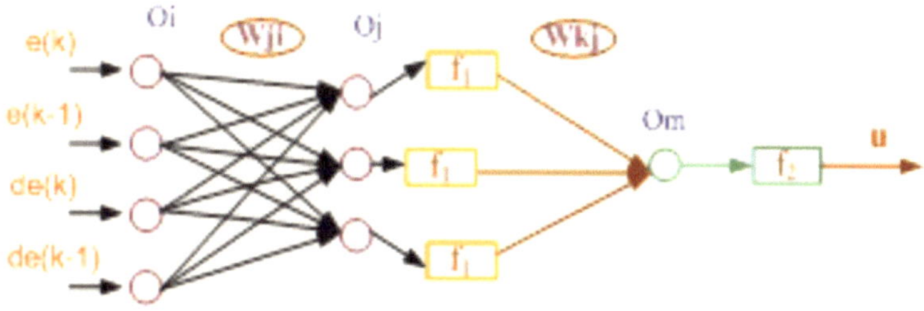

Fig. 23. Architecture of neural corrector

As it has been noted, a corrector type PI (Proportional Integral), for the reference model, which parameters are adapted by a fuzzy inference system, is used (Figure 24).

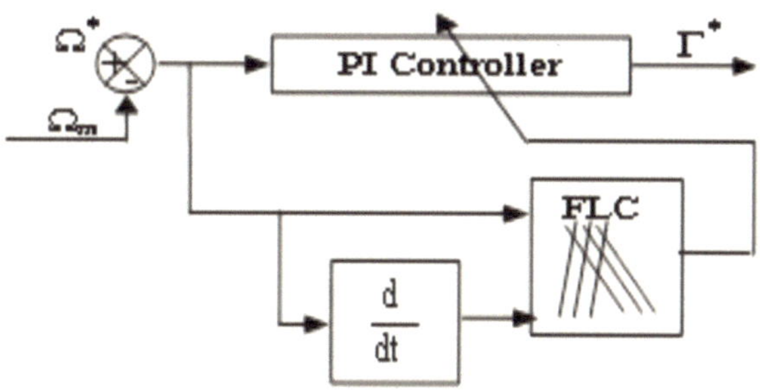

Fig. 24. Controller with PI structure adapted by fuzzy inference system

The PI parameters (Kp, Ki) are calculated by using the intermediate values (K'p and K'i) given by the fuzzy controller as follows:

$$\begin{aligned} K_p &= (K_{p_{\max}} - K_{p_{\min}})K_p^{'} + K_{p_{\min}} \\ K_i &= (K_{i_{\max}} - K_{i_{\min}})K_i^{'} + K_{i_{\min}} \end{aligned} \quad (8)$$

where the gains values are defined by using the Ziegler-Nichols method.
Both parameters (K'p, K'i), corresponding to the output of the system based on fuzzy logic, are meanwhile normalised in the range [0 1].

5.2 Simulation results

Simulations were performed to show the performances of the technique used in this section and based on fuzzy / neural corrector for the control of the IM speed. The following figure presents the model structure tested in the Matlab / Simulink environment.
We have chosen to present the results corresponding to the rotation speed evolution, the electromagnetic torque, the flux evolution in $\alpha\beta$ phase plane and the stator current temporal evolution.

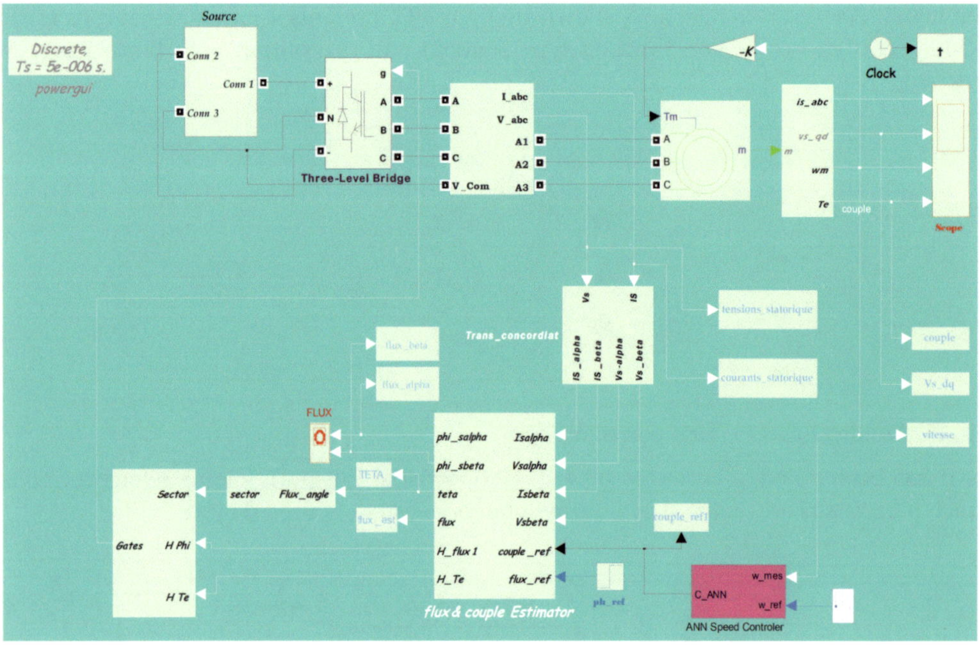

Fig. 25. Stator flux in the αβ phase plane and stator current time evolution

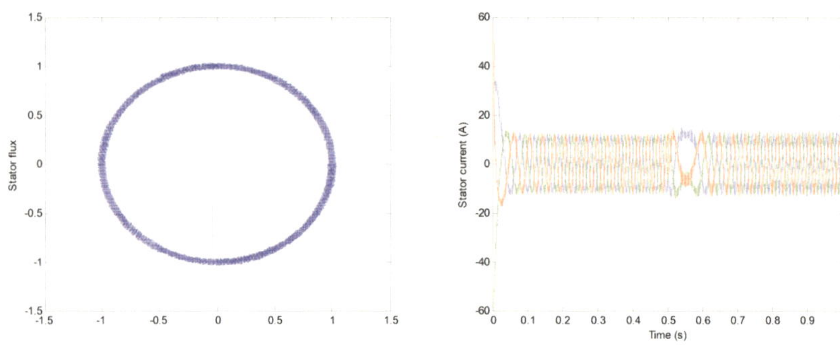

Fig. 26. Stator flux in the ((phase plane and stator current time evolution

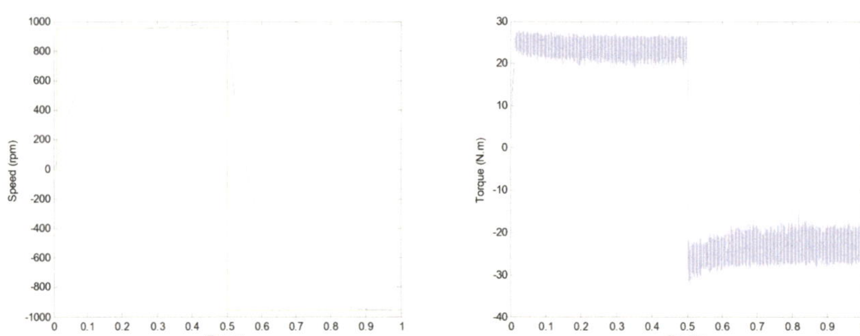

Fig. 27. Time evolution of speed and electromagnetic torque

We can note that mechanical speed track the reference values imposed and corresponding to two senses of the motor rotation (Figure 27). Also the currents are sinusoidal with acceptable value of total harmonic distortion and the flux components trajectory is circular and limited to the reference value in the hysteresis bound.

6. Conclusion

The Direct Torque Control (DTC) is an important alternative method for the induction motor drive, with its high performance and simplicity. The DTC applied to induction machine fed by a 3-level diode clamped inverter presents good performance and undulations reduction. In this case, some techniques were developed in order to replace the conventional DTC switching table adapted for a NPC inverter.

Another issue is concerned with the application of artificial intelligence techniques (fuzzy system, and neural network) to the induction motor control with DTC. Particularly, the application of these techniques for the selection of optimal voltage vectors is presented.

Also, the control of the speed induction motor is realized by the technique of the artificial neural networks (ANN) with reference model. A controller based PI adapted by a fuzzy inference is used as reference model. Attention is focused on the dynamic performance of ANN speed control. The effectiveness of the proposed scheme control is demonstrated by simulation using the blocks PSB of Matlab / Simulink.

Finally, in this chapter, we can conclude that the DTC method applied to an induction motor fed by a three level NPC inverter and based artificial intelligence techniques present most interest and contribute to improvement of system response performances.

The first investigations, presented here, of the induction machine control prove its effectiveness and its high dynamics. It will be completed in a future work by considering others control techniques.

7. References

Takahachi, I. & Noguchi, T. (1986). A new quick response and high efficiency control strategy of an induction motor. IEEE *Trans on Industry Application,* Vol.IA-22.N°5, pp 820-827, September/October 1986.

Baader, U. & Depenbroch, M. (1992). Direct Self Control (DSC) of inverter fed induction machine -A basis for speed control without speed measurement. IEEE. *Trans on Industry Application.* Vol.IA-288.N°3, pp 581-588, May/June 1992.

PUJOL, A.A. (2000). Improvement in direct torque control of induction motors. *Thesis of doctorate of the university polytechnic of Catalonia,* Spain, 2000.

Ozkop, E. & Okumus, H.I. (2008). Direct Torque Control of Induction Motor using Space Vector Modulation (SVM-DTC). MEPCON *12th International Middle-East, Power System Conference,* pp 368-372, March 2008.

Cirrinciane, M., Pucci, M. & Vitale, G. (2003). A Novel Direct Torque Control of Induction Motor Drive with a Three-Level Inverter. IEEE *Power Tech Conference Proceeding,* Vol.3, 7pp, Bologna, June 2003.

Xiying Ding, Qiang Liu, Xiaona Ma, Xiaona. He & Qing Hu (2007). The Fuzzy Direct Torque Control of Induction Motor based on Space Vector Modulation. *Third International Conference on, Natural Computation* ICNC 2007, Vol.4, pp 260-264, Aug.2007.

Guohan Lin & Zhiwei Xu (2009). Direct Torque Control of an Induction Motor using Neural Network. *1st International Conference on, Information Science and Engineering* (ICISE), pp 4827-4830, 28 December.2009.

Martins, A.C., Roboam, X., Meynard, T.A. & Carvaiho, A.C. (2002). Switching Frequency Imposition and Ripple Reduction in DTC Drives by using Multilevel Converter. IEEE *Trans, on Power Electronics*, Vol.17 N°2, March 2002.

Yang Xia & Oghanna, W. (1997). Study on Fuzzy control of induction machine with direct torque control approach. Industrial Electronics. ISIE 97, *Proceeding of the International Symposium*, Vol.2, pp 625-630, Jul.1997.

Yang, J., Ryan, M. & Power, J. (1994). Using Fuzzy Logic," Prentice Hall, 1994.

Kumar, R., Gupta, R.A., Bhangale, S.V. & Gothwal, H. (2008). Artificial Neural Network based Direct Torque Control of Induction Motor Drives. IETECH *Journal of Electrical Analysis*, Vol.2, N°3, pp 159-165, 2008.

Toufouti, R., Mezian, S. & Benalla, H. (2007). Direct Torque Control for Induction Motor using Intelligent Technique. *Journal of Theoretical and Applied Information Technology*, Vol.3, N°3, pp 35-44, 2007.

Dreyfus, G., Martinez, J., Samuelides, M., Gordon, M.B., Badran, F., Thiria, S. & Hérault, L. (2002). Réseaux de neurons : Méthodologie et applications. Editions Eyrolles, 2002.

Grabowski, P.Z., Kazmierkowski, M.P., Bose, B.K. & Blaabjerg, F. (2000). A simple Direct torque Neuro Fuzzy control of PWM Inverter fed Induction motor drive. IEEE *Trans. Electron.* 47 N° 4, pp 863-870, Aug 2000.

Viljamaa, P. (2000). Fuzzy gain scheduling and tuning of multivariable fuzzy control methods of fuzzy computing in control systems. *Thesis for the degree of doctor of technology, Temper University of technology*, Finland, 2000.

Barbara H. K. (2001). Stator and Rotor Flux Based Deadbeat Direct Torque Control of Induction Machines. IEEE *Industry Applications Society, Annual Meeting*, Chicago, September 30-October 4, 2001.

Casadei, D., Profumo, Serra, G. & Tani, A. (2002). FOC And DTC:Tox Viable Schemes For Induction Motors Torque Control. IEEE *trans.Power Electronics. On PE*, Vol.17, N°.5, Sept 2002.

Schibili, N., Nguyen, T. & Rufer, A. (1998). Three-Phase Multilevel Converter for High-Power Induction Motors. IEEE *trans. On Power Elect*. Vol. 13 N°.5, 1998.

Roboan, X. (1991). Variateur de vitesse pour machine asynchrone, Contrôle de la vitesse sans capteur mécanique. Thèse Doctorat de L'INPT, Toulouse, 1991.

Ould Abdeslam, D., Wira, P., Mercklé, J., Chapuis, Y.A. & Flieller, D. (2006). Stratégie neuromimétique d'identification et de commande d'un filtre actif parallèle. *Revue des Systèmes, Série Revue Internationale de Génie Electrique* (RS-RIGE), vol. 9, no. 1, pp 35-64, 2006.

Ould abdeslam, D. (2005). Techniques neuromimétiques pour la commande dans les systèmes électriques: application au filtrage actif parallèle. *Thèse de doctorat d'état en Electronique, Electrotechnique et Automatique*, Université de Batna, 2005.

3

Direct Torque Control using Space Vector Modulation and Dynamic Performance of the Drive, via a Fuzzy Logic Controller for Speed Regulation

Adamidis Georgios, and Zisis Koutsogiannis
Democritus University of Thrace
Greece

1. Introduction

During the last decade, a lot of modifications in classic Direct Torque Control scheme (Takahashi & Noguchi, 1986) have been made (Casadei et al., 2000), (Reddy et al., 2006), (Chen et al., 2005), (Grabowski et al., 2005), (Romeral et al., 2003), (Ortega et al., 2005). The objective of these modifications was to improve the start up of the motor, the operation in overload conditions and low speed region. The modifications also aimed to reduce the torque and current ripple, the noise level and to avoid the variable switching frequency by using switching methods with constant switching frequency.

The basic disadvantages of DTC scheme using hysteresis controllers are the variable switching frequency, the current and torque ripple. The movement of stator flux vector during the changes of cyclic sectors is responsible for creating notable edge oscillations of electromagnetic torque. Another great issue is the implementation of hysteresis controllers which requires a high sampling frequency. When an hysteresis controller is implemented using a digital signal processor (DSP) its operation is quite different to the analogue one.

In the analogue operation the value of the electromagnetic torque and the magnitude of the stator flux are limited in the exact desirable hysteresis band. That means, the inverter can change state each time the torque or the flux magnitude are throwing the specified limits. On the other way, the digital implementation uses specific sample time on which the magnitudes of torque and flux are checked to be in the desirable limits. That means, very often, torque and flux can be out of the desirable limits until the next sampling period. For this reason, an undesirable torque and flux ripple is occurred.

Many researchers are oriented to combine the principles of DTC with a constant switching frequency method for driving the inverter by using space vector modulation. This requires the calculation in the control schemes of the reference voltage vector which must be modulated in the inverter output. Therefore, the Direct Torque Control with Space Vector Modulation method (DTC-SVM) is applied (Koutsogiannis & Adamidis 2007). Since we know the reference voltage vector it is easy to perform the modulation by applying specific switching pattern to the inverter (Koutsogiannis & Adamidis 2006). In the DTC scheme a speed estimation and a torque control are applied using fuzzy logic (Koutsogiannis & Adamidis 2006). An improvement of DTC with a parallel control FOC is observed (Casadei

et al., 2002). The use of the rotor flux magnitude instead of the stator flux magnitude, improves the overload ability of the motor. This control is sensitive to the machine's parameters during transient operations.
Also, the DTC-SVM can be applied using closed loop torque control, for minimization of torque ripple (Wei et. al., 2004). In this case estimation of stator and rotor flux is required. Therefore, all the parameters of the induction motor must be known (Reddy et al., 2006). A new method was developed that allows sensorless field-oriented control of machines with multiple non-separable or single saliencies without the introduction of an additional sensor (Zatocil, 2008). In this paper, the closed loop torque control method is applied which improves the torque response during dynamic and steady state performance. A lot of papers for the speed control of electrical drives, which uses different strategies based on artificial intelligence like neural network and fuzzy logic controller, have presented. For the fuzzy PI speed controller its robustness and disturbance rejection ability Gadou et. Al., 2009) is demonstrated. In this paper fuzzy logic for the speed estimation of the motor and the method DTC-SVM with closed loop torque control will be applied. This paper is further extended through a further improvement of the system control by controlling the magnitudes of torque and flux using closed loop control. The simulation results were validated by experimental results.

2. Overview of the classic DTC scheme

The classic DTC scheme is shown in figure 1.

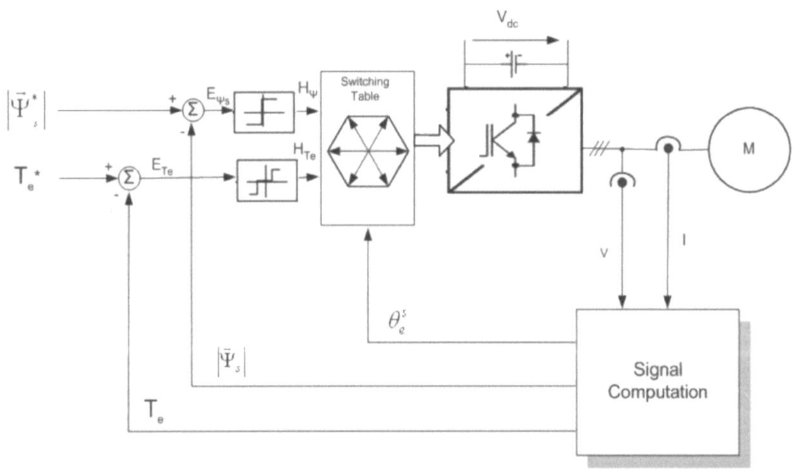

Fig. 1. Classic DTC scheme.

DTC based drives require only the knowledge of the stator resistance R_s. Measuring the stator voltage and current, stator flux vector can be estimated by the following equation:

$$\vec{\psi}_s = \int \left(\vec{V}_s - R_s \vec{I}_s \right) dt \tag{1}$$

the stator flux magnitude is given by,

$$\left| \vec{\Psi}_s \right| = \sqrt{\Psi_{\alpha s}^2 + \Psi_{\beta s}^2} \tag{2}$$

where the indicators *a,β* indicates the *a-β* stationary reference frame. The stator flux angle is given by,

$$\theta_e = \sin^{-1} \frac{\Psi_{\beta s}}{\left|\vec{\Psi}_s\right|} \tag{3}$$

and the electromagnetic torque T_e is calculated by,

$$T_e = \frac{3}{2}\left(\frac{P}{2}\right)\left(\Psi_{\alpha s} i_{\beta s} - \Psi_{\beta s} i_{\alpha s}\right) \tag{4}$$

where P is the number of machine poles.

In the DTC scheme the electromagnetic torque and stator flux error signals are delivered to two hysteresis controllers as shown in figure 1. The stator flux controller imposes the time duration of the active voltage vectors, which move the stator flux along the reference trajectory, and the torque controller determinates the time duration of the zero voltage vectors, which keep the motor torque in the defined-by-hysteresis tolerance band. The corresponding output variables H_{Te}, H_{Ψ} and the stator flux position sector $\theta_{\psi s}$ are used to select the appropriate voltage vector from a switching table scheme (Takahashi & Noguchi, 1986), which generates pulses to control the power switches in the inverter. At every sampling time the voltage vector selection block chooses the inverter switching state, which reduces the instantaneous flux and torque errors.

In practice the hysteresis controllers are digitally implemented. This means that they function within discrete time Ts. Consequently, the control of whether the torque or the flux is within the tolerance limits, often delays depending on the duration of the sampling period. This results in large ripples in the torque and the current of the motor. In conclusion, the abrupt and undesirable ripples in the electromagnetic quantities appear when the control of the values of the torque and the flux takes place at times when their values are near the allowed limits. This means that a voltage vector will be chosen which will continue to modify these quantities in a time Ts, even though these limits have been practically achieved. Accordingly, in the next control which will be carried out after time Ts, these quantities will be quite different from the desirable values. Another reason why the electromagnetic torque of the motor presents undesirable ripples is the position of the $\vec{\psi}_s$ in each of the six sectors of its transition. In general, an undesired ripple of the torque is observed when the $\vec{\psi}_s$ moves towards the limits of the cyclic sectors and generally during the sectors' change. Furthermore, the torque ripple does not depend solely on the systems conditions but on the position of $\vec{\psi}_s$ in the sector as well. Therefore, we can establish that there are more control parameters which could affect the result of the motor's behavior.

3. DTC-SVM with closed-loop torque control

In this section, the DTC-SVM scheme will be presented which uses a closed loop torque control. The block diagram of this scheme is shown in figure 2.

The objective of the DTC-SVM scheme, and the main difference between the classic DTC, is to estimate a reference stator voltage vector V^*_S in order to drive the power gates of the inerter with a constant switching frequency. Although, the basic principle of the DTC is that the electromagnetic torque of the motor can be adjusted by controlling the angle δ_{Ψ} between

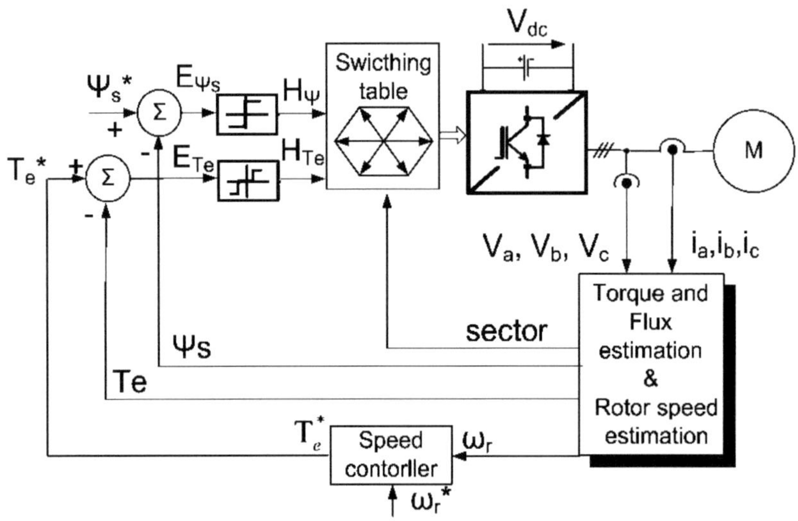

Fig. 2. DTC-SVM with closed-loop torque control

the stator and rotor magnetic flux vectors. Generally, the torque of an asynchronous motor can be calculated by the following equation.

$$T_e = \frac{3}{2}\left(\frac{P}{2}\right)\frac{L_m}{L_r L_s^{'}}|\Psi_r||\Psi_s|\sin\delta_\psi \tag{5}$$

Where $L_s^{'} = L_s L_r - L_m^2$. The change in torque can be given by the following formula,

$$\Delta T_e = \frac{3}{2}\left(\frac{P}{2}\right)\frac{L_m}{L_r L_s^{'}}|\Psi_r|\left|\vec{\Psi}_s + \Delta\vec{\Psi}_s\right|\sin\Delta\delta_\psi \tag{6}$$

where the change in the stator flux vector, if we neglect the voltage drop in the stator resistance, can be given by the following equation,

$$\Delta\vec{\Psi}_s = \vec{V}_s \Delta t \tag{7}$$

where $\Delta t = Ts$, is the sampling period.

Generally, the classic DTC employs a specific switching pattern by using a standard switching table. That means the changes in the stator flux vector and consequently the changes in torque would be quite standard because of the discrete states of the inverter. That happens because the inverter produces standard voltage vectors.

The objective of the DTC-SVM scheme, and the main difference between the classic DTC, is to estimate a reference stator voltage vector V^*_S and modulate it by SVM technique, in order to drive the power gates of the inerter with a constant switching frequency. Now, in every sampling time, inverter can produce a voltage vector of any direction and magnitude. That means the changes in stator flux would be of any direction and magnitude and consequently the changes in torque would be smoother.

According to above observations, and bearing in mind figure 2, we can see that torque controller produces a desirable change in angle $\Delta\delta_\Psi$ between stator and rotor flux vectors.

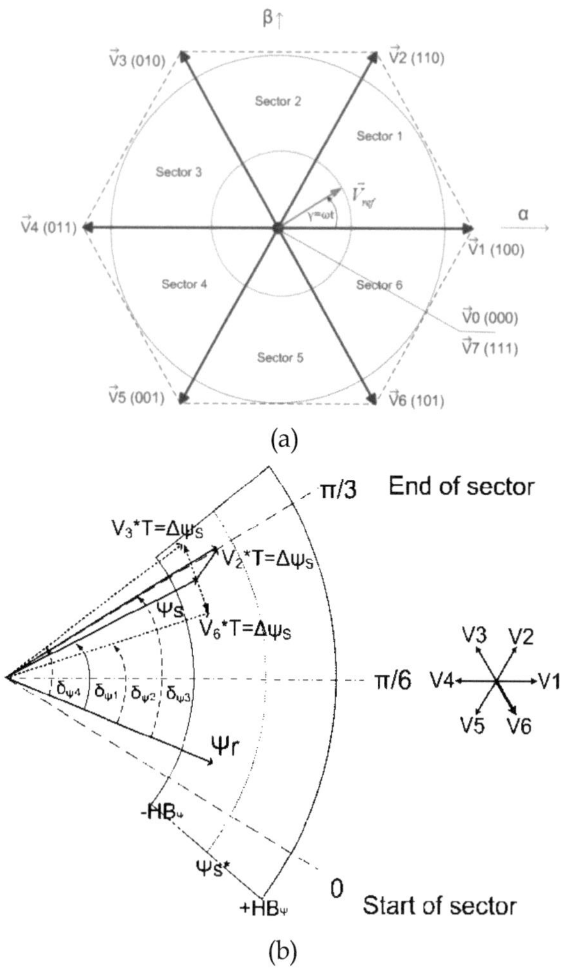

Fig. 3. Principle of Space Vector Modulation (SVPWM)
(a)reference stator vector
(b) modulation of space vector during one switching period which is equal to sampling time of the DTC-SVM method.

The change in angle $\Delta\delta_\Psi$ is added in the actual angle of stator flux vector, so we can estimate the reference stator flux vector by using the following formula, in stationary reference frame.

$$\vec{\psi}_s^* = \left|\vec{\psi}_s^*\right| e^{j\left(\omega_e t + \Delta\delta_\psi\right)} \tag{8}$$

Applying a phasor abstraction between the reference and the actual stator flux vector we can estimate the desirable change in stator flux $\Delta\Psi_S$. Having the desirable change in stator flux, it is easy to estimate the reference stator voltage vector:

$$\vec{V}_s^* = \frac{\Delta\vec{\Psi}_s}{T_S} + R_s\vec{I}_s \tag{9}$$

If the reference stator voltage vector is available, it is easy to drive the inverter by using the SV-PWM technique. So, it is possible to produce any stator voltage space vector (figure 3). As it mentioned before, in the classic DTC scheme, the direction of stator flux vector changes $\Delta\vec{\psi}_S$ are discrete and are almost in the same direction with the discrete state vectors of the inverter. Consequently, in DTC-SVM, stator flux vector changes $\Delta\vec{\psi}_S$ can be of any direction, which means the oscillations of $\vec{\psi}_S$ would be more smoother.

4. Simulation results of DTC and DTC-SVM

The DTC schemes, that are presented so far, are designed and simulated using Matlab/Simulink (figure 4). The proposed scheme is simulated and compared to the classic one. The dynamic and also the steady state behavior is examined in a wide range of motor speed and operating points.

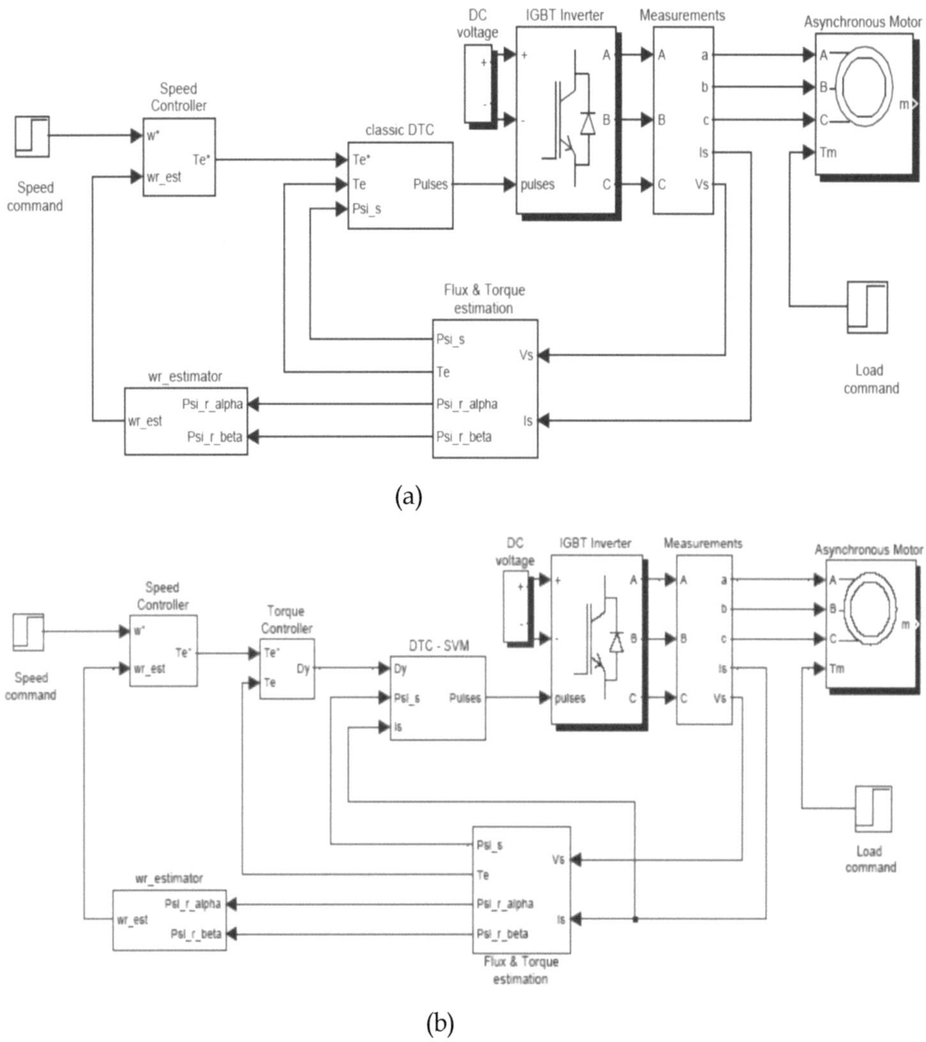

Fig. 4. Simulink models of (a) classic DTC and (b) DTC-SVM.

For simulation purposes, an asynchronous motor is used and its datasheets are shown in the following table I. The nominal values of the asynchronous motor in the simulation system are the same with the nominal values of the asynchronous motor in the experimental electrical system.

P = 4 (2 pair of poles), f = 50 Hz	R_s = 2,81 Ω	L_s = 8,4 mH
230V/ 400V	R'_r = 2,78 Ω	L'_r = 8,4 mH
P = 2,2 kW, N_r = 1420 rpm	L_m = 222,6 mH	
J = 0.0131 kgm2		

Table I. Nominal values of motor.

For the simulations a particular sampling period T_{S_DTC} for torque and flux was chosen as well as the proper limits HB_{ψ} and HB_{Te} for the hysteresis controllers, in order to achieve an average switching frequency which shall be the same with the constant switching frequency produced by the DTC-SVM control. During the simulation, the dynamic behavior of the system has been studied using both the DTC and the DTC-SVM method.

4.1 Steady state operation of the system

The results of the simulations are presented in the figure 5, where the electromechanical magnitudes of the drive system are shown, for both control schemes in various operation points. In more detail, in figure 5 the operation of the system for low speed and low load is shown and figure 6 shows the motor operation in normal mode. All the electromechanical quantities are referred to one electrical period based on the output frequency of the inverter. The average number of switching for the semiconducting components of the inverter during the classic DTC is almost the same with the number of switching of the DTC-SVM method where the switching frequency is constant. In fact, for the classic DTC flux variation of the hysteresis band equal to HB_{Ψ}=0.015 was chosen, which is almost 2% of the nominal flux and for the torque the hysteresis band controller was chosen to be HB_{Te}=0,65, which means 3% of the nominal torque. These adjustments led to an average switching number of inverter states equal to 17540 per second, for the classic DTC, while for the DTC-SVM a switching frequency equal to 2.5kHz was chosen, namely 15000 switching states per second.
The classic DTC has some disadvantages, mainly in the low speed region with low mechanical load in the shaft, where the current ripple is very high, compared to DTC-SVM (figure 5). Also, the classic DTC has variable switching frequency, where it is observed that the switching frequency is high in low speed area and low in high speeds. In practice, it is not easy to change the sampling period of the hysteresis controllers with respect to the operation point of the drive system. For this reason, a value of the sampling period is chosen from the beginning, which shall satisfy the system operation in the complete speed range. The high ripple observed in the classic DTC electrical magnitudes during the operation in low speed area, is due to the fact that many times, instead of choosing the zero voltage vector for the inverter state, in order to reduce the torque, the backwards voltage vector is chosen, which changes the torque value more rapidly.
Figure 6 shows the motor operation in normal mode. The switching frequency is also at the same value in order to have a right comparison. Current ripple has also a notable reduction in DTC-SVM compared to classic DTC. Also, at this operating point it can be seen that in classic

DTC the torque ripple of the electromagnetic torque which is resulted by the cyclic sector changes of stator flux vector and produces sharp edges, is now eliminated by using DTC-SVM.

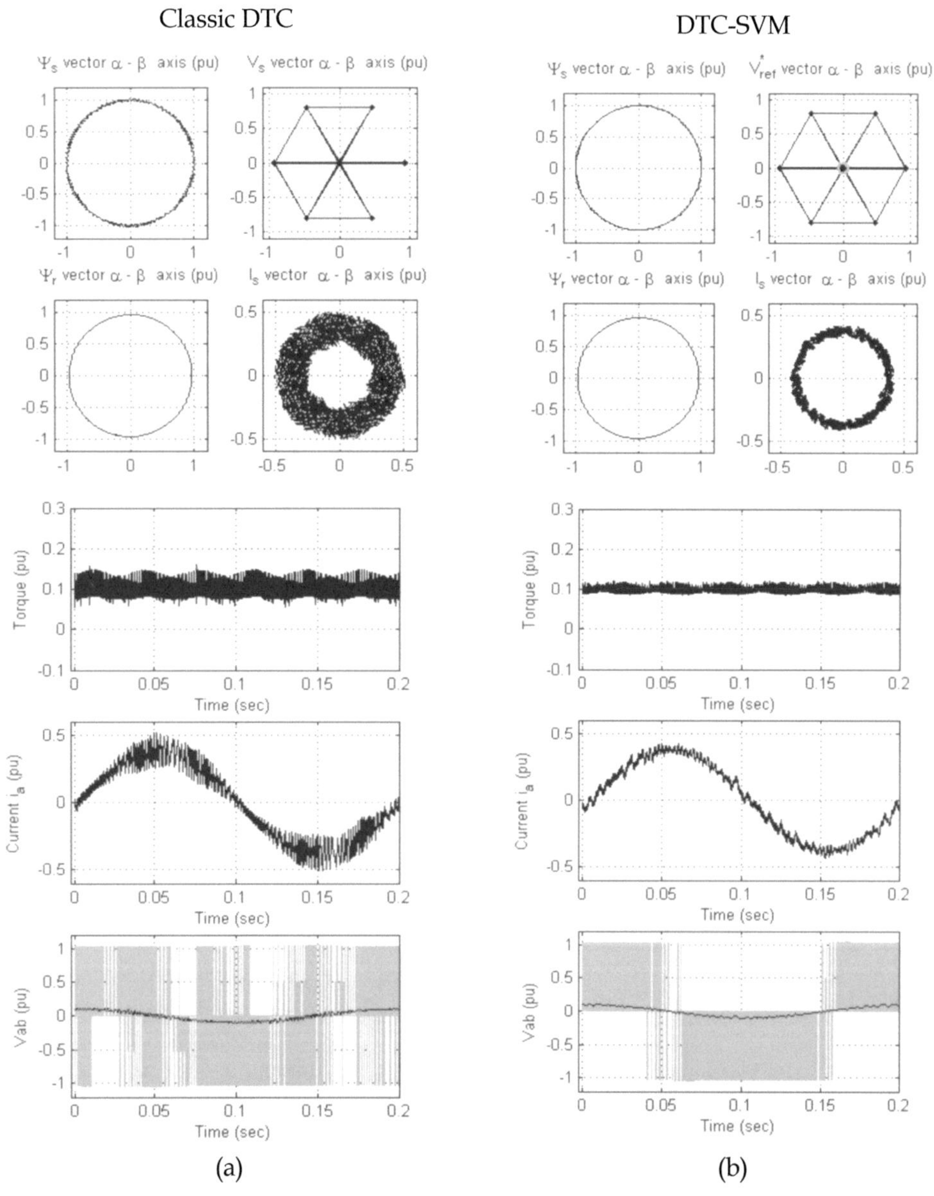

Fig. 5. Steady state of the motor in an operation point where the motor has the 10% of the nominal speed and 10% of nominal load, with $HB_{\psi} = \pm 0.015$, $HB_{Te} = \pm 0.65$

(a) Classic DTC with hysteresis band controllers and $T_{S\_DTC} = 12\mu\text{sec}$ the sampling time for discrete implementation. Inverter produces 16780 states/sec.

(b) DTC with space vector modulation. Switching frequency is equal to 2.5kH and inverter produces 15000 states/sec.

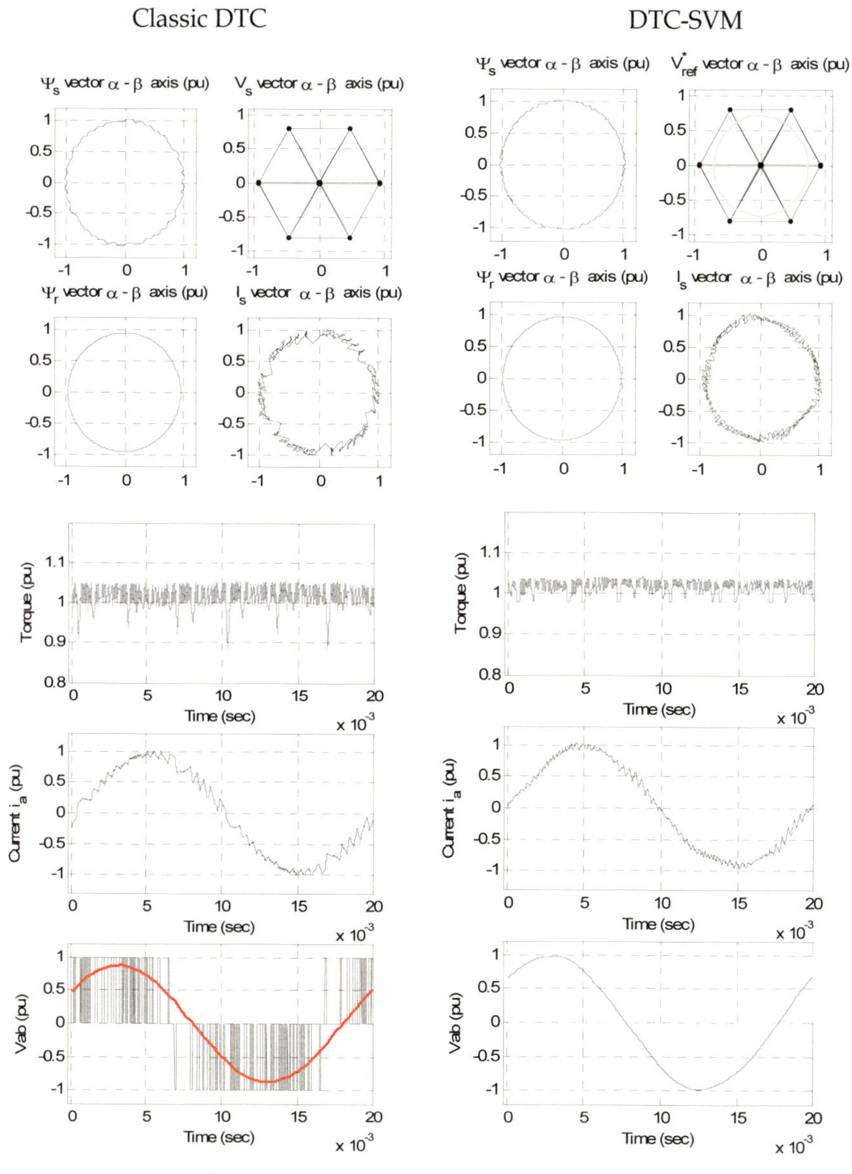

Fig. 6. Steady state of the motor in an operation point where the motor has the 100% of the nominal speed and 100% of nominal load, with $HB_{\psi} = \pm 0.015$, $HB_{Te} = \pm 0.65$ for,

(a) Classic DTC. (b) DTC with space vector modulation.

4.2 Dynamic performance of the system

In figure 7 the simulation results are presented for the dynamic case where the mechanical load is changing while the reference speed must remain constant. The case of this simulation is very rare and extreme where the motor suddenly loses the 80% of its load (from 100% to 20%) while the speed must remain constant.

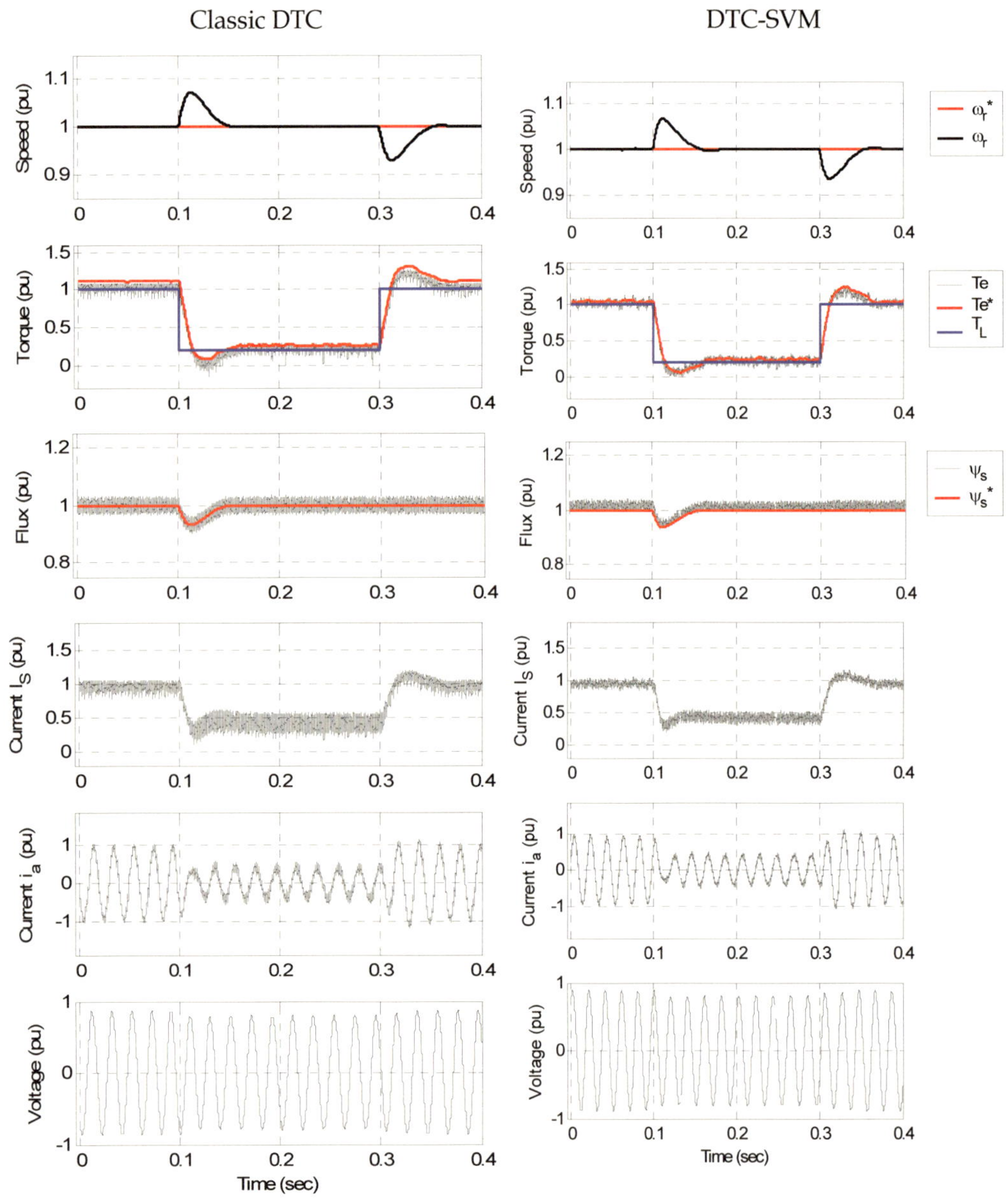

(a) (b)

Fig. 7. Load change: (a) Classic DTC, (b) DTC-SVM.

Figure 8 shows the dynamic performance of the drive system due to the reference speed step commands operation. During the transient operation of the drive system, in both cases, the ripple in electromechanical magnitudes is shown. It must be noted at this point that the speed controller, which is used for the simulations, is a fuzzy PI controller. As it is shown in figure 8 the ripple of the electromechanical magnitudes is higher in the classic DTC method in comparison to the DTC-SVM method.

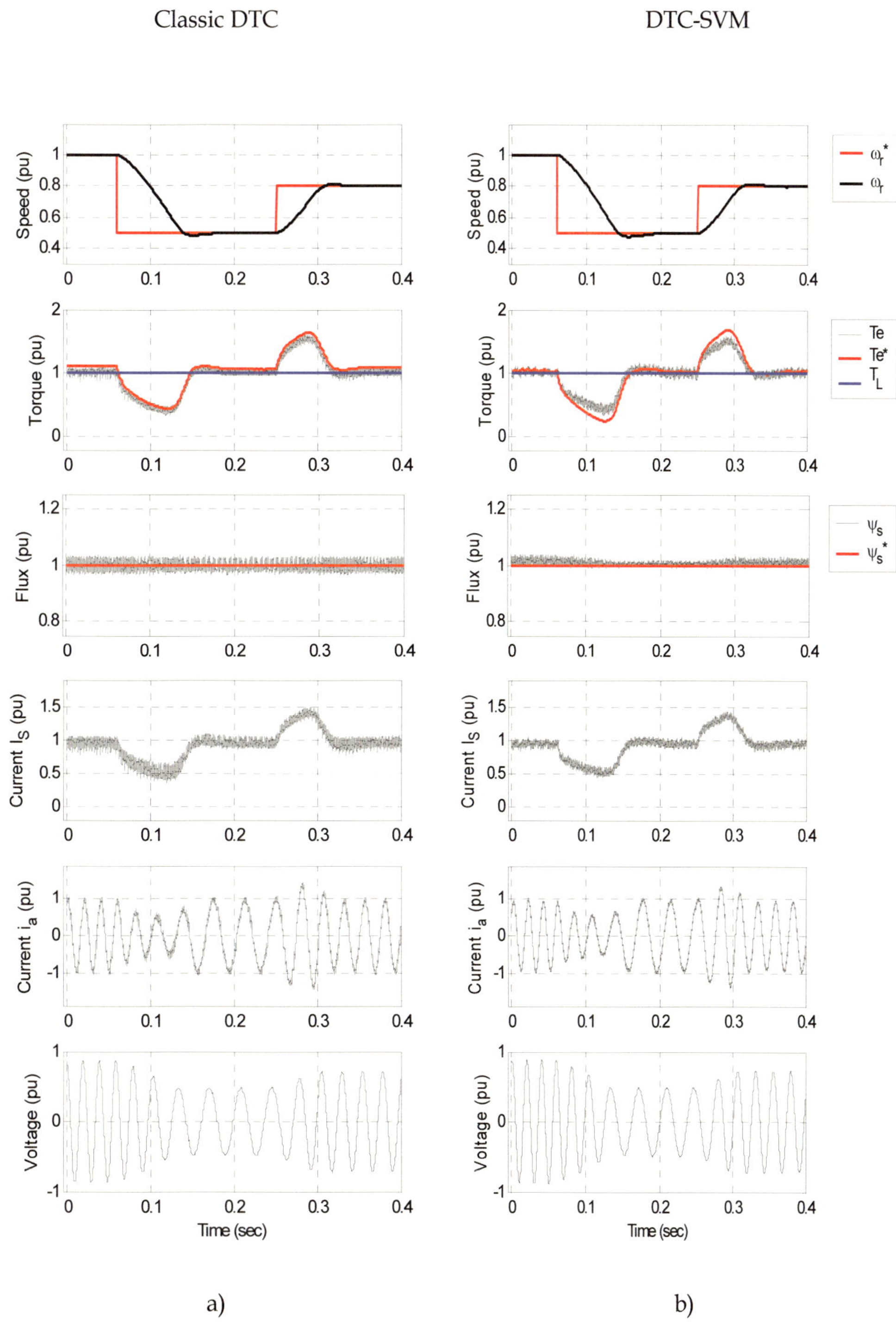

Fig. 8. Speed control response: (a) Classic DTC (b) DTC – SVM.

5. Experimental results

The implementation of the system is carried out with the development system dSPACE and the control panel R&D DS 1104 and the software package Matlab/Simulink. Also the experimental model, consists of one AC motor feed controlled converter and one DC motor feed converter which operates as a load for the system. In table II the datasheet of the experimental system kit is shown.

In figures 9a and 9b the oscillograms of the speed, torque, flux and the current components i_d and i_q of the machine are shown for reference speed variation, for both cases of control method classic DTC and DTC-SVM. In figures 10a and 10b we see the oscillograms of the electromechanical quantities of the system during load loss with classic DTC method and DTC-SVM method respectively. Where **wmegaM** is the actual value, **wref** is the reference value and the **wmega estim** is the estimated value of the angular frequency of the motor. Also the **Torque** is the actual value of the torque and the **Torque calc** is the calculated value of the electromagnetic torque. In figure 10 the **Ia ref** is the reference and **Ia** is the actual value of the DC motor's current, which is performed as a load in the experimental model. From the oscillograms it is shown that the control has more advantages in case of DTC-SVM method compared to the classic DTC.

	Asynchronous Motor	DC-Motor	Converter
Nominal Power	$P_N = 2{,}2$ kW	$P_N = 4{,}2$ kW	$P_N = 4$ kW
Nominal Voltage	$U_N = 400$ V	$U_{AN} = 420$ V	$I_N = 7$ A
Nominal Current	$I_N = 4{,}85$ A	$I_{AN} = 12{,}5$ A	
Nominal Speed	$n_N = 1420$ min^{-1}	$n_N = 2370$ min^{-1}	
Nominal power factor	$\cos\varphi_N = 0{,}82$		
Number of poles	$p = 4$		
Stator ohmic resistance Rotor ohmic resistance	$R_1 = 2{,}82\ \Omega$ $R' = 2{,}78\ \Omega$		
Stator inductance Rotor inductance of stator	$L_s = 8.4$ mH $L'_r = 8.4$ mH		
Excitation voltage		$U_{EN} = 310$ V	
Excitation Current		$I_{EN} = 0{,}93$ A	
Nominal Frequency			$f_N = 4$ kHz

Table II. Datasheets of the asynchronous motor, DC-motor and converter during the implementation

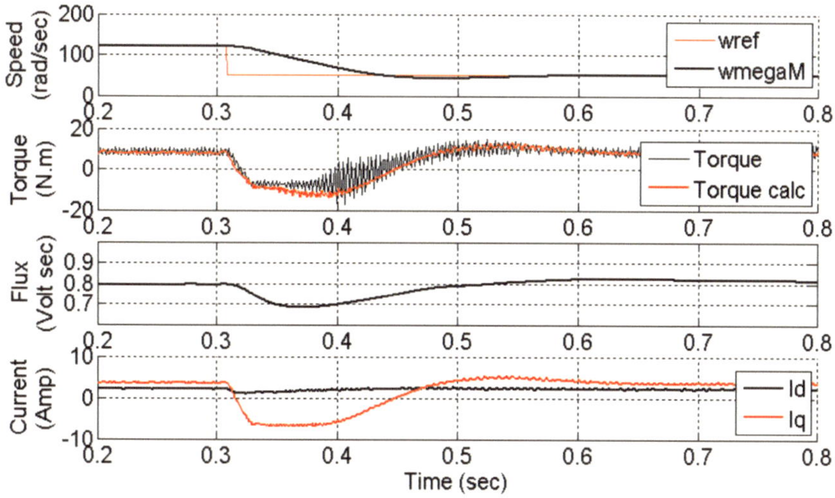

(a)

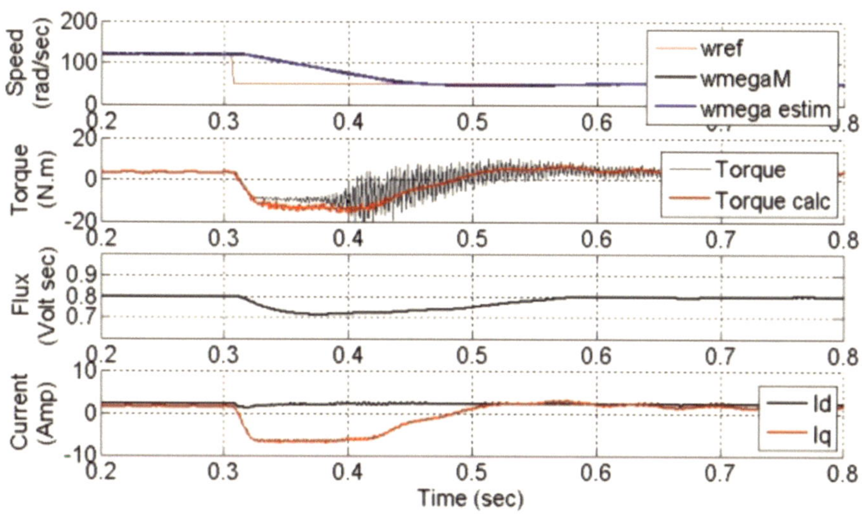

(b)

Fig. 9. Electromechanical quantities in transient operation of the system using, a) classic DTC, b) DTC-SVPWM.

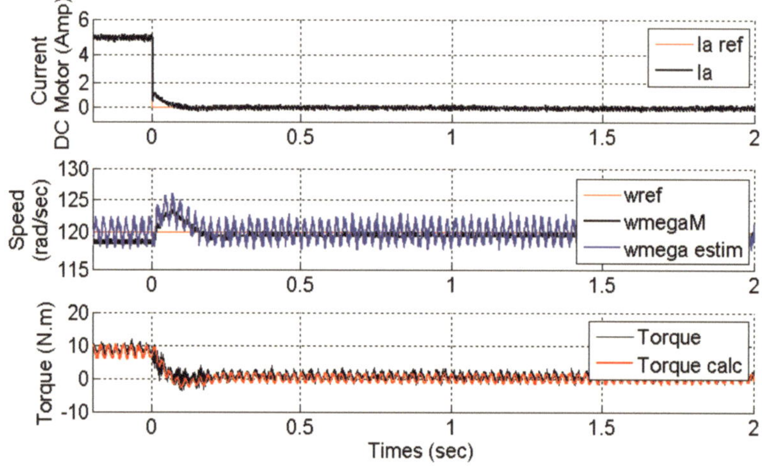

(a)

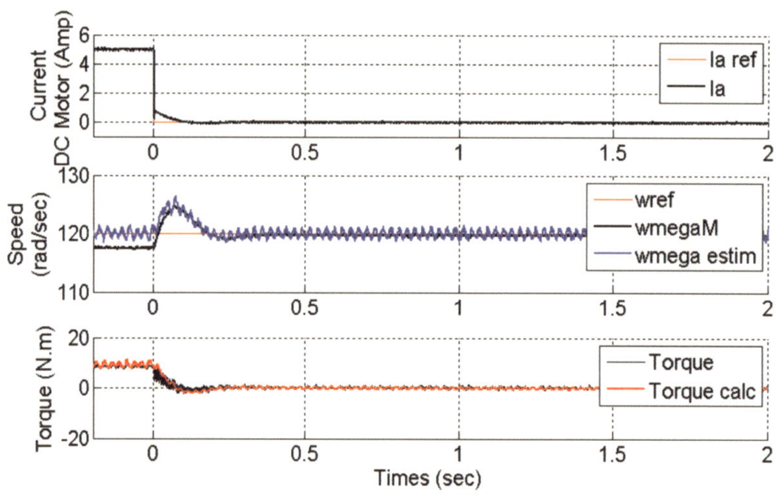

(b)

Fig. 10. Electromechanical quantities in transient operation of the system using, a) classic DTC, b) DTC-SVPWM.

6. Speed regulation using a fuzzy logic controller

So far, two methods were described for controlling the electromagnetic torque of an asynchronous motor drive. When we need to regulate the speed of such a drive a speed controller is needed. The speed controller takes the error signal between the reference and the actual speed and produces the appropriate reference torque value. That means, the drive changes mode from torque control to speed control. So, now the mechanical load on motor shaft defines the electromagnetic torque of the motor. In torque control mode the mechanical load on motor shaft defines the rotor speed. In figure 11 we can see the block diagram of the proposed drive, in speed control mode. A reference speed signal ω_r^* or, in other words, the speed command is given. The actual speed ω_r is estimated or measured with a speed encoder. This depends on the precision requirements of each application. The speed is estimated directly from state equations. The dynamic *a-b* frame state equations of a machine can be operated to compute speed signal directly [2], [4]. Consequently, the speed computation is given by:

$$\omega_r = \frac{1}{\psi_r^2}\left[\left(\psi_{ar}\frac{d}{dt}\psi_{\beta r} - \psi_{\beta r}\frac{d}{dt}\psi_{ar}\right) - \frac{L_m}{T_r}\left(\psi_{ar} i_{\beta s} - \psi_{\beta r} i_{as}\right)\right] \tag{10}$$

Where: $T_r = L_r / R_r$

This method of speed computation requires the knowledge of the machine parameters L_r, L_m, and R_r which are the rotor inductance, the magnetizing inductance and rotor resistance respectively.

The speed controller can be a classic PI controller or a fuzzy PI controller. In [Koutsogiannis], a detailed presentation and comparison of the two controllers is presented and operates with a classic DTC drive. In this paper the fuzzy PI controller is also used for the comparison between the classic DTC and DTC-SVM.

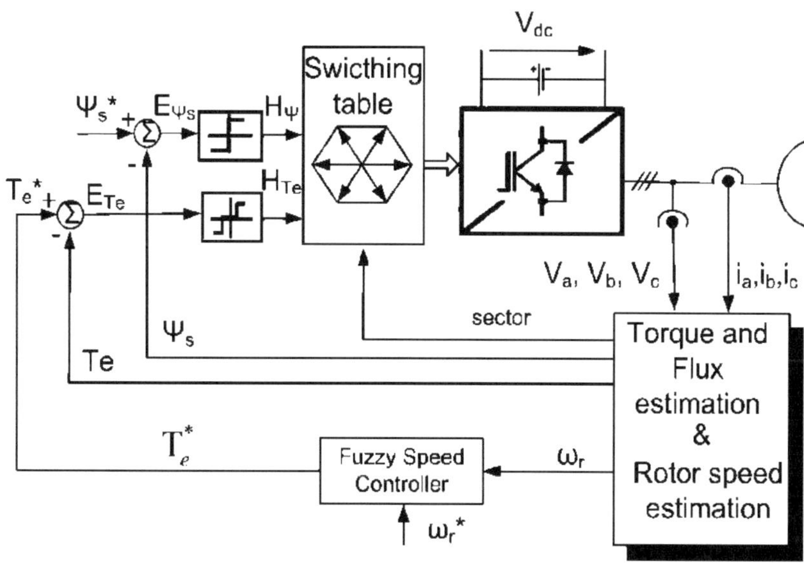

Fig. 11. Speed regulation using a speed controller.

As it will be described in the next section, the error between the estimated speed ω_r and the reference command speed ω_r^* is delivered to the speed controller, which calculates the reference electromagnetic torque T_e^*.

The corresponding output variables H_{Te}, H_{ψ} and the stator flux position sector $\theta_{\psi s}$ are used to select the appropriate voltage vector from a switching table, which generates pulses to control the power switches in the inverter. At every sampling time the voltage vector selection block chooses the inverter switching state, which reduces the instantaneous flux and torque errors.

6.1 Classic PI controller

A classic Proportional plus Integral (PI) controller is suitable enough to adjust the reference torque value T_e^*. Nevertheless, its response depends on the gains K_p and K_i, which are responsible for the sensitivity of speed error and for the speed error in steady state. During computer analysis we use a controller in a discrete system in order to simulate a digital signal processor (DSP) drive system. Its block diagram is shown in Fig.12, where T is the sampling time of the controller.

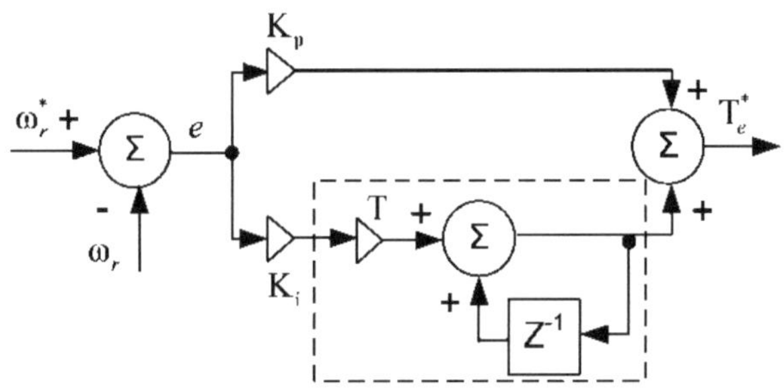

Fig. 12. Block diagram of a discrete classic PI speed controller.

The response of the PI speed controller, in a wide range area of motor speed, is very sensitive to gains *Kp* and *Ki* and it needs good tuning for optimal performance. High values of the PI gains are needed for speeding-up the motor and for rapid load disturbance rejection. This results to an undesired overshoot of motor speed. A solution is to use a variable gain PI speed controller [Giuseppe]. However, in the case of using a variable gain PI speed controller, it is also necessary to know the behaviour of the motor during start up and during load disturbance rejection in several operation points, in order to determine the appropriate time functions for PI gains. This method is also time-consuming and depends on the control system philosophy every time. To overcome this problem, we propose the use of a fuzzy logic PI controller.

6.2 Fuzzy PI controller

Fuzzy control is basically an adaptive and nonlinear control, which gives robust performance for a linear or nonlinear plant with parameter variation. The fuzzy PI speed controller has almost the same operation principles with the classic PI controller. The basic difference of the two controllers is that the output of the fuzzy PI controller gives the change in the reference torque value dT_e^*, which has to be summed or intergraded, to give the T_e^* value (Fig.13). The FL controller has two inputs, the speed error $E = \omega_r^* - \omega_r$ and the change in the speed error CE, which is related to the derivative dE/dt of error. In a discrete system, assuming $dt = T$, where T is the constant sampling time of the controller, $CE = \Delta E$. Fuzzy controller computes, for a specific input condition of the variables, the output signal.

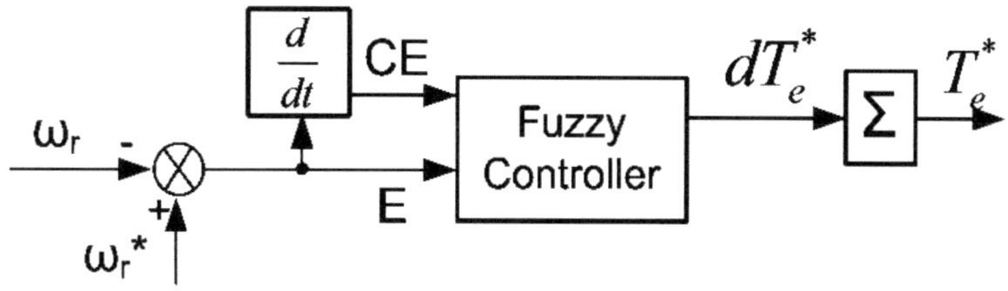

Fig. 13. Basic block diagram of a fuzzy PI speed controller.

All in all, the fuzzy controller is an input/output static nonlinear system which maps the pair values of E and CE according to fuzzy rules (Table III) and gives the following form:

$$K_1E + K_2CE = dT_e^* \tag{11}$$

Where, K_1 and K_2 are nonlinear gain factors.
The analytical block diagram of the fuzzy PI controller is shown in Fig. 14. The input variables E and CE are expressed in per unit values. This is achieved by dividing the variables by specific scale factors. The output will also be expressed in per unit values. The advantage of fuzzy control in terms of per unit variables is that the same control algorithm can be applied to all the plants of the same family. Generally, the scale factors can be constant or programmable. Programmable scale factors control the sensitivity of operation in different regions of control.
The fuzzy rules of the controller are based in linguistic expressions from the physical operation of the system. As mentioned before, the output of the controller is the change in the reference torque value dT_e^*. The fuzzy sets of linguistic expressions of the variables and the membership functions (MFs) of these variables are shown in Fig.15 (a),(b). As mentioned before, the output of the controller is the change in the reference torque value dT_e^*. The MFs of the output variable in per unit values are shown in Fig.15(c). The definition of the MFs depends on the system behavior. All the MFs are asymmetrical because near the origin (steady state), the signals require more precision.
The next step in the analysis of fuzzy speed controller is the definition of fuzzy rules. The fuzzy rules for the speed controller are shown in Table III. We can see that the top row and the left column of the matrix indicate the fuzzy sets of the variables e and ce, respectively, and the MFs of the output variable $dT_e^*(pu)$ are shown in the body of the matrix. There may be 7 x 7 = 49 possible rules in the matrix, where a typical rule reads as:

$$IF\ \ e(pu) = PS,\quad AND\ ce(pu) = NM,\quad THEN\ \ \Delta T_e^*(pu) = NVS \tag{12}$$

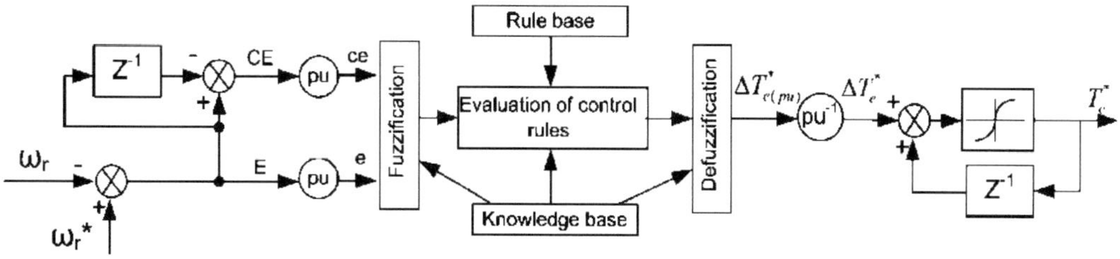

Fig. 14. An analytical discrete block diagram of the fuzzy PI controller.

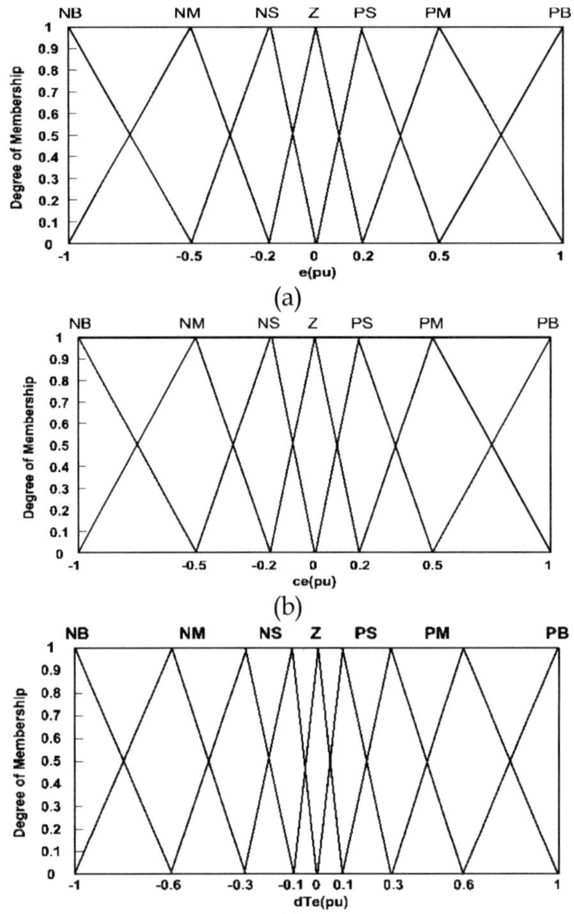

(c)

Fig. 15. Membership functions of the input variables (a) speed error e(pu) (b) change in speed error ce(pu) and of the output variable (c) change in reference torque value ΔTe^* of the fuzzy PI speed controller.

e(pu) / ce (pu)	NB	NM	NS	Z	PS	PM	PB
NB	NB	NB	NB	NM	NS	NVS	Z
NM	NB	NB	NM	NS	NVS	Z	PVS
NS	NB	NM	NS	NVS	Z	PVS	PS
Z	NM	NS	NVS	Z	PVS	PS	PM
PS	NS	NVS	Z	PVS	PS	PM	PB
PM	NVS	Z	PVS	PS	PM	PB	PB
PB	Z	PVS	PS		PB	PB	PB

where, PB = Positive Big, PM = Positive Medium, PS = Positive Small, PVS = Positive Very Small, Z = Zero, NVS = Negative Very Small, NS = Negative Small, NM = Negative Medium, NB = Negative Big.

Table III. Fuzzy Rules

The implication method that we used in simulations is the Mamdani type. There are many types of fuzzy logic controllers, but now the classical structure of Mamdani type is used.

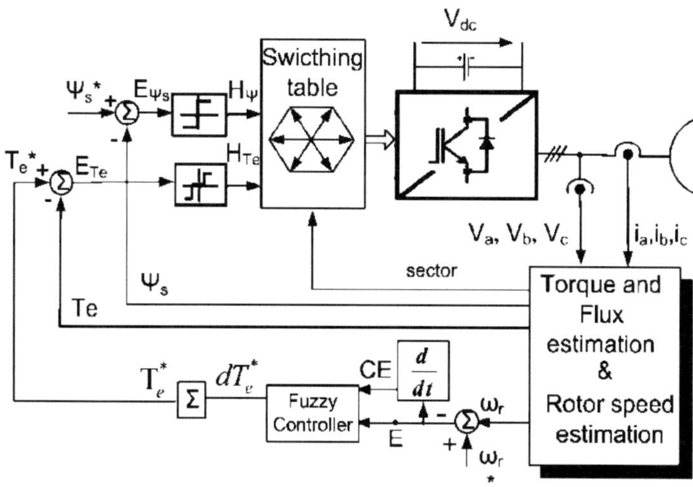

Fig. 16. DTC block diagram with fuzzy logic controller.

The rule matrix and MF description of the variables are based on the knowledge of the system. Summing up, the setting of the fuzzy controller depends on the system requirements for optimal performance. When the fuzzy speed controller is well tuned its performance is excellent in a wide range of motor speed. Fig 16 shows the block diagram of Direct Torque Control Induction Motor Drive using a Fuzzy Speed Controller.

6.3 Simulation results

In this paragraph, the simulation results of a system using the classic PI speed controller (Fig.17.I) and the fuzzy PI speed controller (Fig.17.II) are presented. For the needs of simulation, we used an 160kW, 400V, 50Hz, 1487rpm, $R_s = 0.01379\Omega$ induction motor which is fed by a VS inverter using the DTC method. In more detail the parameters of motor are shown in Table IV,

P=4(2 pair of poles), f=50 Hz	R_s=0,0137 Ω	L_s=0,007705 H
230V/400V	$R_r = 0.00728\Omega$	L_s=0,007705 H
P=160kW, $N_r = 1487$ rpm	L_m=0,00769 H	
J=0.02 Kgm^2	F=0,05658 Nms	

Where *J* is the machine's inertia, and *F* is the friction factor.

Table IV. Induction Machine Parameters

To simulate the drive we used Matllab/Simulink software. The DTC sampling time was 30μs and the speed controller sampling time was 3ms. The reference stator flux magnitude was constant at 1.02 Wb. Fig.17(I) shows the dynamic performance of the DTC-SVM drive using a classic PI speed controller. The results of Fig.17 show that the startup of the motor, until it reaches the command speed value 600 rpm, is made with 400 Nm initial load torque. When the motor runs at 600 rpm/400Nm steady state operation, a step speed command of

1200 rpm is given to the drive and the motor reaches again another operation point (1200rpm/400Nm). Finally, the controllers are tested to step load torque disturbance. It is easy, therefore, to come to the conclusion that fuzzy speed controller has a remarkably better response than the classic PI speed controller.

The system was also investigated during the starting period and its control under different commutative periods. In this fig. 17 it is shown that the torque of the motor has lower ripple when the speed estimation is being carried out using a fuzzy PI controller.

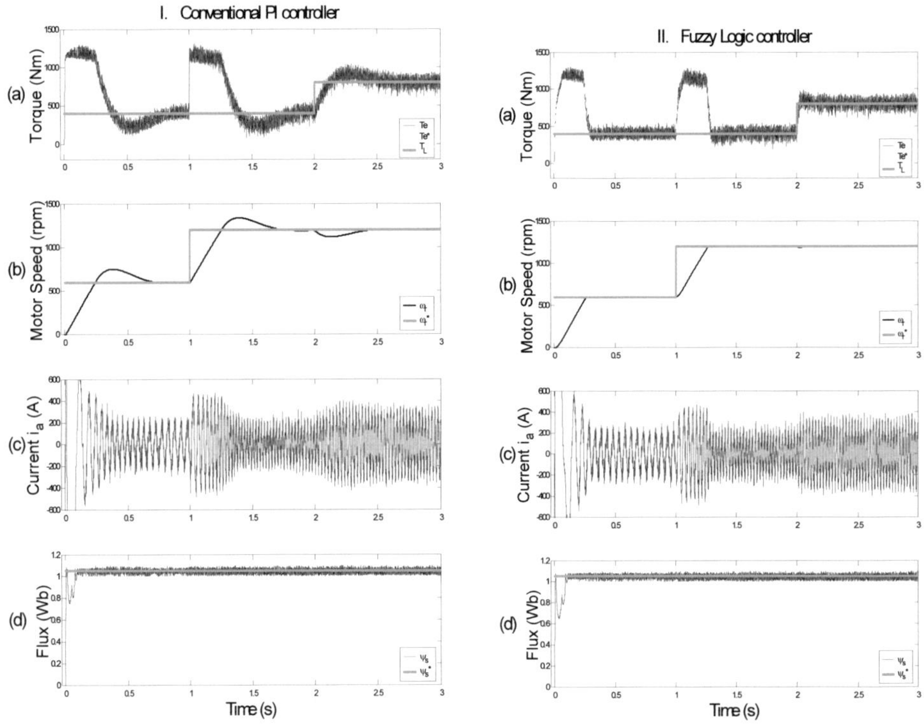

(a) (b)

Fig. 17. Motor control response with steps of speed command and load torque. (a) Electromagnetic torque T_e, speed controller output T_e^*, load torque *TL*, (b) actual motor speed ω_r, reference speed ω_r^*,(c) stator current i_{sa} in phase a (d) stator flux magnitude Ψ_s, *and* reference value Ψ_s^*.

Fig. 18 shows, in more detail, the comparison of the motor speed response using the two different speed controllers, during steps of speed command ω_r^* and load torque. To investigate the system for the classic PI controller more than one pairs of values Kp and Ki have been used. The two controllers were tested in a wide range of engine speed. Extending, namely, from a very low speed to a very high speed of the motor. It was observed, that the fuzzy PI controller has better performance than the classic PI controller.

In fig. 19 we observe that the acceleration of the motor using the classic PI speed controller is almost the same, independently of command step, and generally a linearity is observed, which depends only on the load on the axis of motor. In other words we have the maximum acceleration of the motor under these conditions. This means that when we have a small

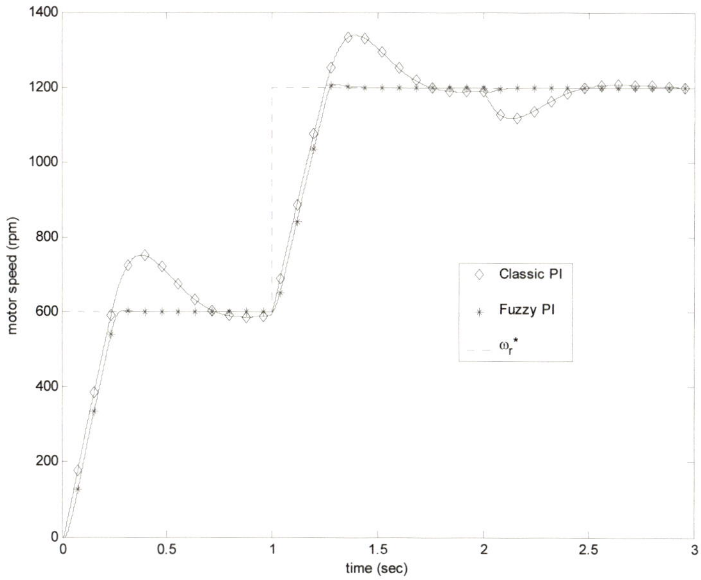

Fig. 18. Motor speed control response with steps of speed command and load torque.

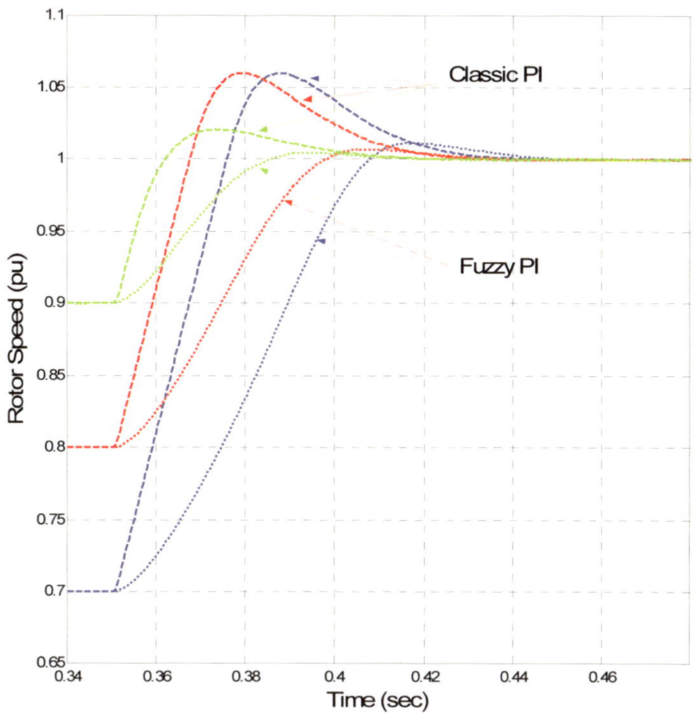

Fig. 19. Dynamic behaviour of classic PI and Fuzzy PI controller during motor startup. Load in the shaft of the motor equal with 50% nominal and various step changes of speed.

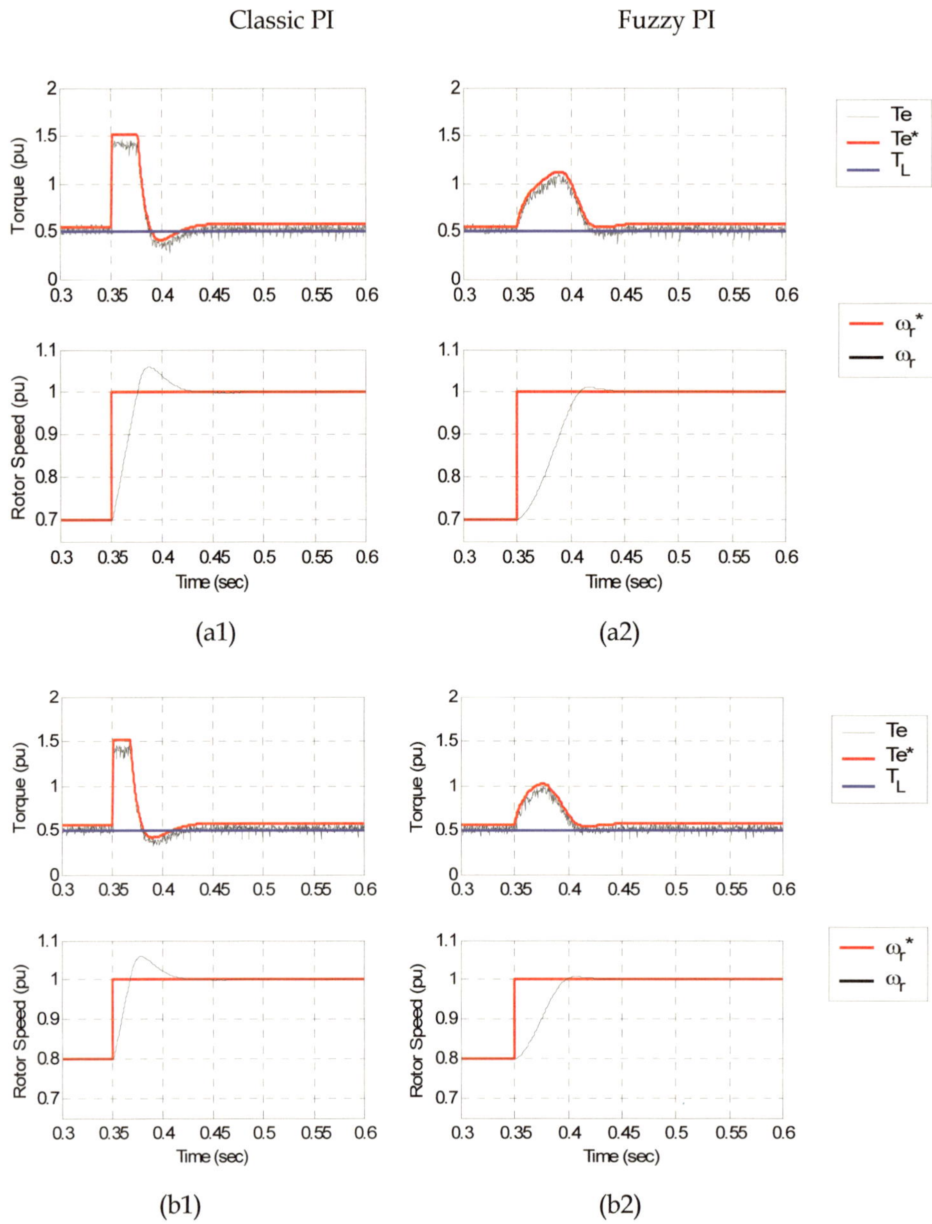

Fig. 20. Simulation results of the speed controller response in various speed step commands. (1) Classic PI controller, (2) Fuzzy PI controller. (a) 30%, (b) 20%

load in the shaft of the motor and the step is small, then an overshoot in the speed of the motor is observed. On the contrary, with the fuzzy PI of controller, we observe an acceleration that depends on the step of command and the load on the shaft. In fig. 20 an analytical comparison of the dynamic performance of the control system is presented. The system behavior can be studied when the motor speed increases, while the load torque in the motor shaft remains constant at 50% of the nominal load. In more detail, the dynamic performance of the two speed controllers, classic PI and fuzzy PI, is presented during increase of the motor speed by 30%, 20% and 10% step commands of the nominal speed respectively. In this figure, the improvement in motor acceleration and the change in motor torque using the fuzzy PI controller can be seen. Classic PI controller shows an undesirable overshoot of the actual speed. On the other hand, fuzzy PI controller has a smoother response. The output of each controller is the value of the reference electromagnetic torque T_e^* . The change in motor speed is the result of applying the produced reference torque to the DTC scheme.

7. Direct torque control for three level inverters

7.1 Control strategy of DTC three-level inverter

The applications of inverter three or multiple level inverters have the advantage of reducing the voltage at the ends of semiconductor that mean the inverters can supply engines with higher voltage at the terminals of the stator. Also, the three level inverters show a bigger number of switching states. A three level inverter shows 3^3=27 switching states. This means an improvement in the higher harmonics in the output voltage of the inverter and hence fewer casualties on the side of the load and less variation of electromagnetic torque. In direct torque control for a two-level inverter there is no difference between large and small errors of torque and flux. The switching states selected by the dynamics of drive system with the corresponding change of desired torque and flux reference is the same as those chosen during the operation in steady state. For the three-level voltage inverter is a quantification of the input variables. In this case, increasing the torque on the control points of the hysteresis comparators in five (Figure 21) and the three magnetic flux (Figure 22). Also divided the cycle recorded by electromagnetic flow of the stator in a rotating, in 12 areas of 30° as shown in Figure 23. This combined with the increased number of operational situations, for three-level inverter is 27 and is expressed in 19 different voltage vectors can be achieved better results. Figure 24a shows the 19 voltage vectors for the three level voltage source inverter of figure 25, and the vector of magnetic flux of the stator Ψs. It should be noted that in Figure 24a vectors V1, V2, V3, V4, V5, V6 shown each for two different operating conditions and the zero vector V0 for three different situations. The angle the vector i in relation to the axis a is less than 30°. The possible changes in magnetic flux stator which can be achieved using the voltage vectors of operating conditions shown in 24b.

From Figures 24.a and 24b seems to change the value of stator flux **Ψs** in a new value should be selected the following voltage vectors. If an increase in the flow can be achieved by applying one of the voltage vectors **V9, V2, V8, V1, V7,** because in this case, the new vector of stator flux will be correspondingly **Ψs+ΔΨ9, Ψs+ΔΨ2, Ψs+ΔΨ8, Ψs+ΔΨ1, Ψs+ΔΨ7.** By the same token if we can achieve a reduction of magnetic flux should implement one of the voltage vectors **V14, V5, V15,** since in this case the new vector of stator flux is είναι **Ψs+ΔΨ14, Ψs+Ψ5, Ψs+ΔΨ15,** which is less than the original **Ψs.** Also for the electromagnetic torque, taking into account the equation 6, if is necessary very sharp increase in torque, then we can apply one from the voltage vectors **V11, V3, V12** because it will grow along with the flow and the angle between the vectors δ of stator flux and the rotor. If a reduction of the torque is needed we can apply one from the voltage vectors **V6, V17, V18.** By the same token if is required large increase in flow and a slight increase in torque can do a combination of the above and apply the vector **V8** or if stator magnetic flux is constant and requires a small reduction of the torque is needed can be chosen one from

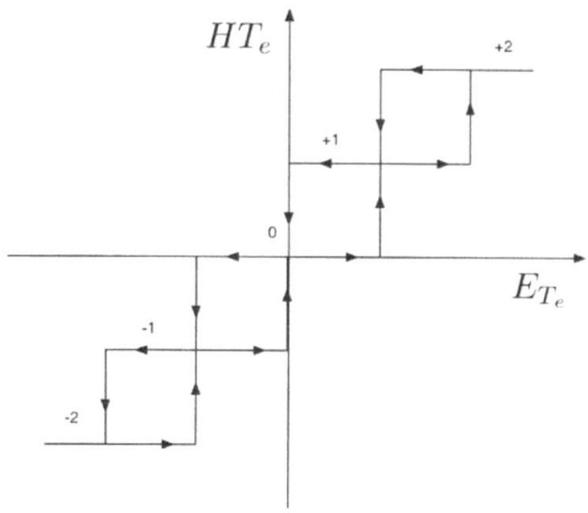

Fig. 21. Hysteresis comparator 5 level for the electromagnetic flux

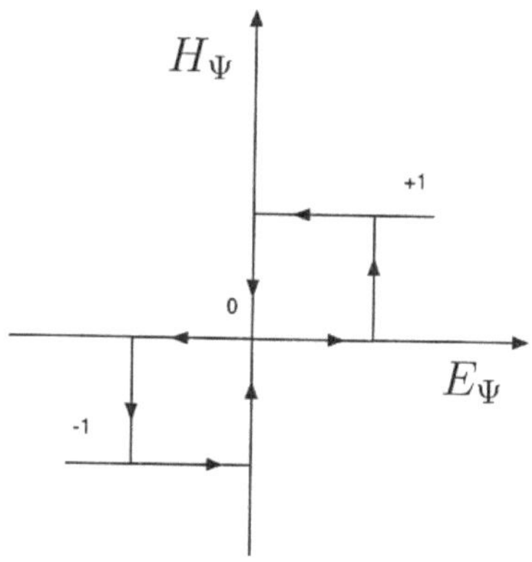

Fig. 22. Hysteresis comparator 3 level for the magnetic flux

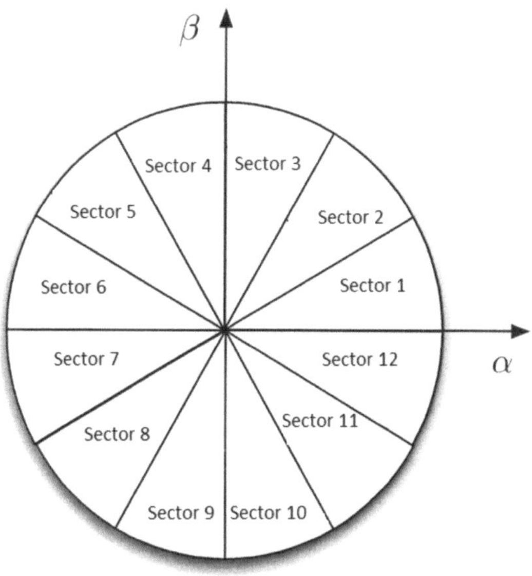

Fig. 23. Sectors of Statorsmagnetic flux.

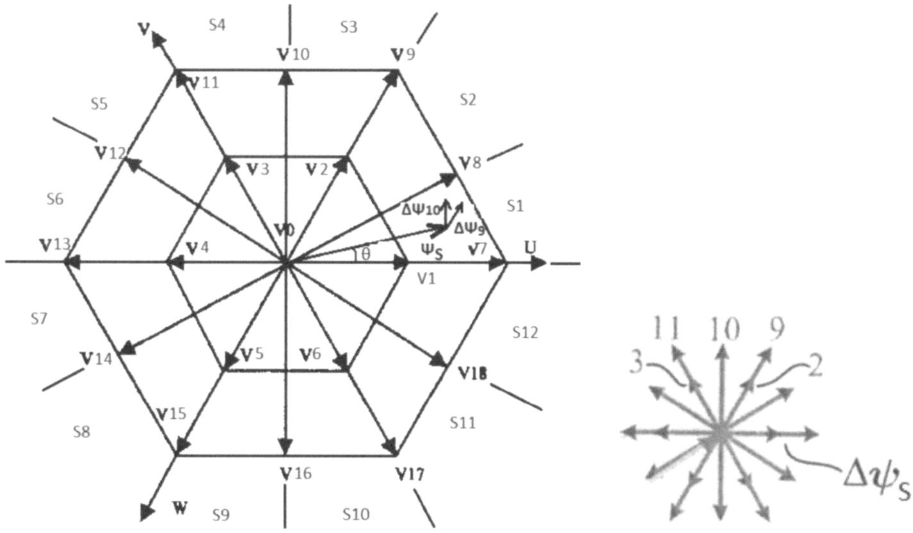

(a) (b)

Fig. 24. a) voltage vectors of 3 level voltage b) changes of the stator's flux with the vector of each switching state.

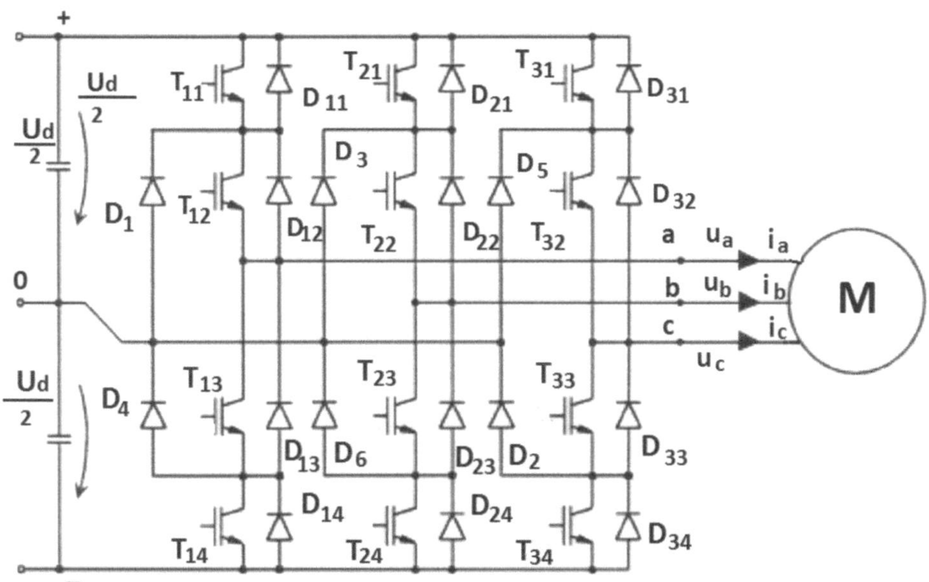

Fig. 25. Three- level voltage source inverter

zero voltage vectors **V0**. Of course the number of vectors that can bring the desired change in magnetic flux in stator and electromagnetic torque varies to the angle the vector of magnetic flux on the axis A. As is natural in such cases there are other suitable candidate voltage vectors. The correct choice of the vectors, depending on the desired change in the flow and torque that we want to do, depending on the sector in which the vector of the flow,

it is the biggest challenge to build such a table in direct torque control for drive systems powered by three-level voltage inverters. So the inverter three-level table is not widely accepted for pulsing as in the case of two-level inverters.
Based on the above logic while taking into account the intersection of Figure 3 in which may be in the vector of the stator magnetic flux, it became the table I.

Flux(ψ_S)	Torque(T_e)	S1	S2	S3	S4	S5	S6	S7	S8	S9	S10	S11	S12
	-2	V0	V0	V0	V0	V0	V0	V0	V0	V0	V0	V0	V0
	-1	V3	V4	V4	V5	V5	V6	V6	V1	V1	V2	V2	V3
-1	0	V3	V4	V4	V5	V5	V6	V6	V1	V1	V2	V2	V3
	1	V11	V12	V13	V14	V15	V16	V17	V18	V7	V8	V9	V10
	2	V11	V12	V13	V14	V15	V16	V17	V18	V7	V8	V9	V10
	-2	V0	V0	V0	V0	V0	V0	V0	V0	V0	V0	V0	V0
	-1	V2	V3	V3	V4	V4	V5	V5	V6	V6	V1	V1	V2
0	0	V2	V3	V3	V4	V4	V5	V5	V6	V6	V1	V1	V2
	1	V10	V11	V12	V13	V14	V15	V16	V17	V18	V7	V8	V9
	2	V10	V11	V12	V13	V14	V15	V16	V17	V18	V7	V8	V9
	-2	V0	V0	V0	V0	V0	V0	V0	V0	V0	V0	V0	V0
	-1	V2	V3	V3	V4	V4	V5	V5	V6	V6	V1	V1	V2
1	0	V2	V3	V3	V4	V4	V5	V5	V6	V6	V1	V1	V2
	1	V9	V11	V11	V13	V13	V15	V15	V17	V17	V7	V7	V9
	2	V9	V11	V11	V13	V13	V15	V15	V17	V17	V7	V7	V9

Table I

7.2 Simulation of the system in the computer

The drive system considered consists of three-phase asynchronous motor, three phase three level voltage inverter and control circuit with hysteresis comparators electromagnetic torque and flux of Figures 21 and 22 respectively. The system design was done by computer simulation with Matlab / Simuling. Figure 26 shows the general block diagram of the simulation.
By simulating the drive system on the computer can pick up traces of electromechanical sizes in both permanent and transition state in the system. From the curves can be drawn for the behavior of both the load response and the response speed. Details of the induction motor and inverter with three levels that will make computer simulations are shown in Tables II and III respectively.

7.3 Simulation resuls

In this text we will present the waveforms of electromechanical changes in the size of the load. To investigate the behavior of the electric drive system in response to load change incrementally load of 25 Nm to 30Nm, then by 30Nm to 25 Nm, maintaining the engine speed steady at 1000 rpm. Figure 27 shown the electromagnetic torque and Figure 8, the engine speed according to the time when the transition state in which they affect the load.

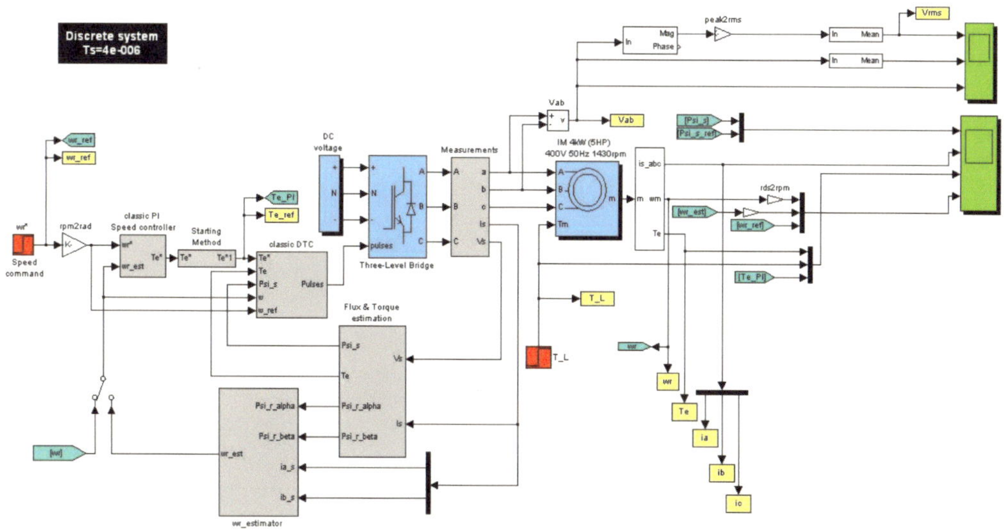

Fig. 26. Block diagram DTC Three-level Inverter in the Simulink with speed estimator.

Nominal power	**P = 4000W**
Stator phase voltage	**V = 460 V**
Ohmic resistance of stator	**R_s = 1.405 Ω**
Ohmic resistance of rotor	**R_r = 1.395 Ω**
Main magnetic induction	**L_m = 172.2x10^{-3} H**
Stator leakage inductance	**L_{ls} = 5.84x10^{-3} H**
Motor leakage inductance	**L_{lr} = 5.84x10^{-3} H**
Leakage torque	**J = 0.0131 Kg.m^2**
Coefficient of friction	**F = 0.002985 Nms**
Number of poles	**P = 4 (two pairs of poles)**

Table II. Nominal details of induction motor

Semiconductor	**IGBT with antiparallel diodes**
Ohmic resistance Snubber	**R_s = 1000 Ω**
Capacitance Snubber	**C_s = infinite**
Internal resistance semiconductor	**Ron = 0.001 Ω**
IGBT voltage crossing	**V_f = 0.8 V**
Diode voltage crossing	**V_f = 0.8 V**

Table III. Nominal details of inverter

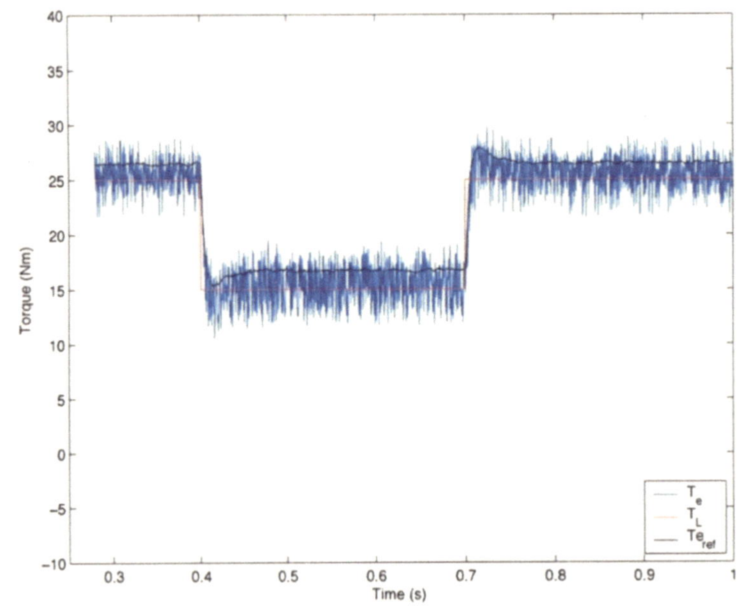

Fig. 27. Electromagnetic flux, reference flux and load flux versus time

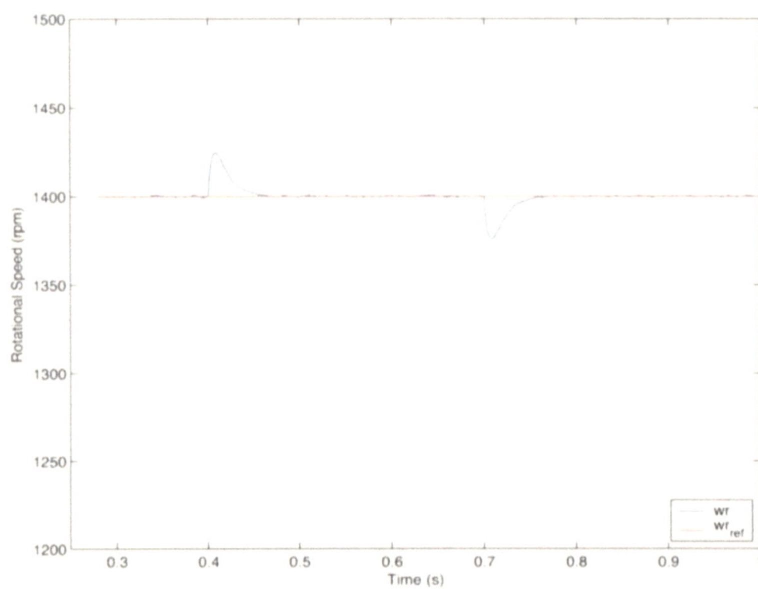

Fig. 28. Speed reference and actual speed versus time

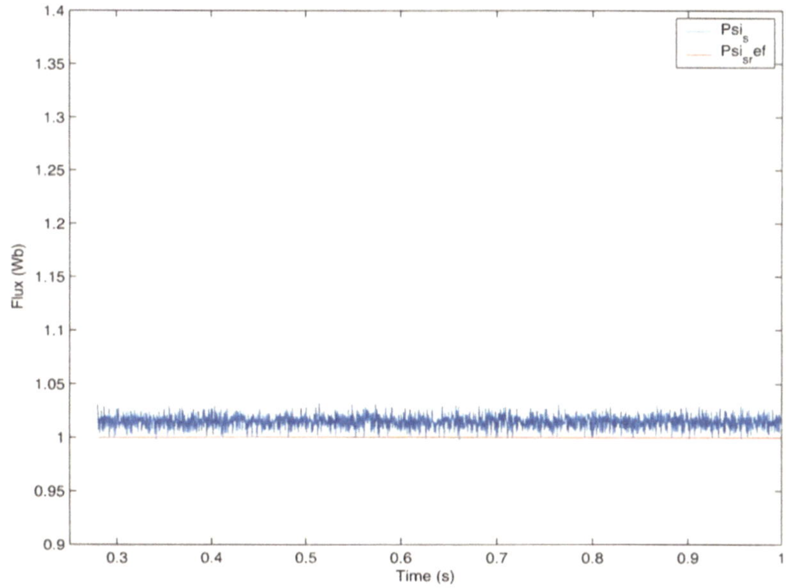

Fig. 29. Electromagnetic stator flow versus time.

By changing the load observed a slight, temporary change of speed. Figure 9 shows the change of the stator flux versus time and Figure 30 is the change of magnetic flux in the stator three-axis system that is α,β system versus time. Figure 31 shows the change of the vector current in the stator system. In this figure shows the change of the modulus of vector power to change the load. When the torque load is reduced and the measure of the vector current and increase the vector of power when the load increases.

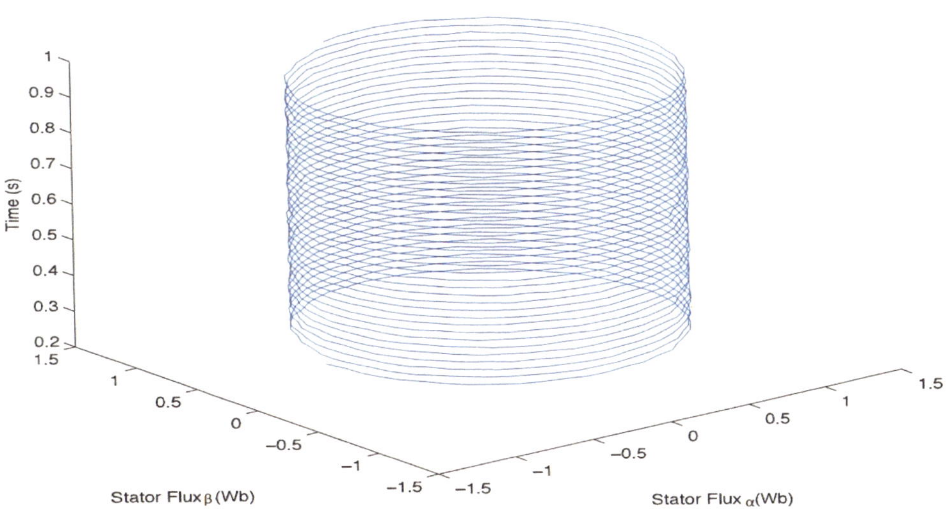

Fig. 30. Electromagnetic flow in the stator ιv α,β system is a function of time

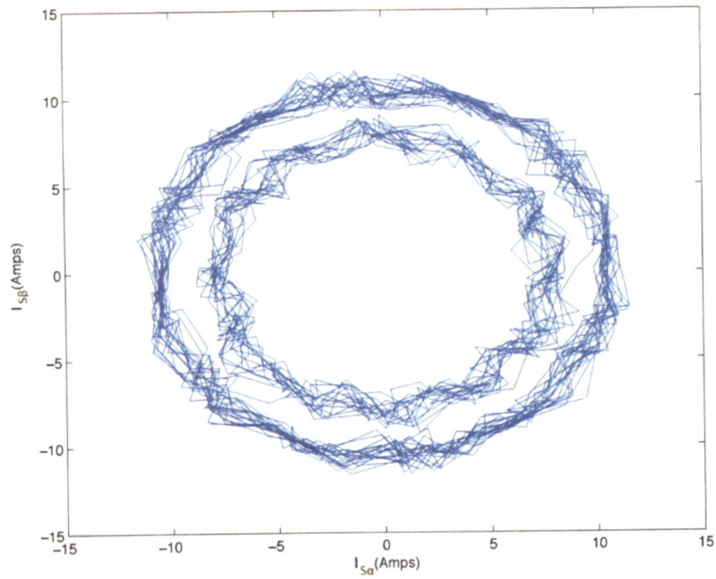

Fig. 31. Current in the stator in α,β reference system

8. Conclusion

This paper has presented a modified Direct Torque Control method for PWM-Inverter fed asynchronous motor drive using constant switching frequency. Constant-switching-frequency is achieved by using space vector modulation and finally, an SVM based DTC system, compared to the classic DTC scheme for torque control. DTC-SVM schemes improve considerably the drive performance in terms of reducing torque and flux pulsations, reliable startup and low-speed operation, well-defined harmonic spectrum, and radiated noise. Therefore, DTC-SVM is an excellent solution for general-purpose asynchronous motor drives. On the contrary, the short sampling time required by the classic DTC schemes makes them suited to very fast torque- and flux-controlled drives because of the simplicity of the control algorithm. When a speed control mode instead of torque control is needed, a speed controller is necessary for producing the reference electromagnetic torque value. For this purpose a fuzzy logic based speed controller is used. Fuzzy PI speed controller has a more robust response, compared to the classic PI controller, in a wide range area of motor speed.

9. References

Takahashi I. & Noguchi T.(1986): *A new quick-response and high efficiency control strategy of an induction machine*, IEEE Trans. Ind. Applicat., vol. 22, pp. 820–827, Sep./Oct.

Bimal K. & .Bose (2002). *Modern Power Electronics And AC Drives*. Prentice Hall.

Andrzej M. & Trzynadlowsky (2002).*Control of Induction Motors*. Academic Press.

Boldea I. &.Nasar S.A. (1998). *Electric Drives*, CRC Press,.

Casadei D., et al.,(2002). *FOC and DTC: Two viable schemes for induction motors torque control*, IEEE Trans. Power Electron., vol. 17, pp. 779–787, Sept..

Casadei D., et al., (2000). *Implementation of a Direct Torque Control Algorithm for Induction Motors Based on Discrete Space Vector Modulation*, IEEE Trans. Ind. Applicat., vol. 15, No.4 pp. 769–777, July.

Giuseppe S. et al. (2004). *Diret Torque Control of PWM Inverter-Fed AC Motors – A survey*, IEEE Trans. Ind. Applicat., vol. 51, pp. 744–757, Aug..

Koutsogiannis Z., et al., (2006). *Computer Analysis of a Direct Torque Control Induction Motor Drive Using a Fuzzy Logic Speed Controller*, XVII International Conference on Electrical Machines, Sept..

Brahmananda T. et al, (2006), *Sensorless Direct Torque Control of Induction Motor based on Hybrid Space Vector Pulsewidth Modulation to Reduce Ripples and Switching Losses – A Variable Structure Controller Approach*, IEEE Power India Conference, 10-12 Apr..

Chen L., et al., (2005). *A scheme of fuzzy direct torque control for induction machine*, IEEE Proceedings of the Fourth International Conference on Machine Learning and Cybernetics, Guangzhou, 18-21 Aug..

Mitronikas E. & Safacas A., (2004). *A Hybrid Sensorless Stator-Flux Oriented Control Method for Induction Motor Drive*, in 35th Annual IEEE Power Electronics Specialists Conference (PESC'04), June 20-25, , Aachen, Germany, pp. 3481-3485.

Grabowski P., (2000). *A Simple Direct Torque Neuro Fuzzy Control of PWM Inverter Fed Induction Motor Drive*, IEEE Trans. Ind. Electron., Vol. 47, No. 4, pp. 863-870, Aug..

Romeral L., et al. (2003). *Novel Direct Torque Control (DTC) Scheme With Fuzzy Adaptive Torque-Ripple Reduction*, IEEE Trans. Ind. Electron., vol.50, pp.487–492,Jun..

Ortega M., et al., (2005). *Direct Torque Control of Induction Motors using Fuzzy Logic with current limitation*, IEEE Industrial Electronics Society, IECON 2005. 32nd Annual Conference.

Mitronikas E. & Safacas A., (2001). *A New Stator Resistance Tuning Method for Stator-Flux-Oriented Vector-Controlled Induction Motor Drive*, IEEE Transaction on Industrial Electronics, Vol. 48, No. 6, December 2001, pp. 1148 - 1157.

Mitronikas E. & Safacas A., (2005) . *An improved Sensorless Vector Control Method for an Induction Motor Drive*, IEEE Transaction on Industrial Electronics, Vol. 52, No. 6, December, pp. 1660-1668.

Miloudi A., et.al., (2004). Simulation and Modelling of a Variable Gain PI Controller For Speed Control of a Direct Torque Neuro Fuzzy Controlled Induction Machine Drive, in: proceedings of 35th Annual IEEE Power Electronics Specialists Conference (PESC'04), June , pp. 3493-3498.

Koutsogiannis Z. &, Adamidis G., (2007). Direct Torque Control using Space Vector Modulation and dynamic performance of the drive via a Fuzzy logic controller for speed regulation, in: proceedings of EPE.

Reddy T. B., at. al. 2006). Sensorless Direct Torque Control of Induction Motor based on Hybrid Space Vector Pulsewidth Modulation to Reduce Ripples and Switching Losses - A Variable Structure Controller Approach, IEEE Power India Conference.

H. Zatocil H., 2008. Sensorless Control of AC Machines using High-Frequency Excitation, in 13th International Power Electronics and Motion Control Conference EPE-PEMC, Poznan, Poland.

Gadoue S. M. at. al., (2009). Artificial intelligence-based speed control of DTC induction motor drives-A comparative study, J. Electric Power Syst. Res. 79, p.p. 210–219.

R. Zaimeddine at. al., (2007). Enhanced Direct Torque Control Using a Three-Level Voltage Source Inverter, Asian Power Electronics Journal, Vol. 1, No. 1, Aug 2007.
Xavier del Toro, at. al, (2005), New DTC Control Scheme for Induction Motors fed with a Three-level Inverter, AUTOMATIKA 46(2005) 1-2, pp. 73–81
R. Zaimeddine, at. al., (2010) DTC Control Schemes for Induction Motorfed by Three-Level NPC-VSI Using Space Vector Modulation, SPEEDAM 2010 International Symposium on Power Electronics, Electrical Drives, Automation and Motion

4

Induction Motor Vector and Direct Torque Control Improvement during the Flux Weakening Phase

Kasmieh Tarek
Higher Institute for Applied Sciences and Technology
Syria

1. Introduction

Some industrial applications, such as spindle, traction, and electric vehicles, need a high speed for the fixed rating power, Fig. 1. To achieve this goal a suitable control method based on the flux weakening is usually applied. This gives an economic solution for the power converter and the motor (Grotstollen, H. & Wiesing, J, 1995).

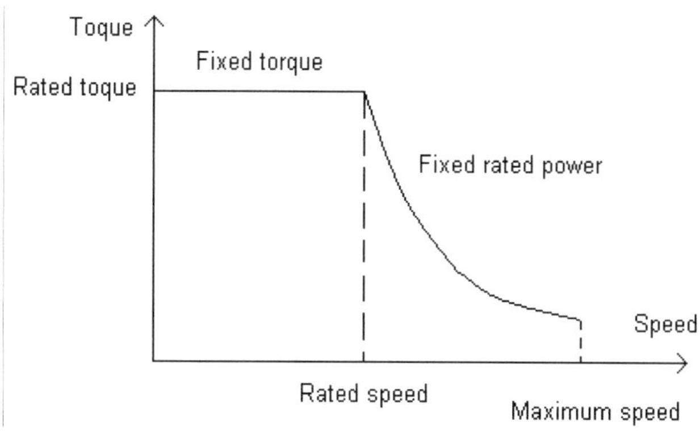

Fig. 1. Motor torque according to speed range

For the induction motor, its magnetic state changes during the flux weakening phase. It goes from the saturation to the linear region, since the rating magnetic point is at the knee of the magnetizing curve of the iron. Therefore, the change of the magnetic state of the motor should be taken into account in the control law.

Field Oriented Control (FOC) and Direct Torque Control (DTC) are based on a linear two-phase model of the induction motor (Vas, P. & Alakula, M. 1990). This model considers that the magnetic state of the motor is fixed, and all control parameters are calculated according to this state. When decreasing the flux level, the motor inductances increase. The change of real inductances values of the motor influences the desired current and speed dynamics.

In this chapter, the linear two-phase model of the induction motor is re-examined, and a new non linear two-phase model of the induction motor is developed. This model takes into account the variation of the saturation level (Kasmieh, T. & Lefevre, Y, 1998). The calculation of this model needs the motor inductances values at each calculation step. The inductances curves, as functions of the magnetic state of the motor, can be obtained using a finite elements calculation program.

Unlike many models developed for the induction motor that take into account the variation of the saturation level (Vas, P, 1981), the model presented in this chapter does not introduce the inductances time derivatives. This leads to an easy computation algorithm, using iteration method at each calculation step. The derived model is validated by comparing its dynamic behavior to the dynamic behavior of a finite elements model.

Based on the new model, a complete sensitivity study of the classical FOC and DTC techniques is presented. The FOC is highly dependent on the motor parameters. During the flux weakening phase, the inductances values increase. This influences the dynamic behavior fixed by the controllers, which is calculated for the rating inductances values. To overcome this problem an adaptive FOC is introduced. At each sampling period, the magnetic state of the motor is calculated by iteration, and then the controllers are tuned to this new magnetic state. Concerning the DTC, this control law is less sensitive to the variation of the saturation level (Kasmieh, T, 2008). The DTC is based on applying the good voltage stator vector in order to achieve the desired stator fluxes and torque variations. The main problem of the DTC lies in is the accuracy of the stator fluxes calculation at each sampling period. Usually, this calculation is easily done by using the stator electric equation. The performance of this estimator is highly dependent on the value of the stator winding resistor, which varies with the motor temperature. A more complicated flux estimator can be derived from the rotor electric equation (Kasmieh, T, 2008). This estimator is less sensitive to the variation of the rotor resistor, but more sensitive to the variation of the saturation level. To overcome this problem, an adaptive flux estimator is presented in this chapter. The estimator parameters are tuned according to the saturation level of the motor. This new estimation method increases the computation time of the DTC, but it remains smaller than the computation time of the FOC.

2. Magnetic state study of the induction motor using finite elements calculation program

The goal is to determine the main variable that influences the magnetic state of the induction motor, and to establish new flux-current relationships in the two-phase reference that take into account the influence of the magnetic saturation level variation. The study is done for a two pole pairs (p=2) 45(KW) induction motor using a finite elements calculation program.

Fig. 2 shows the cross section of the studied 45(KW) induction motor. The motor has two cages of 40 bars each.

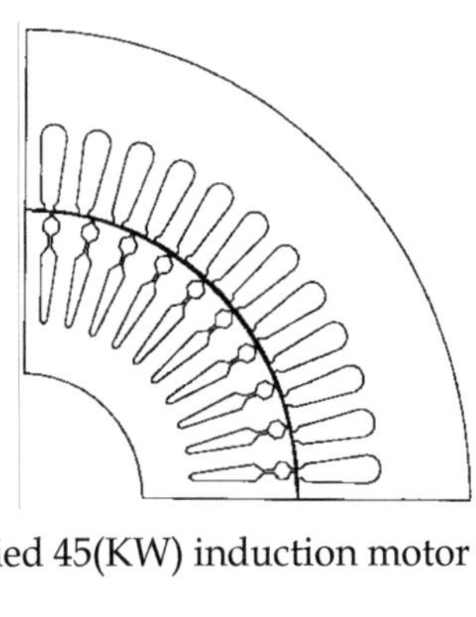

Fig. 2. Cross section of the studied 45(KW) induction motor

The induction motor is modeled as a magnetic circuit of 3 stator phases and m rotor phases, (m=10 in the case of the 45(KW) induction motor), Fig. 3.

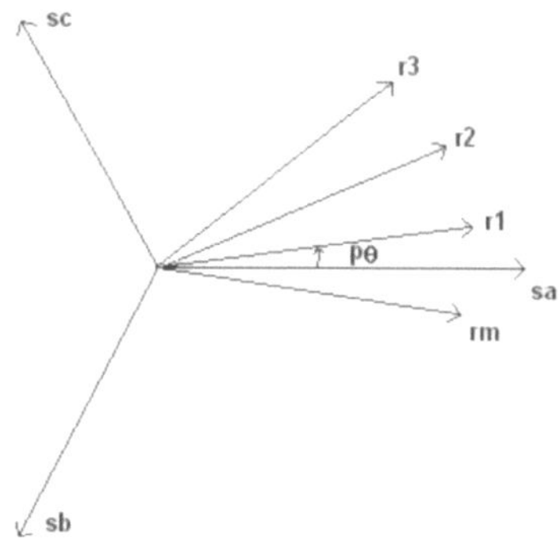

Fig. 3 Axes of multi-phase model of an induction motor

The flux-current relationships can be written as follows:

$$[\phi]_{m+3} = \begin{bmatrix} [L_{ss}]_{3,3} & [M_{sr}(\theta)]_{3,m} \\ [M_{rs}(\theta)]_{m,3} & [L_{rr}]_{m,m} \end{bmatrix} . [I]_{m+3} \quad (1)$$

The elements of the vector $[\phi]_{m+3}$ are the stator and rotor fluxes, the elements of the vector $[I]_{m+3}$ are the stator and rotor currents, $[L_{ss}]_{3,3}$ is the stator inductance matrix, $[L_{rr}]_{m,m}$ is the rotor inductance matrix, $[M_{sr}(\theta)]_{3,m}$ is the stator to rotor mutual inductance matrix, $[M_{rs}(\theta)]_{m,3}$ is the rotor to stator mutual inductance matrix and $[L_{rr}]_{m,m}$ is the rotor inductance matrix.

For the non-linear case where the magnetic saturation effect is taken into account, the stator and rotor fluxes are functions of the motor angle θ, the stator and the rotor currents.

$$[\phi]_{m+3} = [\phi]_{m+3}(\theta, [I]_{m+3}) \quad (2)$$

In this case, it is difficult to find the fluxes analytically, but they can be calculated using a finite elements calculation program, in which the magnetic characteristics of the motor material can be introduced, Fig. 4. The calculation of the fluxes as functions of the motor

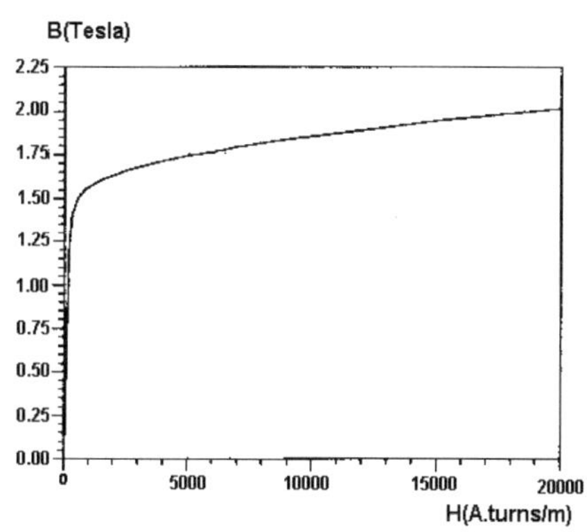

Fig. 4. Iron magnetic characteristics of the 45(KW) motor

angle θ and the currents is possible using the concept of equivalent saturated inductances deduced from the saturation curve of the motor material.
A finite elements calculation program is used to determine the main variable that influences the magnetic state of the induction motor in a two-phase reference related to the rotor, Fig. 5.

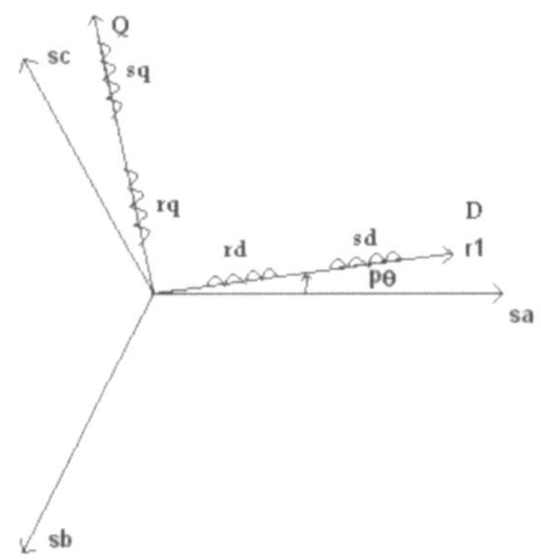

Fig. 5. Two-phase reference related to the rotor

The currents are applied in the two-phase reference, and then the currents $[I]_{m+3}$ are calculated using the inversed Park transformation. These currents are injected in the motor finite elements model, and the program calculates the fluxes $[\phi]_{m+3}$. The fluxes in the two-phase reference are finally calculated by applying the Park transformation. The calculation procedure is shown in Fig. 6.

$$\begin{bmatrix} i_{sd} \\ i_{sq} \\ i_{rd} \\ i_{rq} \end{bmatrix} \xrightarrow[T_{m,2}]{P(-p\theta).T_{3,2}} \begin{bmatrix} [I_s(\theta)]_3 \\ [I_r(\theta)]_m \end{bmatrix} \xrightarrow{\text{Finite Elements Program}} \begin{bmatrix} [\phi_s(\theta)]_3 \\ [\phi_r(\theta)]_m \end{bmatrix} \xrightarrow[T_{2,m}]{T_{2,3}.P(p\theta)} \begin{bmatrix} \phi_{sd} \\ \phi_{sq} \\ \phi_{rd} \\ \phi_{rq} \end{bmatrix}$$

Fig. 6. Calculation procedure of the two-phase fluxes

Where $T_{3,2} = \sqrt{\frac{2}{3}} \begin{bmatrix} 1 & 0 \\ -\frac{1}{2} & \frac{\sqrt{3}}{2} \\ -\frac{1}{2} & -\frac{\sqrt{3}}{2} \end{bmatrix}$, $T_{m,2} = \sqrt{\frac{2}{m}} \begin{bmatrix} 1 & 0 \\ \cos(\frac{2\pi}{m}) & \sin(\frac{2\pi}{m}) \\ . & . \\ . & . \\ \cos((m-1)\frac{2\pi}{m}) & \sin((m-1)\frac{2\pi}{m}) \end{bmatrix}$ are the Concordia matrices, and $P(\psi) = \begin{bmatrix} \cos(\psi) & \sin(\psi) \\ -\sin(\psi) & \cos(\psi) \end{bmatrix}$ is the rotation matrix.

The simulations show that the magnetic state of the induction motor depends on the modulus of the magnetizing current vector $\left|\overline{I_m}\right| = \sqrt{(i_{sd} + i_{rd})^2 + (i_{sq} + i_{rq})^2}$. The magnetizing current vector can be written using complex representation as follows: $\overline{I_m} = (i_{sd} + i_{rd}) + j.(i_{sq} + i_{rq})$. Fig. 7 shows the magnetic state of the motor for different two-phase currents that give the same value of $\left|\overline{I_m}\right|$.

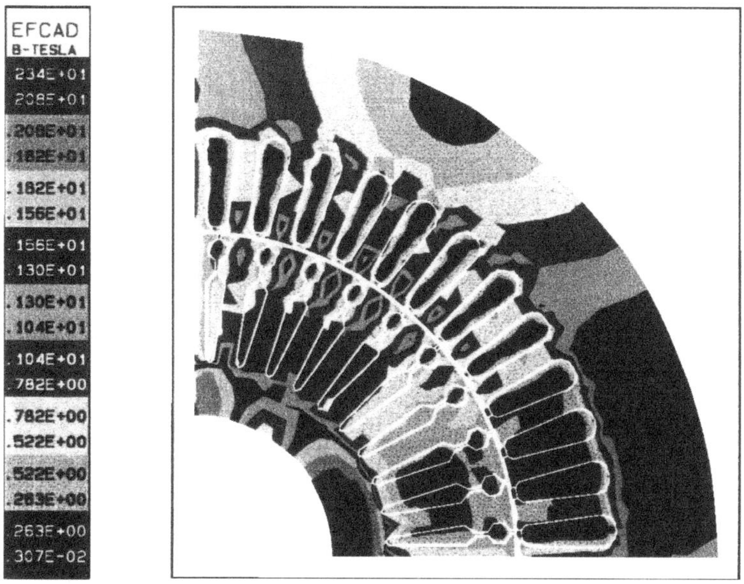

(isd, ird, isq, irq) = (50(A), 380(A), 25(A), 353(A)) |Im| = 100 (A)

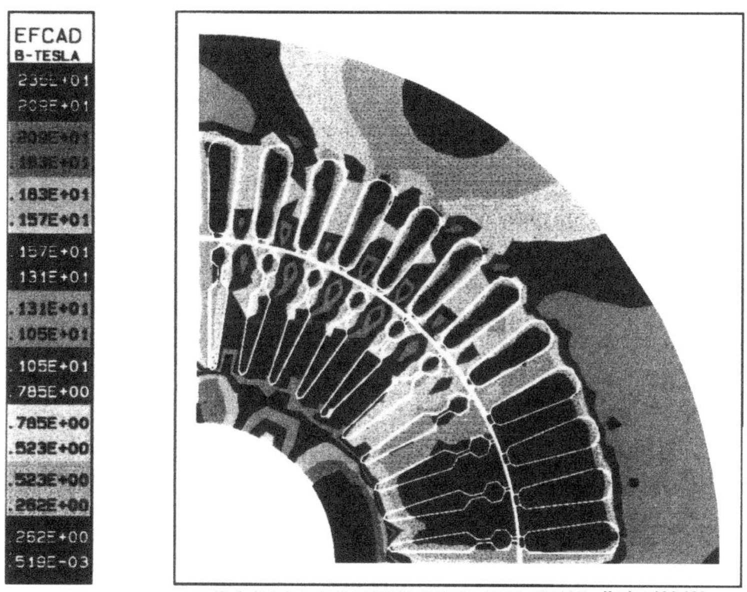

(isd, ird, isq, irq) = (35(A), 300(A), 40(A), 450(A)) |Im| = 100 (A)

Fig. 7. Magnetic state of the motor for different two-phase currents of the same $\left|\overline{I_m}\right|$

The flux-current relationship can then be written, in the two-phase reference, as follows:

$$\begin{bmatrix}\phi_{sd}\\ \phi_{sq}\\ \phi_{rd}\\ \phi_{rq}\end{bmatrix}=\begin{bmatrix}L_s(|\overline{I_m}|) & 0 & M(|\overline{I_m}|) & 0\\ 0 & L_s(|\overline{I_m}|) & 0 & M(|\overline{I_m}|)\\ M(|\overline{I_m}|) & 0 & L_r(|\overline{I_m}|) & 0\\ 0 & M(|\overline{I_m}|) & 0 & L_r(|\overline{I_m}|)\end{bmatrix}\cdot\begin{bmatrix}i_{sd}\\ i_{sq}\\ i_{rd}\\ i_{rq}\end{bmatrix}=[\phi]=\left[M(|\overline{I_m}|)\right].[I] \quad (3)$$

It is possible to obtain the cyclic inductances curves as functions of $|\overline{I_m}|$, by injecting one two-phase current. Fig. 8 shows the cyclic inductances curves as function of this injected current.

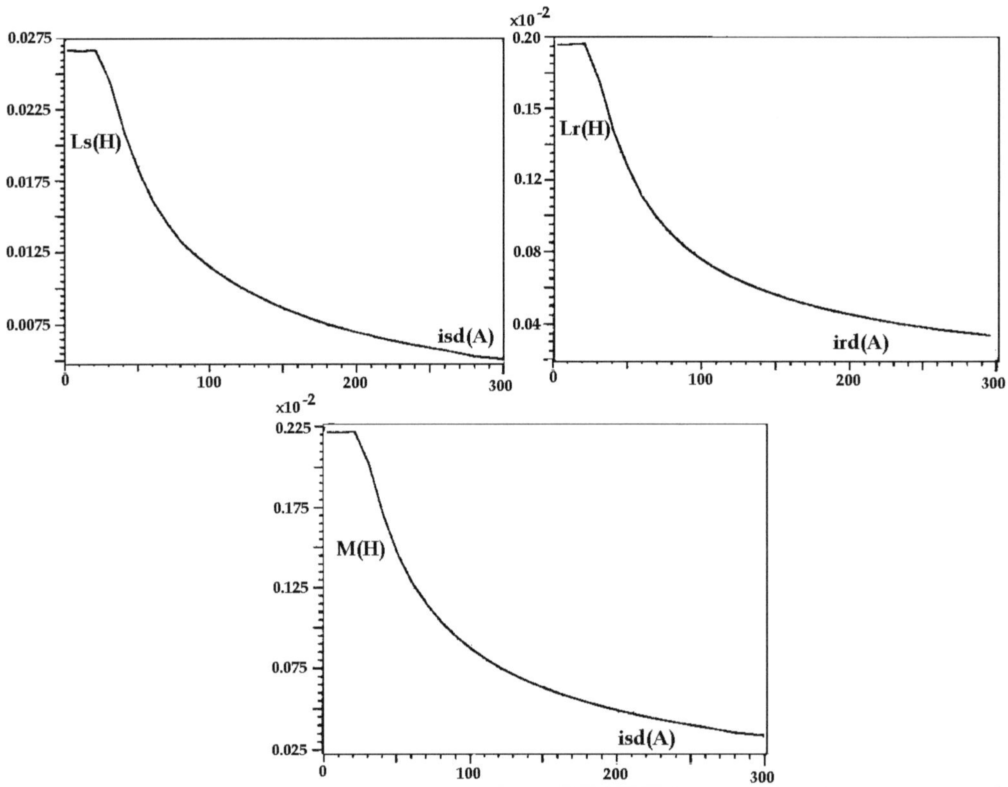

Fig. 8. Cyclic inductances curves

From these curves, inductances lookup tables are established. The values of the injected current can be associated to $|\overline{I_m}|$ (Kasmieh, T. & Lefevre, Y. 1998).
It is important to mention that the saturation harmonics disappear from the two-phase fluxes. This issue can be demonstrated taking into accounts the saturation third harmonics of the fluxes:

$$\begin{aligned}\phi_{sa} &= a.\cos(p\theta) + b.\cos(3p\theta)\\ \phi_{sb} &= a.\cos(p\theta - \frac{2\pi}{3}) + b.\cos(3p\theta - \frac{6\pi}{3})\\ \phi_{sc} &= a.\cos(p\theta - \frac{4\pi}{3}) + b.\cos(3p\theta - \frac{12\pi}{3})\end{aligned}$$

Two-phase stator fluxes can be obtained by applying Park transformation of an angle $p\theta$:

$$\begin{bmatrix}\phi_{sd}\\ \phi_{sq}\end{bmatrix} = \sqrt{\frac{2}{3}}\begin{bmatrix}\cos(p\theta) & \cos(p\theta - \frac{2\pi}{3}) & \cos(p\theta - \frac{4\pi}{3})\\ -\sin(p\theta) & -\sin(p\theta - \frac{2\pi}{3}) & -\sin(p\theta - \frac{4\pi}{3})\end{bmatrix}.\begin{bmatrix}\phi_{sa}\\ \phi_{sb}\\ \phi_{sc}\end{bmatrix}$$

$$\begin{bmatrix}\phi_{sd}\\ \phi_{sq}\end{bmatrix} = \sqrt{\frac{2}{3}}\begin{bmatrix}\frac{3a}{2} + \frac{a+b}{2}(\cos(2p\theta) + \cos(2p\theta - \frac{4\pi}{3}) + \cos(2p\theta - \frac{8\pi}{3})) + \\ \frac{b}{2}(\cos(4p\theta) + \cos(4p\theta - \frac{8\pi}{3}) + \cos(4p\theta - \frac{16\pi}{3}))\\ \frac{b-a}{2}(\sin(2p\theta) + \sin(2p\theta - \frac{4\pi}{3}) + \sin(2p\theta - \frac{8\pi}{3})) + \\ -\frac{b}{2}(\sin(4p\theta) + \sin(4p\theta - \frac{8\pi}{3}) + \sin(4p\theta - \frac{16\pi}{3}))\end{bmatrix} = \begin{bmatrix}\sqrt{\frac{3}{2}}\\ 0\end{bmatrix}$$

The demonstration can be extended to the general expressions of the saturated fluxes.
The next paragraph presents a new dynamic model of the induction motor that takes into account the variation of the saturation level. The resolution of the non-linear equations of the model is done by iteration.

3. Establishment of a saturated two-phase model of the induction motor

Since the magnetic state of the induction motor depends on the modulus of the magnetizing current vector, thus, the new equations that describe the dynamic behavior of a variable saturation level motor in the two-phase reference are:
The electric equations:

$$\begin{aligned}v_{sd} &= R_s.i_{sd} + \frac{d\Phi_{sd}}{dt} - \frac{d\psi}{dt}.\Phi_{sq}\\ v_{sq} &= R_s.i_{sq} + \frac{d\Phi_{sq}}{dt} + \frac{d\psi}{dt}.\Phi_{sd}\\ 0 &= R_r.i_{rd} + \frac{d\Phi_{rd}}{dt} - \frac{d(\psi - p\theta)}{dt}.\Phi_{rq}\\ 0 &= R_r.i_{rq} + \frac{d\Phi_{rq}}{dt} + \frac{d(\psi - p\theta)}{dt}.\Phi_{rd}\end{aligned} \quad \text{or} \quad \begin{aligned}\overline{V_s} &= R_s.\overline{I_s} + \frac{d\overline{\Phi_s}}{dt} + j.\frac{d\psi}{dt}.\overline{\Phi_s}\\ \overline{0} &= R_r.\overline{I_r} + \frac{d\overline{\Phi_r}}{dt} + j.\frac{d(\psi - p\theta)}{dt}.\overline{\Phi_r}\end{aligned} \tag{4}$$

The flux-current relationships:

$$[\phi] = \left[M(\left|\overline{I_m}\right|)\right].[I] \quad \text{or} \quad \begin{aligned}\overline{\Phi_s} &= L_s.\overline{I_s} + M.\overline{I_r}\\ \overline{\Phi_r} &= L_r.\overline{I_r} + M.\overline{I_s}\end{aligned} \tag{5}$$

The mechanical equation: $j\frac{d\Omega}{dt} = T_{em} - T_r$

where

$$T_{em} = p\frac{M}{L_r}(\overline{\Phi_r} \wedge \overline{I_s}) = p\frac{M}{L_r}(\Phi_{rd}.i_{sd} - \Phi_{rq}.i_{sd}) \quad (6)$$

is the electromagnetic torque and T_r is the resistive torque. R_s and R_r are the stator and the rotor windings resistances. V_{sd} and V_{sq} are the stator two-phase voltages and J is the rotor inertia.

The resistive torque is the sum of the viscosity resistive torque, and a resistive torque $T_s : T_r = f.\Omega + T_s$, where f is the viscosity factor. Usually, the variations of T_s are considered smaller than the variation of the velocity when controlling the motor. Note that the complex quantity $\overline{X} = x_d + j.x_q$ is used to represents the vectors in the D, Q reference.

The numeric resolution of the new saturated two-phase model equations is done avoiding the complicated development of the equations as currents deferential equations. The following differential equations can simply be written.

$$\frac{d[\Phi]}{dt} + A(|\overline{I_m}|).[\Phi] = [v] \quad (7)$$

$$A(|\overline{I_m}|) = \begin{bmatrix} \frac{R_s}{\sigma.L_s} & 0 & -\frac{M.R_s}{\sigma.L_s.L_r} & 0 \\ 0 & \frac{R_s}{\sigma.L_s} & 0 & -\frac{M.R_s}{\sigma.L_s.L_r} \\ -\frac{M.R_r}{\sigma.L_s.L_r} & 0 & \frac{R_r}{\sigma.L_r} & \frac{d(p.\theta)}{dt} \\ 0 & -\frac{M.R_r}{\sigma.L_s.L_r} & -\frac{d(p.\theta)}{dt} & \frac{R_r}{\sigma.L_r} \end{bmatrix} (|\overline{I_m}|) \quad (8)$$

The matrix A is written for a two-phase reference related to the stator Ψ=0. $\sigma = 1 - M^2 / (L_s.L_r)$ is the dispersion factor which is never equal to zero because the leakage inductances.

The new non-linear model of the induction motor is described by equations (3), (7) and the expression of the electromagnetic torque. This model is called the saturated two-phase model.

The numeric resolution procedure of these equations starts from an initial state. At each calculation step equation (7) is solved using for example Runge-Kutta 4 (RK4) method. This will give a new flux vector that describes a new magnetic state of the motor. Then, the corresponding current vector must be determined by resolving equation (3). In fact, equation (3) is a non-linear equation. The matrix M depends on the modulus magnetizing current vector. The resolution of this equation can be done by a non-linear iterative resolution method, like substitution method.

Equation (7) can be written as follows:

$$\frac{d[\Phi]}{dt} = [F([\Phi]_t,[I]_t)] = [F]_t \tag{9}$$

where $[F]_t$ is a function of the two-phase fluxes and currents.

The RK4 method gives an approximated numerical solution of equation (9). The fluxes at the instant t+Δt are calculated using equation (10).

$$[\Phi]_{t+\Delta t} = [\Phi]_t + \sum_{i=1}^{4} b_i.\Delta t[F]_i \tag{10}$$

where

$$[\Phi]_1 = [\Phi]_t + \frac{\Delta t}{2}[F]_1 = [\Phi]_t + \frac{\Delta t}{2}[F]_t$$
$$[\Phi]_2 = [\Phi]_1 + \frac{\Delta t}{2}[F]_2 = [\Phi]_1 + \frac{\Delta t}{2}\left[F([\Phi]_1,[I]_1)\right]$$
$$[\Phi]_3 = [\Phi]_2 + \frac{\Delta t}{2}[F]_3 = [\Phi]_2 + \frac{\Delta t}{2}\left[F([\Phi]_2,[I]_2)\right]$$
$$[\Phi]_4 = [\Phi]_3 + \frac{\Delta t}{2}[F]_4 = [\Phi]_3 + \frac{\Delta t}{2}\left[F([\Phi]_3,[I]_3)\right]$$

and $b_1 = \frac{1}{6}$, $b_2 = \frac{1}{3}$, $b_3 = \frac{1}{6}$, $b_4 = \frac{1}{6}$.

To be able to calculate $[\Phi]_{i+1}$, the currents $[I]_i$ must be calculated by solving the non-linear equation $[\phi]_i = \left[M(\left|\overline{I_m}\right|_i)\right].[I]_i$. Finally, Fig. 9 shows the calculation procedure of the saturated two-phase model of the induction motor.

The resolution of the non-linear equations of the flux-current relationships can be done using a non-linear iterative resolution method. The substitution method searches the intersection point between $\left(\left[M(\left|\overline{I_m}\right|)\right].[I]\right)(t)$ and $[\phi]_{t+\Delta t}$ starting from the first iteration $[I]_1 = [I]_t$. The next iteration is calculated from the previous iteration: $[I]_{i+1} = [I]_i + \Delta[I]$, where $\Delta[I] = \left[M(\left|\overline{I_m}\right|_i)\right]^{-1}.\left([\Phi]_{t+\Delta t} - M(\left|\overline{I_m}\right|_i).[I]_i\right)$. In fact the Inductance matrix can be inversed, since the leakage inductances cannot be zero:

$$\left[M(\left|\overline{I_m}\right|)\right]^{-1} = \begin{bmatrix} \frac{1}{\sigma.L_s} & 0 & -\frac{M}{\sigma.L_s.L_r} & 0 \\ 0 & \frac{1}{\sigma.L_s} & 0 & -\frac{M}{\sigma.L_s.L_r} \\ -\frac{M_r}{\sigma.L_s.L_r} & 0 & \frac{1}{\sigma.L_r} & 0 \\ 0 & -\frac{M}{\sigma.L_s.L_r} & 0 & \frac{1}{\sigma.L_r} \end{bmatrix}(\left|\overline{I_m}\right|)$$

Fig. 10. shows the substitution calculation procedure for vectors dimension equal to one.

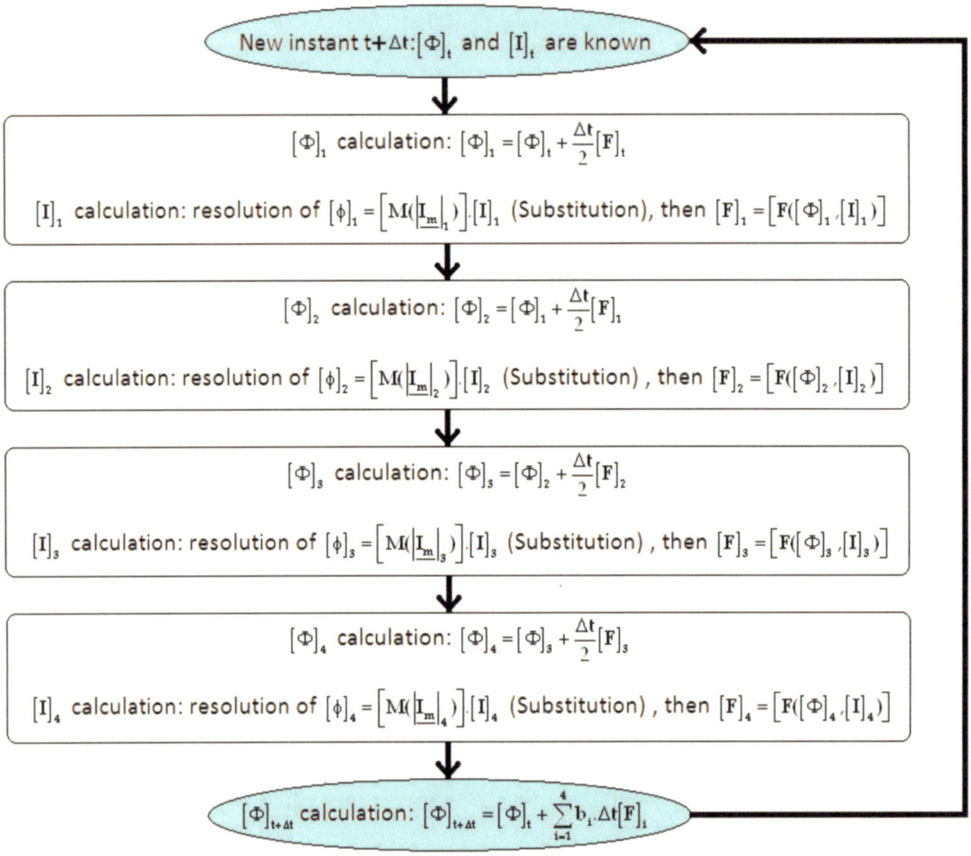

Fig. 9. Calculation procedure of the saturated two-phase model of the induction motor

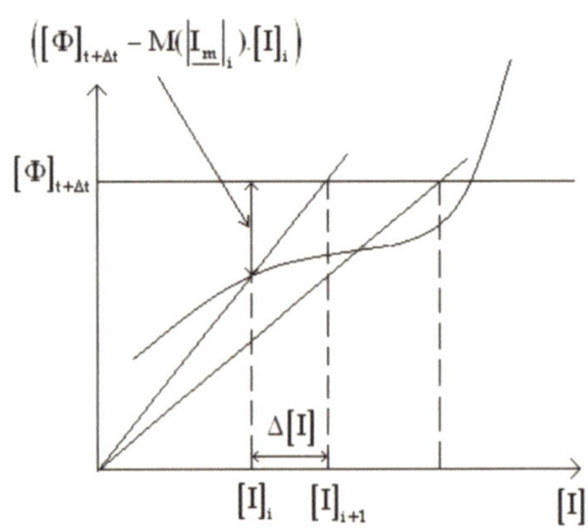

Fig. 10. Substitution calculation procedure

The iteration procedure is stopped when achieving a suitable error of the modulus of the flux vector.

The execution of the calculation procedure of the Fig. 9 gives the results shown in Fig. 11.

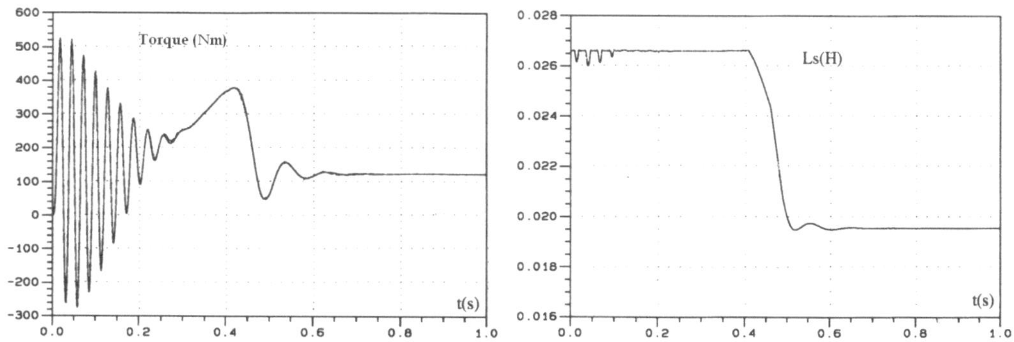

Fig. 11. Dynamic behavior of the saturated two-phase model of the induction motor

The comparison between the saturated two-phase model and the finite elements model is shown in Fig. 12. It is clear that it gives closer results to the finite elements model results than the results of the linear model.

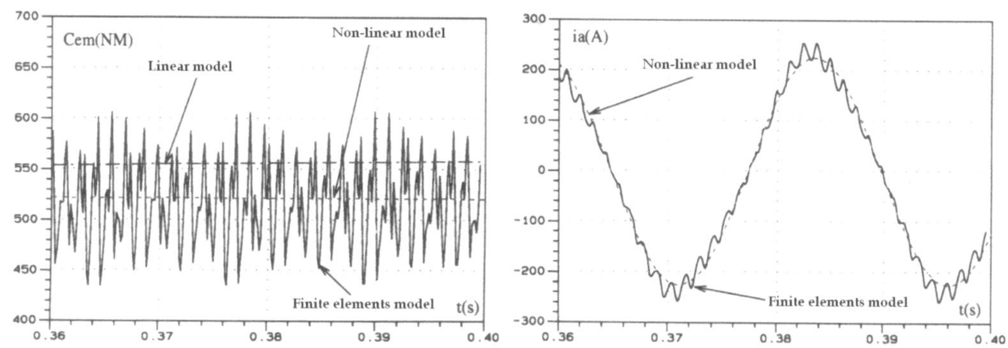

Fig. 12. Saturated two-phase model, linear model and finite element model results comparison

4. Field oriented control law improvement during the flux weakening phase

The vector control law or field-oriented control (FOC) law of an induction motor has become a powerful and frequently adopted technique world-wide. It is based on the two-phase model, Park model. The aim of this control is to give the induction motor a dynamic behavior like the dynamic behavior of a direct current motor. This can be done by controlling separately the modulus and the phase angle of the flux (Blaschke, F. 1972).

Using this control technique, the electrical and mechanical dynamic responses of the induction motor are determined by fixing the coefficients of the current loops controllers, flux loop controller and the velocity loop controller. Usually, these coefficients are calculated for the rating values of the cyclic inductances, which correspond to the rating saturation level. In fact, this level is achieved by applying the rating flux value as a reference value to the flux loop.

Some industrial applications require the induction motor to operate at a high speed over the rating speed. The method used to reach this speed is to decrease the reference value of the flux in order to work at the rating power. This decrease can cause a coupling between the two-phase axes D and Q, so FOC does not work properly (Kasmieh, T. & Lefevre, Y. 1998).

Many published papers have studied the effects of the variation of the saturation level on FOC law (Vas, P. & Alakula, M. 1990) (Vas, P. 1981), but few attempts have been made to develop a FOC law that takes into account this variation.

In this paragraph the sensitivity of the classical FOC law to the variation of saturation level of an induction motor is studied. Then, a new indirect vector control law in accordance to the rotor flux vector that takes into account this variation is developed. This law is based on the saturated two-phase model found in the previous sections.

The simulations are done using an electromechanical simulation program called "A_MOS", Asynchronous Motor Open Simulator, (Kasmieh, T. 2002), Fig. 13.

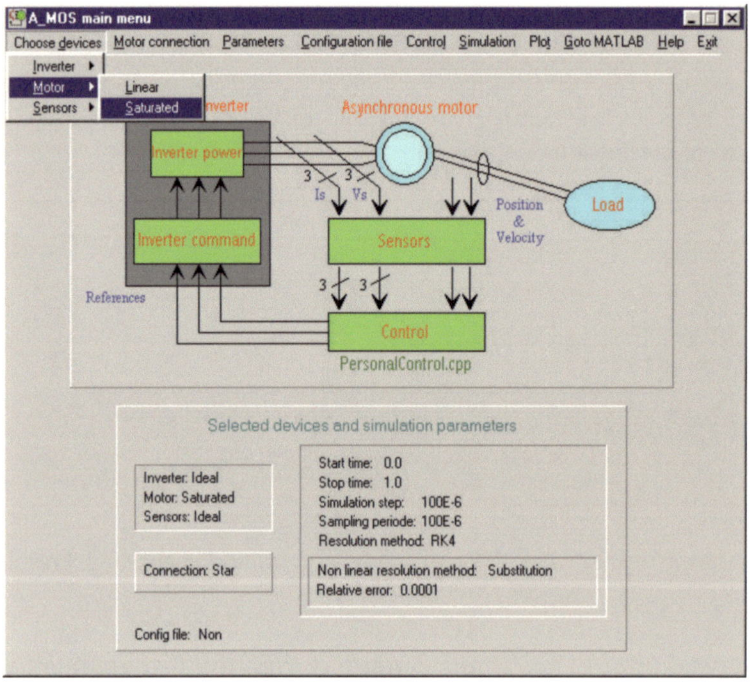

Fig. 13. The main window of "A-MOS" Software

The resolution algorithm of the non-linear model is implemented in this programmed. The user can write his own control algorithm.

4.1 Classical FOC law

The strategy of the FOC in accordance with the rotor flux vector is adopted. This strategy leads to simpler equations than those obtained with the axis D aligned on the stator flux vector or with the magnetizing flux vector (Vas, P. & Alakula, M. 1990).

The development of the FOC equations in accordance to the rotor flux vector can be done by supposing $\underline{\phi}_r = \left[\phi_{rd}, \phi_{rq}\right]^t = \left[\phi_r, 0\right]^t$, Fig. 14. The expression of the motor torque is reduced to:

$$T_{em} = p\frac{M}{L_r}.\Phi_r.i_{sd} \tag{11}$$

Since the rotor flux vector turns at the synchronized speed ω_s, the electric equations become:

$$\begin{aligned}
v_{sd} &= R_s.i_{sd} + \frac{d\Phi_{sd}}{dt} - \omega_s.\Phi_{sq} \\
v_{sq} &= R_s.i_{sq} + \frac{d\Phi_{sq}}{dt} + \omega_s.\Phi_{sd} \\
0 &= R_r.i_{rd} + \frac{d\Phi_r}{dt} \\
0 &= R_r.i_{rq} + (\omega_s - p\frac{d\theta}{dt}).\Phi_r
\end{aligned} \tag{12}$$

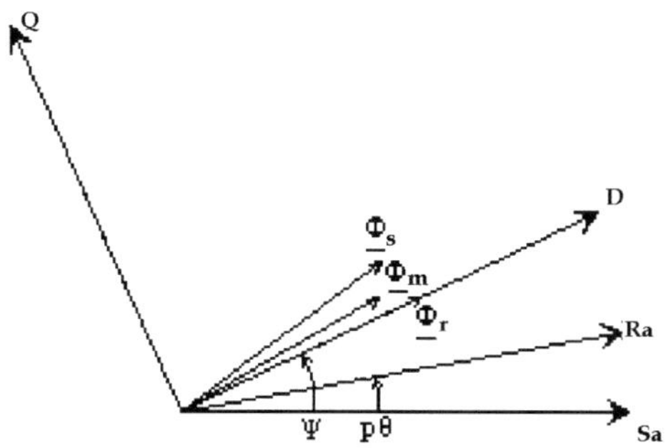

Fig. 14. Two-phase reference in accordance with the rotor flux vector

4.1.1 Stator voltages and stator fluxes equations

The stator voltages of equation (12), and the stator fluxes expressions can be written using complex representation ($\overline{X} = x_d + j.x_q$):

$$\overline{V_s} = R_s.\overline{I_s} + \frac{d\overline{\Phi_s}}{dt} + j.\omega_s.\overline{\Phi_s}$$
$$\overline{\Phi_s} = L_s.\overline{I_s} + M.\overline{I_r}$$

By adding and subtracting the term $\frac{M^2}{L_r}.\overline{I_s}$ in the stator flux vector expression, the magnetizing rotor current vector is introduced $\overline{I_{mr}}$:

$$\overline{\Phi_s} = \sigma L_s.\overline{I_s} + \frac{M^2}{L_r}.(\overline{I_s} + \frac{L_r}{M}.\overline{I_r}) = \sigma L_s.\overline{I_s} + \frac{M^2}{L_r}.(\overline{I_{mr}}).$$

Since the rotor flux vector is aligned on the magnetizing rotor current vector: $\overline{\Phi_r} = \Phi_r = L_r.\overline{I_r} + M.\overline{I_s} = M.\overline{I_{mr}}$, the stator flux vector can be written as a function of the stator current vector and the rotor flux.

$$\overline{\Phi_s} = \sigma L_s.\overline{I_s} + \frac{M}{L_r}.\Phi_r \tag{13}$$

Substituting (13) in the expression of the stator voltage vector:

$$\overline{V_s} = R_s.\overline{I_s} + \sigma L_s.\frac{d\overline{I_s}}{dt} + \frac{M}{L_r}.\frac{d\Phi_r}{dt} + j.\omega_s.\overline{\Phi_s} \qquad (14)$$

4.1.2 Rotor voltages and rotor fluxes equations

The pulsation $(\omega_s - p\frac{d\theta}{dt})$ is the rotor pulsation ω_r, thus the rotor electric equations become:

$$\begin{aligned} 0 &= R_r.i_{rd} + \frac{d\Phi_r}{dt} \\ 0 &= R_r.i_{rq} + \omega_r.\Phi_r \end{aligned} \qquad (15)$$

From the rotor fluxes expressions, the rotor currents are expressed as functions of the rotor flux and the stator currents:

$$\begin{aligned} \phi_{rd} &= L_r.i_{rd} + M.i_{sd} \\ \phi_{rq} &= L_r.i_{rq} + M.i_{sq} \end{aligned} \Rightarrow \begin{aligned} \phi_r &= L_r.i_{rd} + M.i_{sd} \\ 0 &= L_r.i_{rq} + M.i_{sq} \end{aligned} \Rightarrow$$

$$i_{rd} = \frac{\phi_r}{L_r} - \frac{M}{L_r}.i_{sd} \qquad (16)$$

$$i_{rq} = -\frac{M}{L_r}.i_{sq} \qquad (17)$$

4.1.3 Transfer functions of the induction motor

In order to establish the FOC strategy, the transfer functions of the motor are developed. The inputs of the transfer functions are v_{sd} and v_{sq}, and the outputs the variables that determine the motor torque Φ_r and i_{sd}.

Transfer functions on D axis:

It is possible to control the rotor flux via the stator current on the D axis. This can be demonstrated from the rotor electric equation on the D axis and from equation (16):

$$\frac{d\Phi_r}{dt} = -\frac{R_r}{L_r}.\Phi_r + R_r.\frac{M}{L_r}.i_{sd} \qquad (18)$$

Developing equation (14) on the axis D yields to:

$$v_{sd} = R_s.i_{sd} + \sigma L_s.\frac{di_{sd}}{dt} + \frac{M}{L_r}.\frac{d\Phi_r}{dt} - \omega_s.\Phi_{sq}$$

By substituting equation (17) in the expression of Φ_{sq}, the following equation is obtained:

$$\phi_{sq} = L_s.i_{sq} + M.i_{rq} = L_s.i_{sq} - \frac{M^2}{L_r}.i_{sq} = (L_s - \frac{M^2}{L_r}).i_{sq} = L_s.(1 - \frac{M^2}{L_r.L_s}).i_{sq} = L_s.\sigma.i_{sq}\,.$$

The D stator voltage expression becomes:

$$v_{sd} = R_s.i_{sd} + \sigma L_s.\frac{di_{sd}}{dt} + \frac{M}{L_r}.\frac{d\Phi_r}{dt} - \omega_s.L_s.\sigma.i_{sq} \tag{19}$$

By replacing (18) in (19), the stator voltage of the D axis can be written as follows:

$$v_{sd} = R_{sr}.i_{sd} + \sigma L_s.\frac{di_{sd}}{dt} + E_d \tag{20}$$

where $R_{sr} = R_s + R_r.\left(\frac{M}{L_r}\right)^2$, and the electrical force $E_d = -R_r.\frac{M}{L_r^{\ 2}}.\Phi_r - \omega_s.L_s.\sigma.i_{sq}$ represents the coupling between the two axes D and Q.

Transfer functions on Q axis:

By developing equation (14) on the axis Q, the stator voltage of the same axis is obtained:

$$v_{sq} = R_s.i_{sq} + \sigma L_s.\frac{di_{sq}}{dt} + \omega_s.\Phi_{sd}$$

From equation (13) the D stator flux is: $\Phi_{sd} = \sigma L_s.i_{sd} + \frac{M}{L_r}.\Phi_r$. By replacing Φ_{sd} in the previous expression, v_{sq} becomes:

$$v_{sq} = R_s.i_{sq} + \sigma L_s.\frac{di_{sq}}{dt} + \omega_s.\sigma L_s.i_{sd} + \omega_s.\frac{M}{L_r}.\Phi_r \tag{21}$$

Φ_r can be written as a function of the stator current on the Q axis by substituting the expression of i_{rq}, equation (17), in the rotor electric equation on the Q axis:

$$\Phi_r = R_r.\frac{M}{\omega_r.L_r}i_{sq} \tag{22}$$

By replacing (22) in (21) :

$$v_{sq} = R_s.i_{sq} + \sigma L_s.\frac{di_{sq}}{dt} + \omega_s.\sigma L_s.i_{sd} + \frac{\omega_s}{\omega_r}.\left(\frac{M}{L_r}\right)^2.R_r.i_{sq}$$

$$v_{sq} = R_s.i_{sq} + \sigma L_s.\frac{di_{sq}}{dt} + \omega_s.\sigma L_s.i_{sd} + \frac{\omega + \omega_r}{\omega_r}.\left(\frac{M}{L_r}\right)^2.R_r.i_{sq}$$

Finally v_{sq} can be written as follows:

$$v_{sq} = R_{sr}.i_{sq} + \sigma L_s.\frac{di_{sq}}{dt} + \omega_s.\sigma L_s.i_{sd} + \omega.\left(\frac{M}{L_r}\right)^2.R_r.i_{sq} = R_{sr}.i_{sq} + \sigma L_s.\frac{di_{sq}}{dt} + E_q \tag{23}$$

The electrical force E_q represents the coupling between the two axes D and Q.
The equations (18), (20) and (23) describe the transfer functions of the induction motor if the D axis is aligned on the rotor flux vector, Fig. 15.

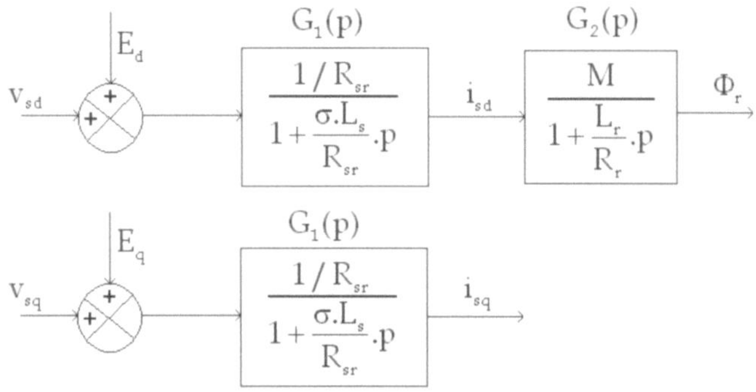

Fig. 15. Transfer functions of the induction motor (D axis is aligned on the rotor flux vector)

4.1.4 Establishment of the classical FOC law

It is important to mention that the transfer functions shown on Fig. 15 are valid if the axis D is rotating with the rotor flux vector. Taking into account this hypothesis the control scheme of Fig. 16 can be built.

The two axes D and Q are decoupled by estimating the electric forces E_d and E_q:

$E_d^e = -R_r.\frac{M}{L_r^2}.\Phi_r^e - \omega_s^e.L_s.\sigma.i_{sq}^m$ and $E_q^e = \omega_s^e.\sigma L_s.i_{sd}^m + \omega^m.\left(\frac{M}{L_r}\right)^2.R_r.i_{sq}^m$. The index e is for the estimated variables, and the index m is for the measured variables.

Φ_r^e is calculated by solving numerically the equation (18). The value of Φ_r^e is also used as a feedback for the rotor flux control closed loop.

ω_s^e is calculated from equation (18): $\omega_s^e = \omega^m + R_r.\frac{M}{\Phi_r^e.L_r}i_{sq}^m$. $\omega^m = p.\Omega^m = p.d\theta / dt$ is the electric speed of the motor that can be measured using a speed sensor, and p is the pole pairs number.

For the induction motor, L_r / R_r is ten times bigger than $\sigma.L_s / R_{sr}$, so it is possible to do poles separation by doing an inner closed loop for the current and an outer closed loop for the rotor flux.

From Fig. 16, it is clear that the D axis closed loops are for controlling the amplitude of the rotor flux, and the closed loop of the Q axis is for controlling the stator current, thus for controlling the motor torque, equation (11).

In practice, the three phase currents are measured, and then the two phase currents are calculated using Park transformation of an angle Ψ. The angle Ψ is estimated by integrating $\omega_s^e = \omega^m + R_r.\frac{M}{\Phi_r^e.L_r}i_{sq}^m$. After calculating the control variables v_{sd} and v_{sq}, the three phase control variables v_{sa}, v_{sb} and v_{sc} are found using the inversed Park transformation.

4.2 Sensitivity of the classical FOC law to the variation of the saturation level

the FOC algorithm is implemented in "A_MOS" program. The controller parameters are fixed according to rating values of the induction motor cyclic inductances. The simulation results of fig. 17 show that during the flux weakening phase, the rotor flux does not follow its reference and the dynamic response of the speed is disturbed. This due to the fact that the

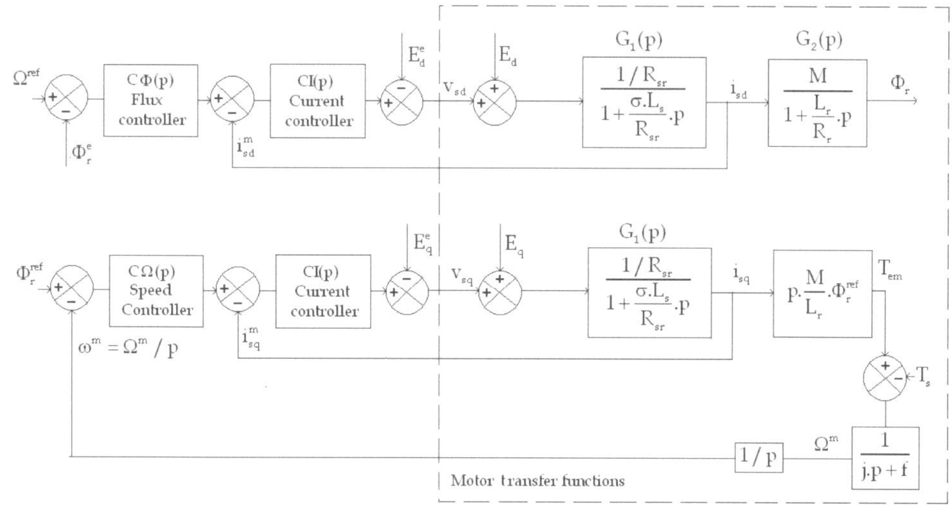

Fig. 16. FOC law scheme

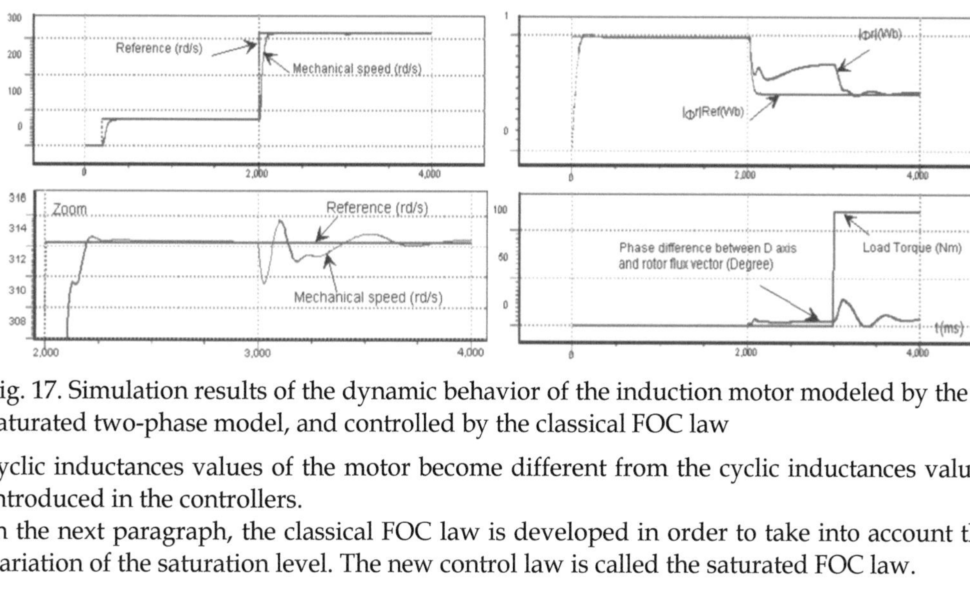

Fig. 17. Simulation results of the dynamic behavior of the induction motor modeled by the saturated two-phase model, and controlled by the classical FOC law

cyclic inductances values of the motor become different from the cyclic inductances values introduced in the controllers.

In the next paragraph, the classical FOC law is developed in order to take into account the variation of the saturation level. The new control law is called the saturated FOC law.

4.3 New saturated FOC law

To simplify the study, stator and rotor leakage inductances (L_{sf} and L_{rf}), are supposed to be constant. Only the mutual cyclic inductance M is considered to be variable with the modulus of magnetizing current vector, where $L_s = M + L_{sf}$ and $L_r = M + L_{rf}$.

From expression (13), The derivative of the stator flux vector is:

$$\frac{d\overline{\Phi_s}}{dt} = \sigma.L_s.\frac{d\overline{I_s}}{dt} + \frac{M}{L_r}.\frac{d\overline{\Phi_r}}{dt} + \overline{I_s}.\frac{d(\sigma L_s)}{dt} + \overline{\Phi_r}.\frac{d(\frac{M}{L_r})}{dt} = \sigma.L_s.\frac{d\overline{I_s}}{dt} + \frac{M}{L_r}.\frac{d\overline{\Phi_r}}{dt} + \overline{I_s}.\frac{dM}{dt}.(\frac{L_{rf}^2}{L_r^2}) + \overline{\Phi_r}.\frac{dM}{dt}.\frac{L_{rf}}{L_r^2}$$

Finally the expression of the stator flux vector derivative is:

$$\frac{d\overline{\Phi_s}}{dt} = \sigma.L_s.\frac{d\overline{I_s}}{dt} + \frac{M}{L_r}.\frac{d\Phi_r}{dt} + (\overline{I_s}.L_{rf} + \Phi_r)\frac{L_{rf}}{L_r^2}.\frac{dM}{dt} \tag{24}$$

The stator voltage vector is then modified to:

$$\overline{V_s} = R_s.\overline{I_s} + \sigma.L_s.\frac{d\overline{I_s}}{dt} + \frac{M}{L_r}.\frac{d\Phi_r}{dt} + (\overline{I_s}.L_{rf} + \Phi_r)\frac{L_{rf}}{L_r^2}.\frac{dM}{dt} + j.\omega_s.\overline{\Phi_s} \tag{25}$$

As previous, the resistance R_{sr} can be introduced. The stator voltages on the D and Q axes are:

$$\begin{aligned} v_{sd} &= R_{sr}.i_{sd} + \sigma.L_s.\frac{di_{sd}}{dt} - R_r.\frac{M}{L_r^2}.\left|\overline{\Phi_r}\right| - \sigma.L_s.\omega_s.i_{sq} + (i_{sd}.L_{rf} + \left|\overline{\Phi_r}\right|)\frac{L_{rf}}{L_r^2}.\frac{dM}{dt} \\ &= R_{sr}.i_{sd} + \sigma.L_s.\frac{di_{sd}}{dt} + E_d \end{aligned} \tag{26}$$

$$v_{sq} = R_{sr}.i_{sq} + \sigma.L_s.\frac{di_{sq}}{dt} + \omega.\frac{M}{L_r}.\left|\Phi_r\right| + i_{sq}.\frac{L_{rf}^2}{L_r^2}.\frac{dM}{dt} + \sigma.L_s.\omega_s.i_{sd} = R_{sr}.i_{sq} + \sigma.L_s.\frac{di_{sq}}{dt} + E_q \tag{27}$$

where E_d and E_q are electrical forces and equal to:

$$E_d = R_r.\frac{M}{L_r^2}.\left|\overline{\Phi_r}\right| + \sigma.L_s.\omega_s.i_{sq} - (i_{sd}.L_{rf} + \left|\overline{\Phi_r}\right|).\frac{L_{rf}}{L_r^2}.\frac{dM}{dt},$$

$$E_q = -\omega.\frac{M}{L_r}.\left|\overline{\Phi_r}\right| - \sigma.L_s.\omega_s.i_{sd} - i_{sq}.\frac{L_{rf}^2}{L_r^2}.\frac{dM}{dt}.$$

The obtained transfer functions are approximately the same as in the linear case. The main difference is that the parameters of these transfer functions are time variant. Terms containing $\frac{dM}{dt}$ appear in the expressions of E_d and E_q. Anyhow, this term can be neglected since $\frac{L_r}{R_r}$ is bigger than $10\frac{\sigma L_s}{R_s}$ for induction machines, so the expressions of E_d and E_q become:

$$E_d \approx R_r.\frac{M}{L_r^2}.\left|\overline{\Phi_r}\right| + \sigma.L_s.\omega_s.i_{sq},\quad E_q \approx -\omega.\frac{M}{L_r}.\left|\overline{\Phi_r}\right| - \sigma.L_s.\omega_s.i_{sd}.$$

The idea of the saturated FOC is to tune the coefficients of the controllers according to the value of $\left|\overline{I_m}\right|$. At each sampling period $\left|\underline{I_m}\right|$ is calculated, and the corresponding cyclic inductances are found from look up tables to update the controller's coefficients.
The expression of $\left|\overline{I_m}\right|$ is $\left|\overline{I_m}\right| = \sqrt{(i_{sd} + i_{rd})^2 + (i_{sq} + i_{rq})^2}$. i_{sd} and i_{sq} can be measured at each sampling period. i_{rd} can be calculated from the first rotor equation (15), and i_{rq} from the equation (17) using a non-linear resolution method as the substitution method.

Fig. 18 shows the strategy of the new FOC law. The blocks with dashed lines are the blocks necessary for calculating the modulus of magnetizing current vector. At each sampling period the controller's coefficients are updated according to the new values of the cyclic inductances.

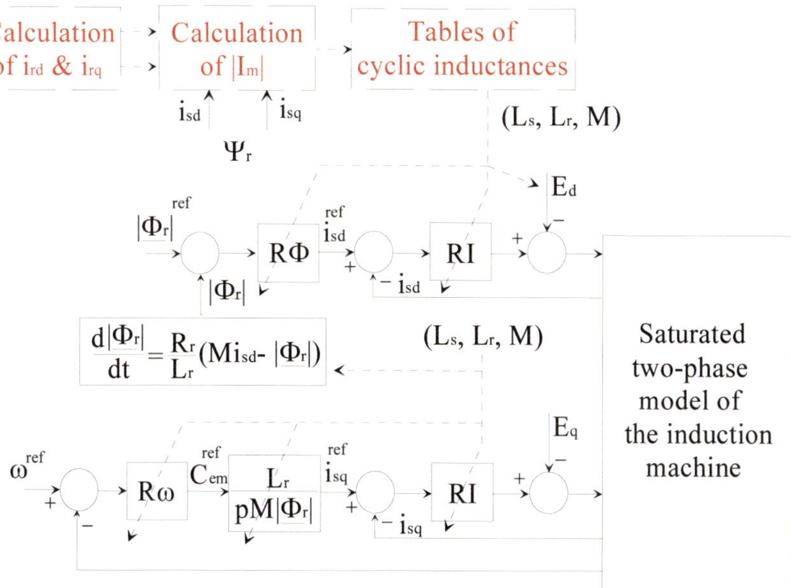

Fig. 18. Saturated FOC law

Fig. 19 presents simulation results of the dynamic response of the 45KW induction motor controlled by the new saturated FOC control. This simulation is done for the same inputs of figure 5. It is clear that the performance of the machine is clearly improved.

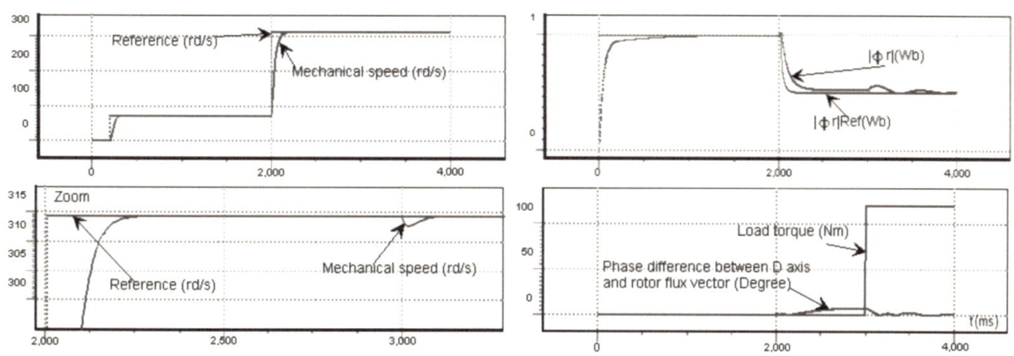

Fig. 20. Simulation results with saturated FOC

5. Stator flux estimation improvement during the flux weakening phase for the Direct Torque Control Law

Thirteen years after developing the FOC law by F. Blaschke in 1971 (Blaschke, F. 1972), I. Takahashi and M. Depenbrock presented a new technique for the induction motor torque

control called Direct Torque Control (DTC), (Noguchi, T. & Takahashi, I. 1984), Depenbrock, M. & Steimel A. 1990). DTC is based on applying the appropriate voltage space vector in order to achieve the desired flux and torque variations.
DTC permits to have very fast dynamics without any intermediate current control loops.
The DTC is based on the fact that the variations of the stator flux vector are directly controlled by the stator voltage vector for high speed:

$$\overline{V_s} = R_s.\overline{I_s} + \frac{d\overline{\Phi_s}}{dt} \approx \frac{d\overline{\Phi_s}}{dt} \tag{28}$$

5.1 Direct Torque Control Law for an induction machine with a voltage source inverter drive

A small variation of the stator flux vector is in fact the product of the stator voltage vector and the sampling period ΔT :

$$\Delta\overline{\Phi_s} = \overline{V_s}.\Delta T \tag{29}$$

Usually, the motor is driven by a voltage source inverter. The stator voltage vector for such an inverted has only 8 positions, Fig. 21. From Fig. 21 If the stator flux vector is in sector i, then its magnitude is increased when applying $\overline{V_i}$, $\overline{V_{i+1}}$ or $\overline{V_{i-1}}$. To decrease $\overline{\Phi_s}$, the vector $\overline{V_{i+2}}$, $\overline{V_{i-2}}$ or $\overline{V_{i+3}}$ can be applied.

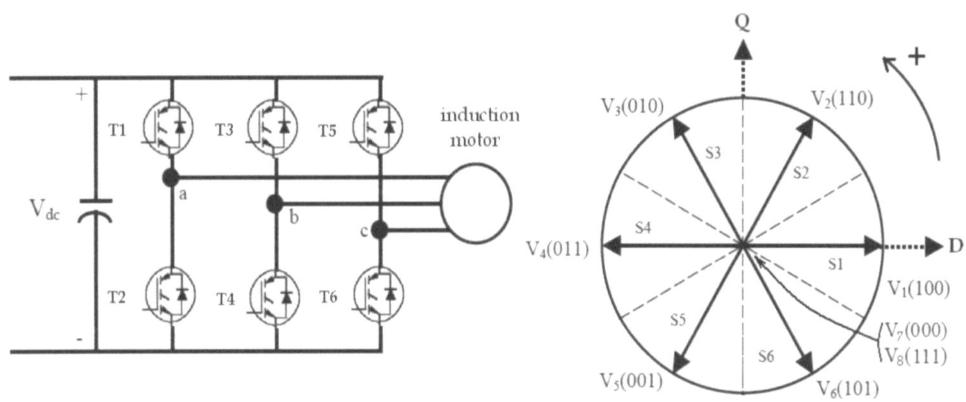

Fig. 21. Stator Voltage space vector for a voltage source inverter

In order to search what does the stator voltage space vector act on the motor torque, its expression can be rewritten starting from equation (6) and taking into account the flux-current relationships as follows:

$$T_{em} = p.\overline{\Phi_s} \wedge \overline{I_s} \tag{30}$$

$$T_{em} = p.\frac{M}{Ls.Lr - M2}.\overline{\Phi_s} \wedge \overline{\Phi_r} = p.\frac{M}{Ls.Lr - M2}.\left|\overline{\Phi_s}\right|.\left|\overline{\Phi_r}\right|.\sin\theta_{sr} \tag{31}$$

where θ_{sr} is the angle difference between $\overline{\Phi_s}$ and $\overline{\Phi_r}$.

It is important to mention that the rotor flux vector time constant is bigger than the time constant of the stator flux vector. This can be demonstrated by writing the transfer function from the stator flux vector to the rotor flux vector. For a two-phase reference related to the rotor: $\psi = p\theta$, the rotor electric equation becomes: $\overline{0} = R_r.\overline{I}_r + \frac{d\overline{\Phi}_r}{dt}$. From the flux-current relationships: $\overline{I}_r = \frac{\overline{\Phi}_r}{\sigma.L_r} - \frac{M}{\sigma.L_s.L_r}.\overline{\Phi}_s$. By substituting the expression of $\overline{I}_r$ in the rotor electric equation, the following transfer function is obtained:

$$\frac{\overline{\Phi}_r}{\overline{\Phi}_s} = \frac{M/L_r}{1+\sigma.\tau_r.p} \tag{32}$$

where $\tau_r = R_r/L_r$ is the rotor time constant. From equation (32), it is clear that the stator flux vector changes slowly compared to the stator flux vector.

Going back to the expression of the motor torque, equation (31), if the stator flux vector modulus is maintained constant, then the motor torque can be rapidly changed and controlled by changing the angle θ_{sr}. Thus the tangential component of $\Delta\overline{\Phi}_s = \overline{V}_s.\Delta T$ is for controlling the torque, and its radial component is for controlling $\left|\overline{\Phi}_s\right|$.

For a stator flux vector existing in sector i, the following stator voltage vector can is applied in order to have the desired variations of the stator flux vector modulus and the motor torque.

$\overline{V}_s$	Increase	Decrease
$\left\|\overline{\Phi}_s\right\|$	$\overline{V}_i$, $\overline{V}_{i+1}$ or $\overline{V}_{i-1}$	$\overline{V}_{i+2}$, $\overline{V}_{i-2}$ or $\overline{V}_{i+3}$
T_{em}	$\overline{V}_{i+1}$ or $\overline{V}_{i+2}$	$\overline{V}_{i-1}$ or $\overline{V}_{i-2}$

Table 1. Stator voltage vector for the desired variations of $\left|\overline{\Phi}_s\right|$ and T_{em}

The vectors $\overline{V}_i$ and $\overline{V}_{i+3}$ are not considered for controlling the torque because they increase the torque for the positive 30 degree half sector, and decrease it for the negative 30 degree half sector. They can be used if 12 sectors are considered for dividing the total locus.

By analyzing Table 1, it is possible to do a decoupled control of $\left|\overline{\Phi}_s\right|$ and T_{em}. For all the six sectors, Table 2 shows the good stator voltage vector that gives the desired variations of $\left|\overline{\Phi}_s\right|$ and T_{em}.

Fig. 22 shows the scheme of the DTC.

There are two different loops for controlling the stator flux vector modulus and the motor torque. The reference values of $\left|\overline{\Phi}_s\right|$ and T_{em} are compared with the estimated values. The resulting errors are fed into the two-level and three-level hysteresis comparators respectively. The outputs of the hysteresis comparators and the position of the stator flux vector are used as inputs for the look up table (selection table of Table 2).

$\lvert\overline{\Phi_s}\rvert$	T_{em}	S1	S2	S3	S4	S5	S6
FI	TI	V_2	V_3	V_4	V_5	V_6	V_1
	=	V_0	V_7	V_0	V_7	V_0	V_7
	TD	V_6	V_1	V_2	V_3	V_4	V_5
FD	TI	V_3	V_4	V_5	V_6	V_1	V_2
	=	V_7	V_0	V_7	V_0	V_7	V_0
	TD	V_5	V_6	V_1	V_2	V_3	V_4

Table 2. Stator voltage vector for the desired variations of $\lvert\overline{\Phi_s}\rvert$ and T_{em} in all sectors

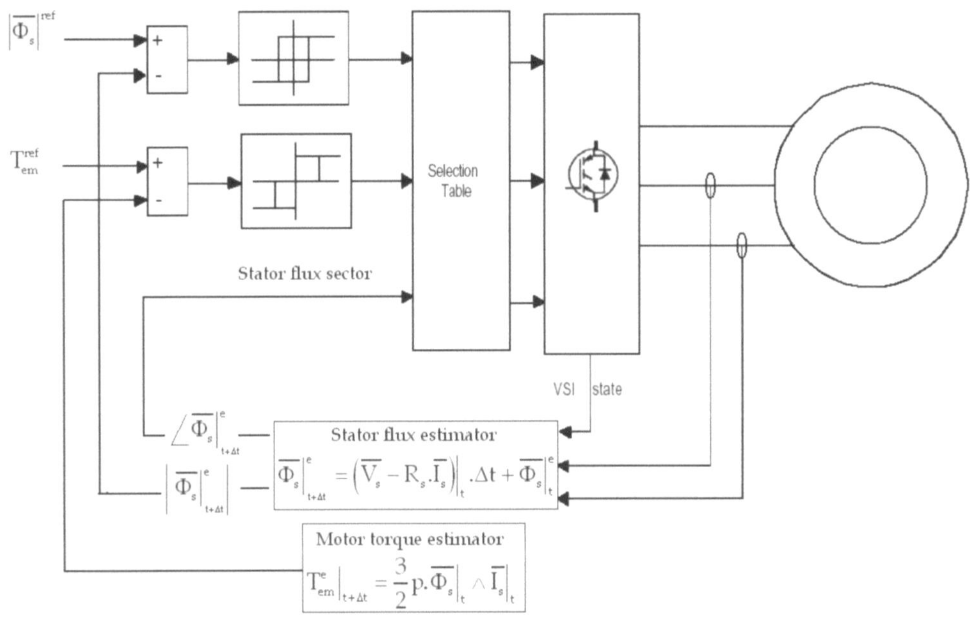

Fig. 22. Scheme of the DTC law

Usually, the estimation of the stator flux vector is done using the stator electric equation:

$$\overline{\Phi_s}\Big|_{t+\Delta t}^{e} = \left(\overline{V_s} - R_s.\overline{I_s}\right)\Big|_t .\Delta t + \overline{\Phi_s}\Big|_t^{e} \tag{33}$$

The accuracy of this flux estimator is highly dependent on the value of the stator winding resistor, which varies with the motor temperature.

This chapter proposes a new estimation technique that uses the rotor electric equation. It shows that it is less sensitive to the variation of the rotor resistor, but more sensitive to the variation of the saturation level. To overcome this problem, an adaptive estimator is proposed, based on a previous saturation phenomenon study.

5.2 Direct Torque Control Law for an induction machine for a fixed chopping frequency voltage source inverter

It is possible to develop the expression of a continuous optimal stator voltage vector that gives the desired variations of $\left|\overline{\Phi_s}\right|$ and T_{em} (C.A, Martins.; T.A, Meynard.; X, Roboam. & A.S, Carvalho2, 1999). The control voltages v_{sd}^{opt} and v_{sq}^{opt} that give the desired $\Delta T_{em}^{Des}/\Delta t$ and $\Delta\left|\overline{\Phi_s}\right|^{Des}/\Delta t$ are searched.

The expression of the motor torque derivative is:

$$\frac{dT_{em}}{dt} = p(\frac{d\Phi_{sd}}{dt}.i_{sq} + \Phi_{sd}.\frac{di_{sq}}{dt} - \frac{d\Phi_{sq}}{dt}.i_{sd} - \Phi_{sq}.\frac{di_{sd}}{dt}) \tag{34}$$

The expressions of $\frac{d\Phi_{sd}}{dt}$ and $\frac{d\Phi_{sq}}{dt}$ can be found from the stator electric equations in the fixed reference:

$$\begin{aligned} \frac{d\Phi_{sd}}{dt} &= v_{sd} - R_s.i_{sd} \\ \frac{d\Phi_{sq}}{dt} &= v_{sq} - R_s.i_{sq} \end{aligned} \tag{35}$$

By writing the expressions of i_{sd} and i_{sq} from the flux-current relationships, the derivatives of these currents versus time are:

$$\begin{aligned} \frac{di_{sd}}{dt} &= \frac{1}{\sigma.L_s}.\left(\frac{d\Phi_{sd}}{dt} - \frac{M}{Lr}.\frac{d\Phi_{rd}}{dt}\right) \\ \frac{di_{sq}}{dt} &= \frac{1}{\sigma.L_s}.\left(\frac{d\Phi_{sq}}{dt} - \frac{M}{Lr}.\frac{d\Phi_{rq}}{dt}\right) \end{aligned} \tag{36}$$

The rotor electric equations give the expressions of the rotor fluxes derivatives versus time:

$$\begin{aligned} \frac{d\Phi_{rd}}{dt} &= -\frac{d(p\theta)}{dt}.\Phi_{rq} - R_r.i_{rd} = -\frac{d(p\theta)}{dt}.\Phi_{rq} - \frac{R_r}{M}.(\Phi_{sd} - L_s.i_{sd}) \\ \frac{d\Phi_{rq}}{dt} &= \frac{d(p\theta)}{dt}.\Phi_{rd} - R_r.i_{rq} = \frac{d(p\theta)}{dt}.\Phi_{rd} - \frac{R_r}{M}.(\Phi_{sq} - L_s.i_{sq}) \end{aligned} \tag{37}$$

The final expressions of the stator fluxes derivatives can be obtained by substituting Φ_{rd} and Φ_{rq} by their expressions using stator variables:

$$\begin{aligned} \frac{d\Phi_{rd}}{dt} &= -\frac{d(p\theta)}{dt}.\Phi_{rq} - R_r.i_{rd} = -\frac{d(p\theta)}{dt}.\frac{L_r}{M}(\Phi_{sq} - \sigma.L_s.i_{sq}) - \frac{R_r}{M}.(\Phi_{sd} - L_s.i_{sd}) \\ \frac{d\Phi_{rq}}{dt} &= \frac{d(p\theta)}{dt}.\Phi_{rd} - R_r.i_{rq} = \frac{d(p\theta)}{dt}.\frac{L_r}{M}(\Phi_{sd} - \sigma.L_s.i_{sd}) - \frac{R_r}{M}.(\Phi_{sq} - L_s.i_{sq}) \end{aligned} \tag{38}$$

By replacing (38) in (36), the stator currents derivatives become:

$$\begin{aligned}\frac{di_{sd}}{dt} &= \frac{1}{\sigma.L_s}.\left(\frac{d\Phi_{sd}}{dt} - \frac{M}{Lr}.\left(-\frac{d(p\theta)}{dt}.\frac{L_r}{M}(\Phi_{sq} - \sigma.L_s.i_{sq}) - \frac{R_r}{M}.(\Phi_{sd} - L_s.i_{sd})\right)\right)\\ \frac{di_{sq}}{dt} &= \frac{1}{\sigma.L_s}.\left(\frac{d\Phi_{sq}}{dt} - \frac{M}{Lr}.\left(\frac{d(p\theta)}{dt}.\frac{L_r}{M}(\Phi_{sd} - \sigma.L_s.i_{sd}) - \frac{R_r}{M}.\left(\Phi_{sq} - L_s.i_{sq}\right)\right)\right)\end{aligned} \quad (39)$$

The motor torque derivative is finally obtained as a function of stator voltage, stator current and stator flux components.

$$\frac{dT_{em}}{dt} = p(v_{sd}.K_{sq} - v_{sq}.K_{sd} + K_1) \quad (40)$$

with

$$K_{sd} = i_{sd} - \frac{\Phi_{sd}}{\sigma.L_s},\ K_{sq} = i_{sq} - \frac{\Phi_{sq}}{\sigma.L_s},\ R_s^{'} = R_s + \frac{L_s}{L_r}.R_r,\ \left|\overline{\Phi_s}\right| = \sqrt{\frac{2}{3}}\sqrt{\left(\Phi_{sd}\right)^2 + \left(\Phi_{sq}\right)^2} \text{ and}$$

$$\begin{aligned}K_1 &= -\frac{R_s^{'}}{\sigma.L_s.p}T_{em} - \frac{3.p.\frac{d\theta}{dt}}{2.\sigma.L_s}.\left|\overline{\Phi_s}\right|^2 + p.\frac{d\theta}{dt}(\Phi_{sd}.i_{sd} - \Phi_{sq}.i_{sq}) =\\ &-\frac{R_s^{'}}{\sigma.L_s.p}T_{em} - \frac{3.p.\frac{d\theta}{dt}}{2.\sigma.L_s}.\left|\overline{\Phi_s}\right|^2 + p.\frac{d\theta}{dt}(\Phi_{sd}.(\Phi_{sd} - \sigma.L_s.i_{sd}) - \Phi_{sq}.(\Phi_{sq} - \sigma.L_s.i_{sq}))\end{aligned}$$

Using the stator electric equations, the derivative of $\left|\overline{\Phi_s}\right| = \sqrt{\frac{2}{3}}\sqrt{\left(\Phi_{sd}\right)^2 + \left(\Phi_{sq}\right)^2}$ can be found:

$$\frac{d\left|\overline{\Phi_s}\right|}{dt} = \frac{2}{3.\left|\overline{\Phi_s}\right|}\left(\Phi_{sd}.v_{sd} + \Phi_{sq}.v_{sq} - R_s.(\Phi_{sd}.i_{sd} + \Phi_{sq}.i_{sq})\right) \quad (41)$$

Finally, the optimal control v_{sd}^{opt} and v_{sq}^{opt} are obtained by replacing the desired variations during the sampling period $\Delta T_{em}^{Des}/\Delta t$ and $\Delta\left|\overline{\Phi_s}\right|^{Des}/\Delta t$ in equations (40) and (41) instead of the derivatives $\frac{dT_{em}}{dt}$ and $\frac{d\left|\overline{\Phi_s}\right|}{dt}$.

$$v_{sd}^{opt} = \frac{\frac{3}{2}.\Phi_s.\left(\Delta\left|\overline{\Phi_s}\right|^{Des}/\Delta t\right).K_{sd} + R_s.K_{sd}.(\Phi_{sd}.i_{sd} + \Phi_{sq}.i_{sq}) + \Phi_{sq}.\left(\left(\Delta T_{em}^{Des}/\Delta t\right)/p - K_1\right)}{\Phi_{sd}.K_{sd} + \Phi_{sq}.K_{sq}} \quad (42)$$

At each sampling period the stator currents are measured and the stator fluxes are estimated from the stator electric equations. Actual values of $\left|\overline{\Phi_s}\right|$ and T_{em} are then calculated. Using the reference values for the motor torque and for the modulus of the stator flux vector, the desired variations during a period of Δt are calculated and used in equation (42) to find the optimal values of the control v_{sd}^{opt} and v_{sq}^{opt}.

This control strategy can be implemented using a fixed chopping frequency source voltage inverter.

5.3 Sensitivity study of the DTC stator flux estimator to the variation of the stator resistor

The classical stator flux estimator used generally for the DTC is based on the stator electric equation written in a fixed two-phase reference: $\Psi=0$, $\underline{V_s} = R_s \underline{I_s} + \frac{d\underline{\Phi_s}}{dt}$. It is clear that this estimator is highly affected by the stator resistor variations, due to the motor temperature variations, especially for low speed applications.

The DTC for fixed chopping frequency of the voltage source inverter is implemented in A_MOS program. Fig. 23 shows simulation results of a 45(KW) induction machine controlled by the DTC law with the previous estimator. A difference of 15% between the motor stator resistor and its value implemented in the control estimator is considered.

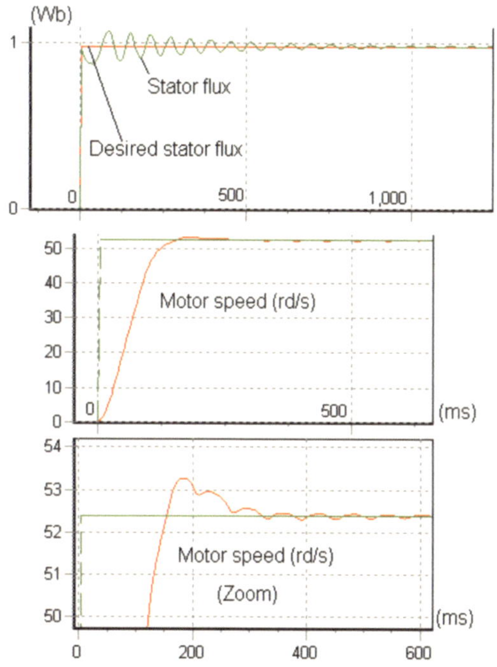

Fig. 23. Stator electric equation estimator results with 15% increase for the stator resistor

The difference may cause oscillations to the motor speed, and this problem is more important for low speed.

6. New stator flux estimator for the DTC

If the motor speed is available, the stator fluxes can be calculated from the flux currents relationships:

$$\Phi_{sd} = \sigma L_s.i_{sd} + \frac{M}{L_r}.\Phi_{rd}$$
$$\Phi_{sq} = \sigma L_s.i_{sq} + \frac{M}{L_r}.\Phi_{rq} \qquad (43)$$

At each sampling period the stator currents are measured and the rotor fluxes are calculated using the rotor electric equations:

$$\frac{d\Phi_{rd}}{dt} = -R_r.i_{rd} - \frac{d(p\theta)}{dt}\Phi_{rq} = -R_r.\left(\frac{\Phi_{sd}}{M} - \frac{L_s}{M}.i_{sd}\right) - \frac{d(p\theta)}{dt}\Phi_{rq}$$
$$\frac{d\Phi_{rq}}{dt} = -R_r.i_{rq} + \frac{d(p\theta)}{dt}\Phi_{rd} = -R_r.\left(\frac{\Phi_{sq}}{M} - \frac{L_s}{M}.i_{sq}\right) + \frac{d(p\theta)}{dt}\Phi_{rd} \qquad (44)$$

The calculation of the stator fluxes using equations (43) and (44) does not require the stator resistor, thus any change in its value has no influence. In fact, the estimator uses the value of the rotor resistor which determines the time constant of the rotor flux. It is obvious that the accuracy in measuring the rotor resistor has no big effect on estimating the stator flux vector using the two previous equations. This is due to the fact that the stator fluxes time constant is smaller than the time constant of the rotor fluxes, as it was shown previously. Fig. 24 shows that for an increase of 15% in the rotor resistor value, the DTC with the new estimator gives better results.

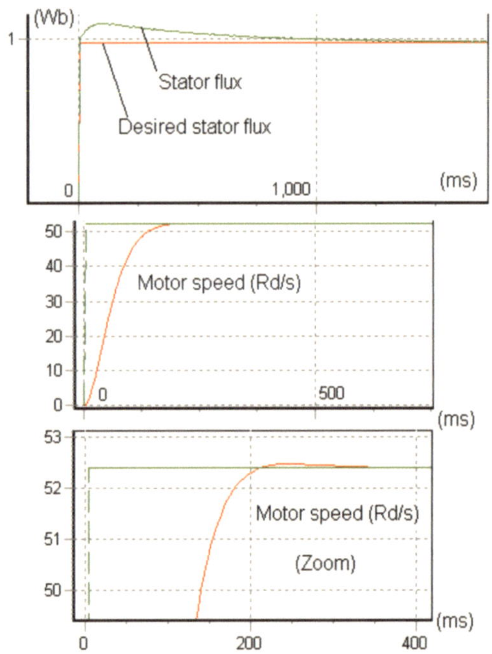

Fig. 24. New estimator results with 15% increase for the rotor resistor

The new method of estimating the stator fluxes requires the knowledge of a greater number of motor parameters. It is clear that the new estimator will be more sensitive to the variation of the induction motor saturation level, since it uses the cyclic inductances.
Fig. 25 shows simulation results of the induction motor controlled by the DTC with the new stator fluxes estimator. During the flux weakening phase a big difference between the desired stator flux and the real one was obtained. This influences the dynamic behavior of the speed when applying a load torque of 110(Nm) at 3(s).

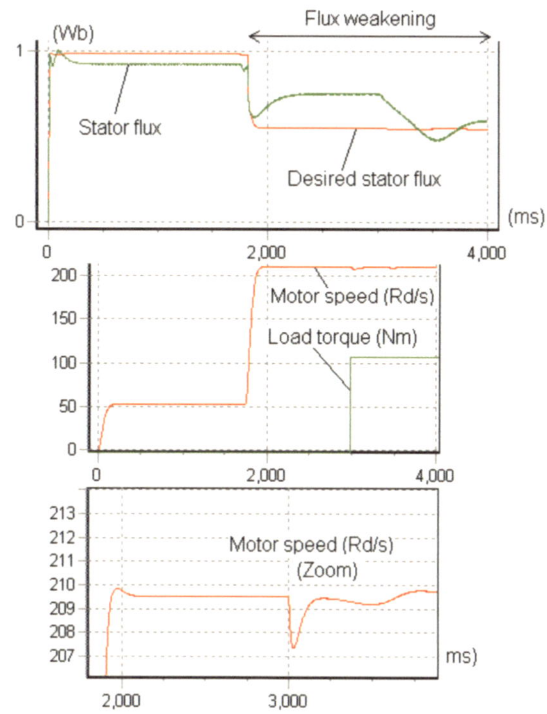

Fig. 25. New estimator sensitivity to the variation of saturation level

It is possible to overcome this problem by tuning the new estimator parameters at each sampling period according to the magnetizing current modulus that can be estimated as described in the section of the saturated FOC.
The scheme of the DTC using this estimator is presented in Fig. 26.
Fig. 27 shows the improvement of dynamic behavior of the induction motor after implementing the adaptive estimator of Fig. 26.

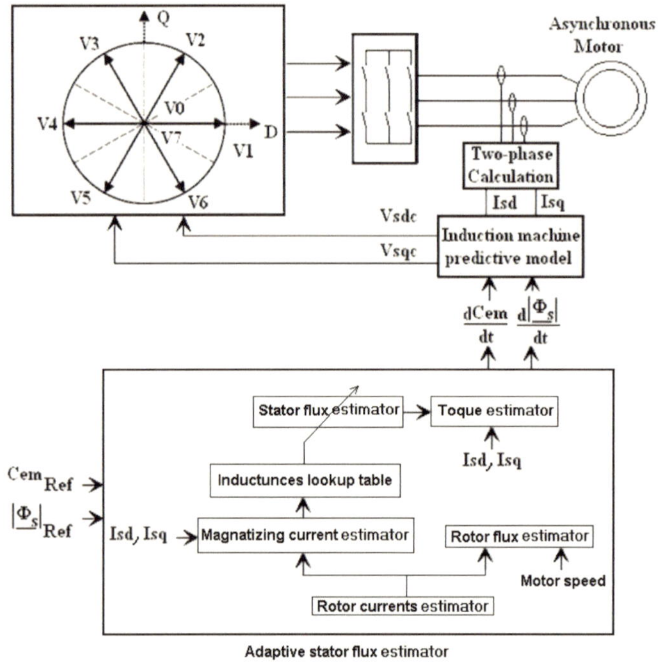

Fig. 26. Adaptive stator fluxes estimator

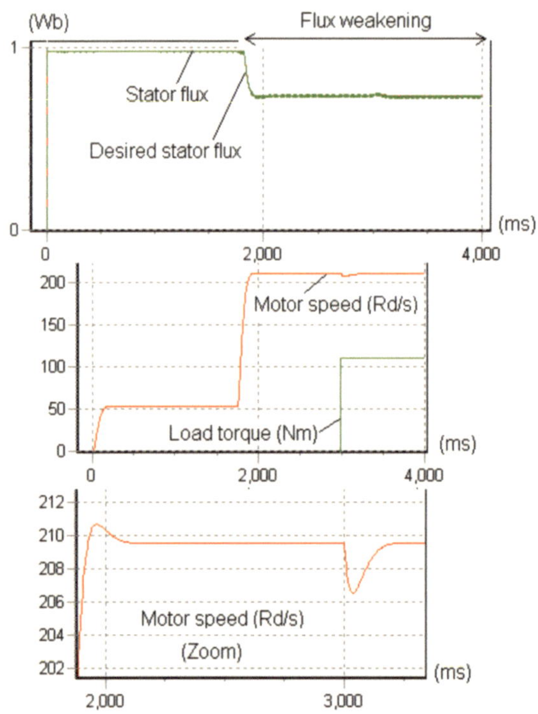

Fig. 27. 45(KW) motor behavior with the adaptive stator flux estimator

It is possible to combine the two previous estimators. The first one can be used for high speed, while the second estimator can be used for low speed range. In this case, there is no need to program the adaptive estimator of the stator fluxes, since it will not work during the flux weakening phase.

7. Conclusion

This chapter has presented a full study of the magnetic state variation of the induction motor. Using a finite elements calculation program, it was possible to establish a two-phase model that takes into account the variation of the saturation level. A very simple resolution method of this new model was presented. The dynamic response of the new model was validated by comparing it to the dynamic response of the induction motor given by the finite element calculation program. After establishing the new model it was possible to review the advanced control laws like the FOC and the DTC laws. A new saturated FOC law was developed in order to enhance the dynamic behavior of the motor during the flux weakening phase, because of the difference between the motor cyclic inductances values and the values of the cyclic inductances introduced in the controllers. Concerning the DTC law, it was shown that a small error in the stator resistor value will highly influence the stator flux estimation, which is done using the stator electric equation. A new stator fluxes estimator was developed using rotor electric equations. This estimator is less sensitive to the motor temperature variation, but it is more sensitive to the variation of the saturation level. An adaptive solution was proposed to tune the estimator parameters according to the saturation level of the motor. Nevertheless the adaptive part added to the DTC algorithm, its computation time remains very small comparing to the FOC algorithm that takes into account the variation of the saturation level. It is important to mention that it is possible to combine the classical estimator and the new estimator according to the speed range. The classical estimator can be used at high speed, but at low speed, it is better to use the new stator flux estimator.

8. References

Grotstollen, H. & Wiesing, J. (1995). Torque capability and control of saturated induction motor over a wide range of flux weakening, *Transaction on Industrial Electronics*, Vol. 42, No. 4, (August 1990) page numbers (374-381).

Vas, P. & Alakula, M. (1990). Field oriented control of saturated induction motors, *IEEE Transaction on Energy Conversion*, Vol. 5, No. 1, (March 1990) page numbers (218-224), ISSN 0885-8969.

Kasmieh, T. & Lefevre, Y. (1998). Establishment of two-phase non-linear simulation model of the dynamic operation of the induction motor, *EPJ European Physical Journal*, Vol. 1, No. 1, (January 1998) page numbers (57-66).

Vas, P. (1981). Generalized transient analysis of saturated a.c motors, *Archiv fur Elektrotechnik*, Vol. 64, No. 1-2, (June 1981) page numbers (57-62).

Kasmieh, T. (2008). Adaptive stator flux estimator for the induction motor Direct Torque Control, *Proceedings of SPEEDAM 2008*, pp. 1239-1241, Ischia, June 2008, Italy.

Blaschke, F. (1972). The principal of field orientation as applied to the new trans-vector closed-loop control system for rotating field machines, *Siemens Review*, (May (1972).

Kasmieh, T.(2002), Presentation of a powerful opened simulator for the saturated induction motor traction system, *Proceedings of SPEEDAM 2002*, (June 2002), pp. A1 24-A1 37, Ravello, June 2002, Italy.

Noguchi, T. & Takahashi, I. (1984). Quick torque response control of an induction motor based on a new concept, *IEEE Tech,* Vol. RM84-76, (September 1984) page numbers (61-70).

Depenbrock, M. & Steimel A.(1990). High power traction drives and convertors. *Proc. of Elect. Drives Symp.'90*, pp. 1–9, Capri,1990, Italy.

C.A, Martins.; T.A, Meynard.; X, Roboam. & A.S, Carvalho2. (1999). A predictive sampling scale model for direct torque control of the induction machine fed by multilevel voltage-source inverters. *European Physical Journal-Applied Physics*, AP. 5, (1999) page numbers (51-61).

5

Control of a Double Feed and Double Star Induction Machine Using Direct Torque Control

Leila Benalia
Department of electrical Engineering
Batna University, Rue Chahid Med El Hadi boukhlouf
Algeria

1. Introduction

DTC is an excellent solution for general-purpose induction drives in very wide range The short sampling time required by the TC schemes makes them suited to a very fast torque and flux controlled drives as well the simplicity of the control algorithm. DTC is inherently a motion sensor less control method.

2. Objective of the work

This chapter describes the control of doubly fed induction machine (DFIM) and the control of doubly star asynchronous machine (DSAM), using direct torque control (DTC).

3. Principe du control direct du couple

Direct torque control is based on the flux orientation, using the instantaneous values of voltage vector.
An inverter provides eight voltage vectors, among which two are zeros (Roys & Courtine, 1995), (Carlos et al., 2005). This vector are chosen from a switching table according to the flux and torque errors as well as the stator flux vector position. In this technique, we don't need the rotor position in order to choose the voltage vector. This particularity defines the DTC as an adapted control technique of ac machines and is inherently a motion sensor less control method (Casdei et al., 2001), (Kouang-kyun et al., 2000).

4. Double feed induction machine (DFIM)

In the training of high power as the rolling mill, there is a new and original solution using a double feed induction motor (DFIM). The stator is feed by a fixed network while the rotor by a variable supply which can be either a voltage or current source.
The three phase induction motor with wound rotor is doubly fed when, as well as the stator windings being supplied with three phase power at an angular frequency ω_s, the rotor windings are also fed with three phase power at a frequency ω_{rr}.

Under synchronous operating conditions, as shown in (Prescott & Alii., 1958), (Petersson., 2003) , the shaft turns at an angular velocity ω_r , such that:

$$\omega_r = \omega_s + \omega_{rr}$$

The sign on the right hand side is (+) when the phase sequences of the three phase supplies to the stator and rotor are in opposition and (-) when these supplies have the same phase sequence. The rotational velocity of the shaft, ω_r , is expressed in electric radians per second, to normalize the number of poles.

4.1 Double feed induction machine modelling

Using the frequently adopted assumptions, like sinusoid ally distributed air-gap flux density distribution and linear magnetic conditions and considering the stator voltages ($v_{s\alpha}$, $v_{s\beta}$) and rotor voltages ($v_{r\alpha}$, $v_{r\beta}$) as control inputs, the stator flux ($\Phi_{s\alpha}$, $\Phi_{s\beta}$), and the rotor current ($i_{r\alpha}$, $i_{r\beta}$) as state variables. In the referential axis fixed in relation to the stator, the following electrical equations are deduced:

$$\begin{bmatrix} V_{s\alpha} \\ V_{s\beta} \end{bmatrix} = \begin{bmatrix} R_s & 0 \\ 0 & R_s \end{bmatrix} \begin{bmatrix} I_{s\alpha} \\ I_{s\beta} \end{bmatrix} + \frac{d}{dt} \begin{bmatrix} \Phi_{s\alpha} \\ \Phi_{s\beta} \end{bmatrix} \tag{1}$$

$$\begin{bmatrix} V_{r\alpha} \\ V_{r\beta} \end{bmatrix} = \begin{bmatrix} R_r & 0 \\ 0 & R_r \end{bmatrix} \begin{bmatrix} I_{r\alpha} \\ I_{r\beta} \end{bmatrix} + \frac{d}{dt} \begin{bmatrix} \Phi_{r\alpha} \\ \Phi_{r\beta} \end{bmatrix} + \begin{bmatrix} 0 & \omega \\ -\omega & 0 \end{bmatrix} \begin{bmatrix} \Phi_{r\alpha} \\ \Phi_{r\beta} \end{bmatrix} \tag{2}$$

Expressions of fluxes are given by:

$$\begin{cases} \Phi_{s\alpha} = l_s I_{s\alpha} + M I_{r\alpha} \\ \Phi_{s\beta} = l_s I_{s\beta} + M I_{r\beta} \\ \Phi_{r\alpha} = l_r I_{r\alpha} + M I_{s\alpha} \\ \Phi_{r\beta} = l_r I_{r\beta} + M I_{s\beta} \end{cases} \tag{3}$$

The mathematical model is written as a set of equations of state, both for the electrical and mechanical parts:

$$\frac{dX}{dt} = \dot{X} = AX + BU \tag{4}$$

Where:

$$X = \begin{bmatrix} I_{r\alpha} \\ I_{r\beta} \\ \Phi_{s\alpha} \\ \Phi_{s\beta} \end{bmatrix} \text{ and } U = \begin{bmatrix} V_{s\alpha} \\ V_{s\beta} \\ V_{r\alpha} \\ V_{r\beta} \end{bmatrix} \tag{5}$$

The matrices A and B are given by:

$$A = \begin{bmatrix} \frac{-1}{T'_s\delta} & \omega_r & \frac{1-\delta}{\delta M T_s} & \frac{1-\delta}{\delta M}\omega_r \\ -\omega_r & \frac{-1}{T'_s\delta} & -\frac{1-\delta}{\delta M}\omega_r & \frac{1-\delta}{\delta M T_s} \\ \frac{M}{T_s} & 0 & -\frac{1}{T_s} & 0 \\ 0 & \frac{M}{T_s} & 0 & -\frac{1}{T_s} \end{bmatrix} \tag{6}$$

$$B = \begin{bmatrix} -\frac{1-\delta}{\delta M} & 0 & \frac{1}{L_r\delta} & 0 \\ 0 & -\frac{1-\delta}{\delta M} & 0 & \frac{1}{L_r\delta} \\ 1 & 0 & 0 & 0 \\ 0 & 1 & 0 & 0 \end{bmatrix} \tag{7}$$

$$J\frac{d\Omega}{dt} = C_{em} - C_r - K_f\Omega. \tag{8}$$

Where J is the moment of inertia of the revolving parts, K_f is the coefficient of viscous friction, arising from the bearings and the air flowing over the motor, and C_r is the load couple.
The equation of the electromagnetic torque is:

$$C_e = \frac{3pM}{2L_s}(\Phi_{s\alpha}I_{r\beta} - \Phi_{s\beta}I_{r\alpha}) \tag{9}$$

The block diagram for the direct torque and flux control applied to the double feed induction motor is shown in figure 1.The stator flux Ψ_{ref} and the torque C_{emref} magnitudes are compared with respective estimated values and errors are processed through hysteresis-band controllers.
Stator flux controller imposes the time duration of the active voltage vectors, which move the stator flux along the reference trajectory, and torque controller determinates the time duration of the zero voltage vectors, which keep the motor torque in the defined-by hysteresis tolerance band (Kouang-kyun et al.,2000), (Xu & Cheng.,1995). Finally, in every sampling time the voltage vector selection block chooses the inverter switching state, which reduces the instantaneous flux and torque errors (Presada et al., 1998).

5. Simulation results machine

Figure 2 refer in order, to the variation in magnitude of the following quantities, speed, flux and electromagnetic torque obtained while starting up the induction motor initially under no load then connecting the nominal load. During the starting up with no load the speed reaches rapidly its reference value without overtaking, however when the nominal load is applied a little overtaking is noticed and the command reject the disturbance. The excellent dynamic performance of torque and flux control is evident.

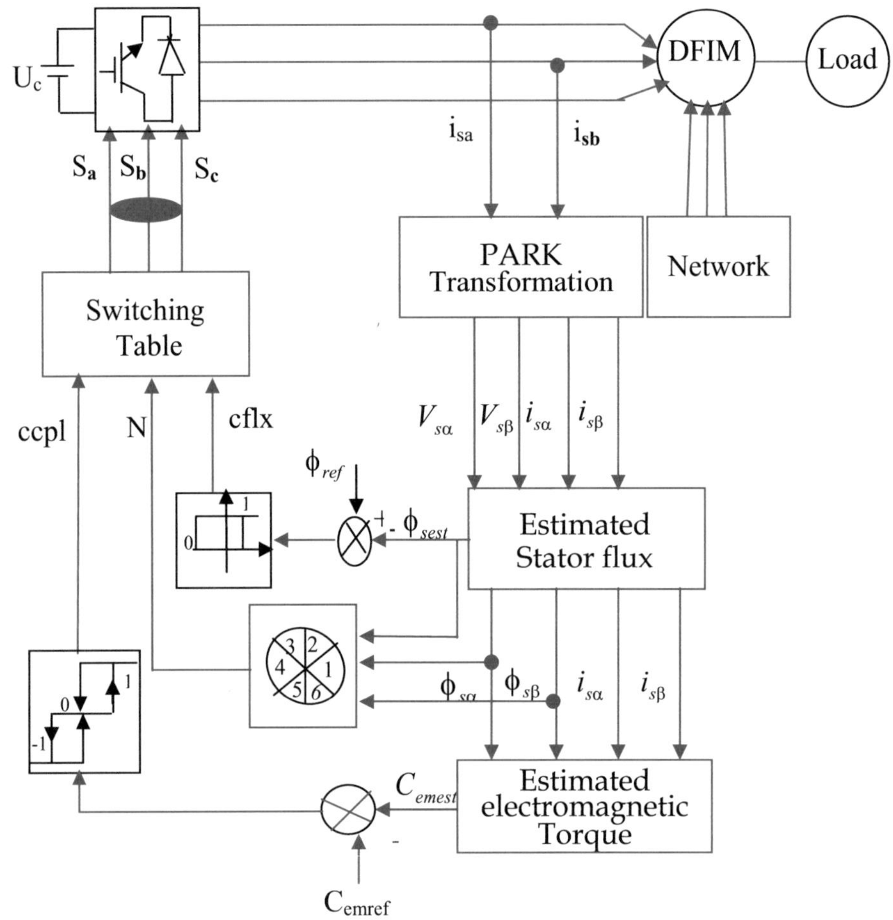

Fig. 1. DTC applied to double feed induction machine

6. Robust control of the IP regulator

a) Speed variation

Figure 3 shows the simulation results obtained for a speed variation for the values: (Ω_{ref} = 157, 100 and 157 rad/s), with the load of 3 N.m applied at t =0.8s. This results show that the variation lead to the variation in flux and the torque. The response of the system is positive, the speed follow its reference value while the torque return to its reference value with a little error.

b) Speed reversal of rated value

The excellent dynamic performance of torque control is evident in figure 4, which shows torque reversal for speed reversal of (157, -157 rad/s), with a load of 5N.m applied at t=1 s. The speed and torque response follow perfectly their reference values with the same response time. The reversal speed leads to a delay in the speed response, to a peak oscillation the current as well as a fall in the flux magnitude which stabilise at its reference value.

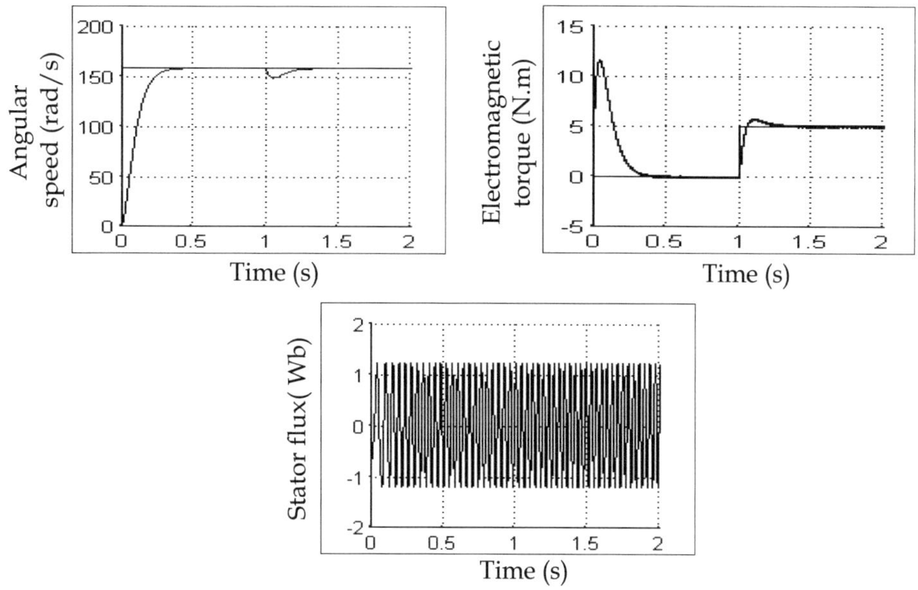

Fig. 2. Simulation results obtained with an IP regulator

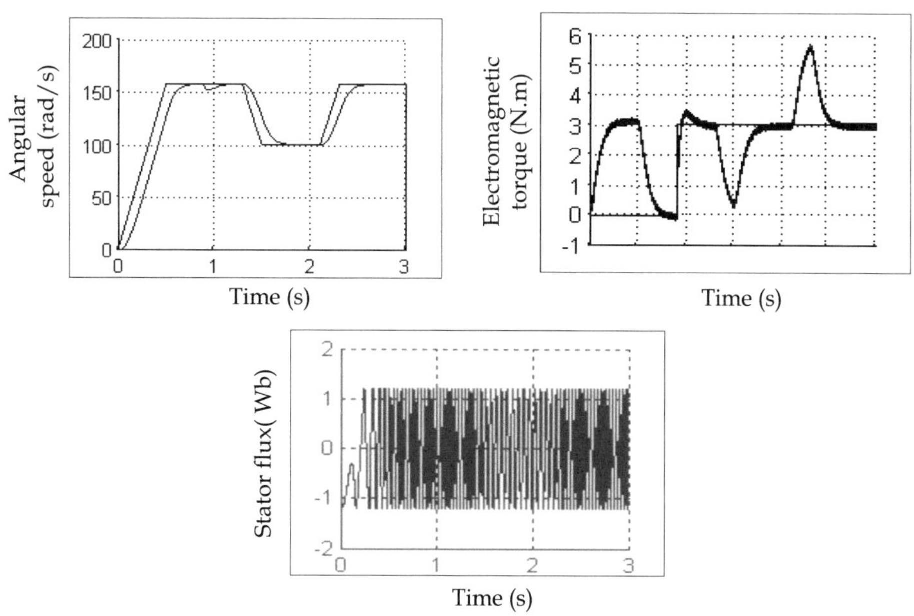

Fig. 3. Robust control for a speed variation

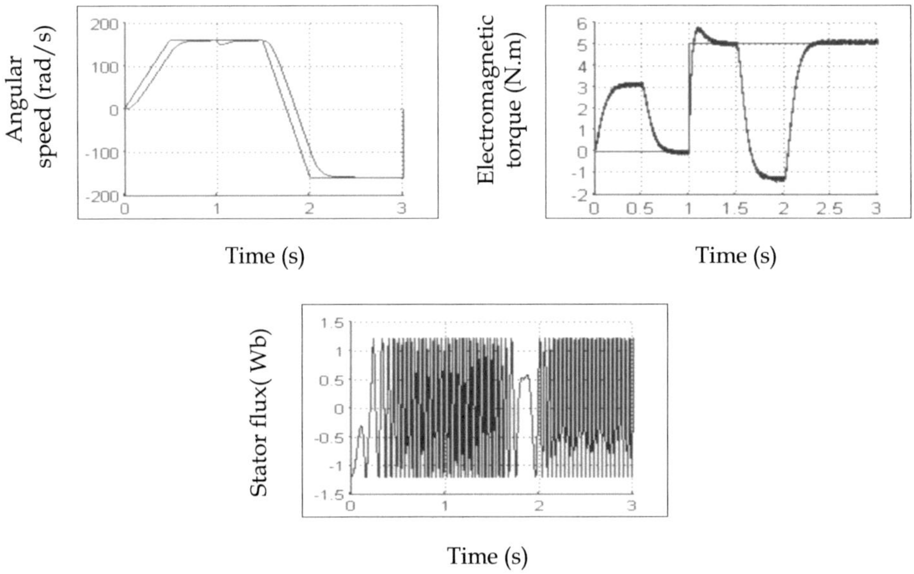

Fig. 4. Robust control under reversal speed

c) Robust control for load variation

The simulation results obtained for a load variation (C_r = 3 N.m, 6 N.m) in figure 5, show that the speed, the torque and the flux are inflated with this variation. Indeed the torque and the speed follow their reference values.

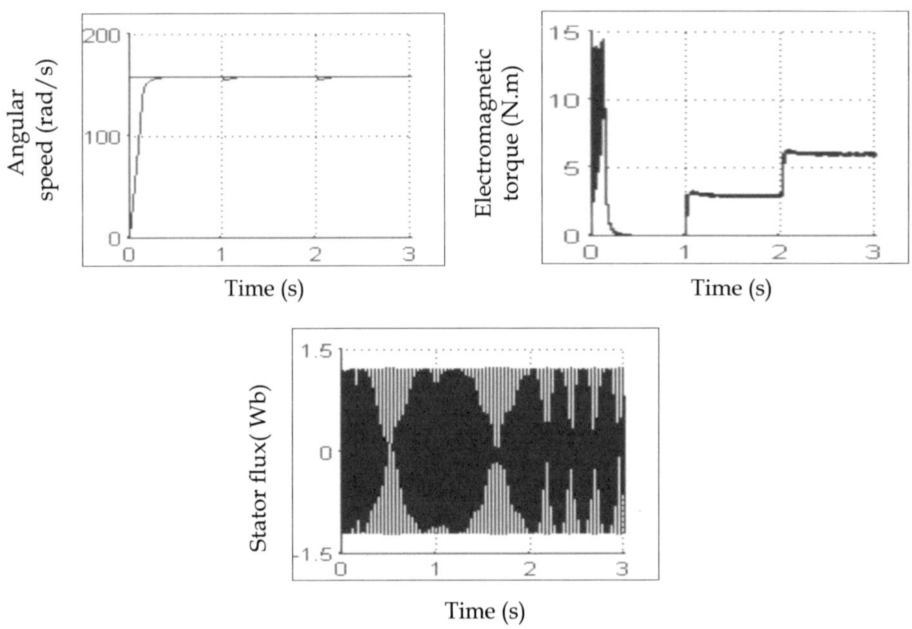

Fig. 5. Robust control under load variation

d) Robust control of the regulator under stator resistance variation

In order to verified the robustess of the regulator under motor parameters variations we carried out a test for a variation of 50% in the value of stator resistance at tile t= 1.5s. The speed is fixed at 157 rad/s and a resistant torque of 5 N.m is applied at t= 1s. Figure 6 shows the in order the torque response, the current, the stator flux and the speed. The results indicate that the regulator is very sensitive to the resistance change which results in the influence on the torque and the stator flux.

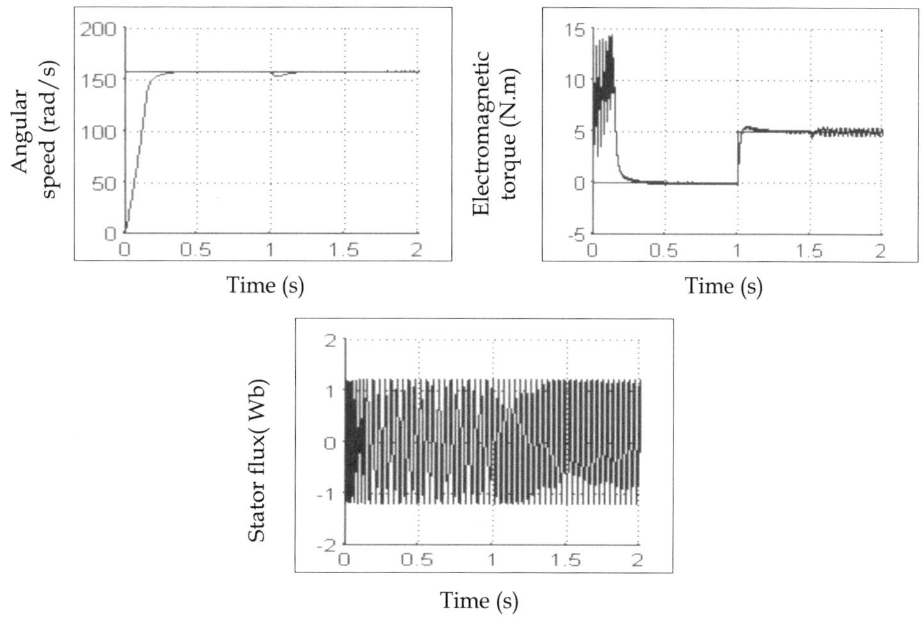

Fig. 6. Robust control under stator resistance variation

7. Double star induction machine (DSIM)

For the last 20 years the induction machines with a double star have been used in many applications for their performances in the power fields because of their reduced pulsation when the torque is minimum (Kalantari et al.,2002).The double stator induction machine needs a double three phase supply which has many advantages. It minimise the torque pulsations and uses a power electronics components which allow a higher commutation frequency compared to the simple machines. However the double stator induction machines supplied by a source inverter generate harmonic which results in supplementary losses (Hadiouche et al., 2000). The double star induction machine is not a simple system, because a number of complicated phenomena's appears in its function, as saturation and skin effects (Hadiouche et al., 2000).

The double star induction machine is based on the principle of a double stators displaced by $\alpha=30^0$ and rotor at the same time. The stators are similar to the stator of a simple induction machine and fed with a 3 phase alternating current and provide a rotating flux.

Each star is composed by three identical windings with their axes spaced by $2\pi/3$ in the space. Therefore, the orthogonality created between the two oriented fluxes, which must be

strictly observed, leads to generate decoupled control with an optimal torque (Petersson., 2003).
This is a maintenance free machine.
The machine studied is represented with two stars windings: $A_{s1}B_{s1}C_{s1}$ et $A_{s2}B_{s2}C_{s2}$ which are displaced by $\alpha = 30°$ and thee rotorical phases: A_r B_r C_r.

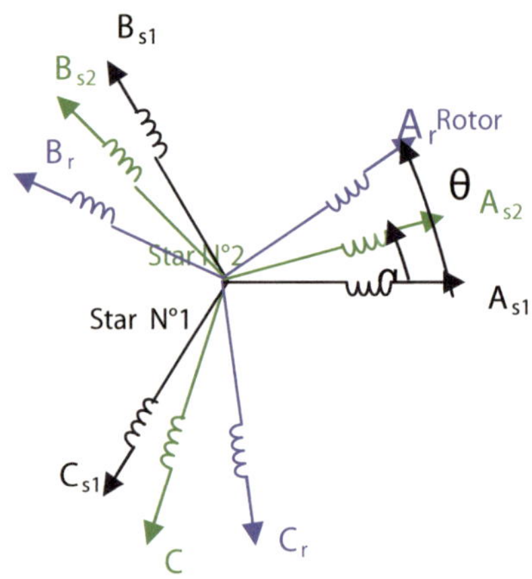

Fig. 7. Double star winding representation

8. Double star induction machine modeling

The mathematical model is written as a set of state equations, both for the electrical and mechanical parts:

$$\begin{aligned}\left[V_{abc,s1}\right]&=\left[R_{s1}\right]\left[I_{abc,s1}\right]+\frac{d}{dt}\left[\Phi_{abc,s1}\right].\\ \left[V_{abc,s2}\right]&=\left[R_{s2}\right]\left[I_{abc,s2}\right]+\frac{d}{dt}\left[\Phi_{abc,s2}\right].\\ \left[V_{abc,r}\right]&=\left[R_{r}\right]\left[I_{abc,r}\right]+\frac{d}{dt}\left[\Phi_{abc,r}\right].\end{aligned} \tag{1}$$

$$J\frac{d\Omega}{dt}=C_{em}-C_r-K_f\Omega. \tag{2}$$

Where:
J is the moment of inertia of the revolving parts.
K_f is the coefficient of viscous friction, arising from the bearings and the air flowing over the motor.
C_{em} is the electromagnetic torque.
The electrical state variables are the flux, transformed into vector $[\Phi]$ by the "dq" transform, while the input are the "dq" transforms of the voltages, in vector [V].

$$\frac{d}{dt}[\Phi]=[A].[\Phi]+[B].[v] \tag{3}$$

$$[\Phi]=\begin{bmatrix}\Phi_{ds1}\\ \Phi_{ds2}\\ \Phi_{qs1}\\ \Phi_{qs2}\\ \Phi_{dr}\\ \Phi_{qr}\end{bmatrix}\quad [V]=\begin{bmatrix}v_{ds1}\\ v_{ds2}\\ v_{qs1}\\ v_{qs2}\end{bmatrix} \tag{4}$$

The equation of the electromagnetic torque is given by

$$Cem = p\,\frac{Lm}{Lm+Lr}\,(\Phi_{dr}\,(i_{qs1}+i_{qs2})-\Phi_{qr}\,(i_{ds1}+i_{ds2})) \tag{5}$$

The flux equation is:

$$\Phi_{md}=L_a\left(\frac{\Phi_{ds1}}{L_{s1}}+\frac{\Phi_{ds2}}{L_{s2}}+\frac{\Phi_{dr}}{L_r}\right) \tag{6}$$

$$\Phi_{mq}=L_a\left(\frac{\Phi_{qs1}}{L_{s1}}+\frac{\Phi_{qs2}}{L_{s2}}+\frac{\Phi_{qr}}{L_r}\right) \tag{7}$$

Given that the "dq"axes are fixed in the synchronous rotating coordinate system, we have:

$$[A]=\begin{bmatrix}a_{11}&a_{12}&a_{13}&a_{14}&a_{15}&a_{16}\\ a_{21}&a_{22}&a_{23}&a_{24}&a_{25}&a_{26}\\ a_{31}&a_{32}&a_{33}&a_{34}&a_{35}&a_{36}\\ a_{41}&a_{42}&a_{43}&a_{44}&a_{45}&a_{46}\\ a_{51}&a_{52}&a_{53}&a_{54}&a_{55}&a_{56}\\ a_{61}&a_{62}&a_{63}&a_{64}&a_{65}&a_{66}\end{bmatrix} \tag{8}$$

$$[B]=\begin{bmatrix}1&0&0&0\\ 0&1&0&0\\ 0&0&1&0\\ 0&0&0&1\\ 0&0&0&0\\ 0&0&0&0\end{bmatrix} \tag{9}$$

Where:

$a_{11}=a_{33}=\dfrac{R_{s1}L_a}{L_{s1}^2}-\dfrac{R_{s1}}{L_{s1}}$

$a_{12}=a_{34}=\dfrac{R_{s1}L_a}{L_{s1}L_{s2}}$

ω_s $a_{13}= a_{24}= - a_{31}= - a_{42}=$

$a_{14}= a_{16} =a_{23} = a_{26}= a_{32}=a_{35}=a_{41}=a_{45} =a_{53}= a_{54}=a_{61}=a_{62} = 0$

$a_{15}= a_{36}=\dfrac{R_{s1}L_a}{L_r L_{s1}}$, $a_{21}= a_{43}=\dfrac{R_{s2}L_a}{L_{s1}L_{s2}}$

$a_{22}= a_{44}=\dfrac{R_{s2}L_a}{L_{s2}^2}-\dfrac{R_{s1}}{L_{s1}}$, $a_{25}= a_{46}=\dfrac{R_{s2}L_a}{L_r L_{s2}}$

$a_{51}= a_{63}=\dfrac{R_r L_a}{L_r L_{s1}}$, $a_{52}= a_{64}=\dfrac{R_r L_a}{L_r L_{s2}}$

$a_{55}= a_{66}=\dfrac{R_r L_a}{L_{r2}^2}-\dfrac{R_r}{L_r}$, $a_{56}= - a_{65}= \omega_{gl}$

Figure 8 shows the block diagram for the direct torque and flux control applied to the double star induction motor shown in.

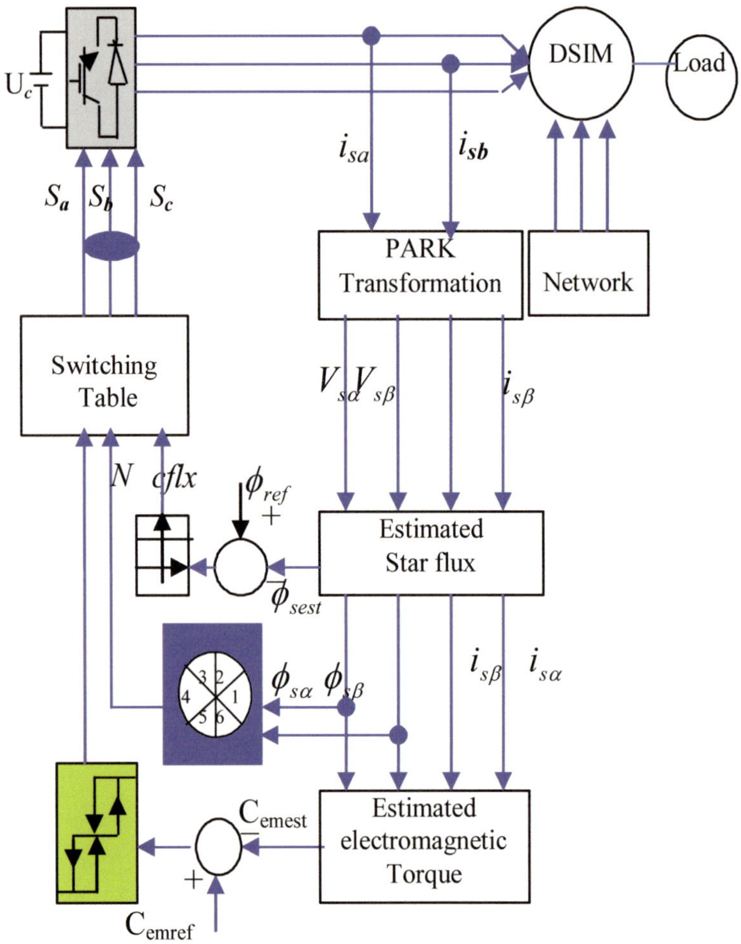

Fig. 8. DTC applied to double star induction machine

9. Simulation results

Figure 9 refer in order, to the variation in magnitude of the following quantities, speed, electromagnetic torque, current and flux obtained while starting up the induction motor initially under no load then connecting the nominal load. During the starting up with no load the speed reaches rapidly its reference value without overtaking, however when the nominal load is applied a little overtaking is noticed and the command reject the disturbance. The excellent dynamic performance of torque and flux control is evident.

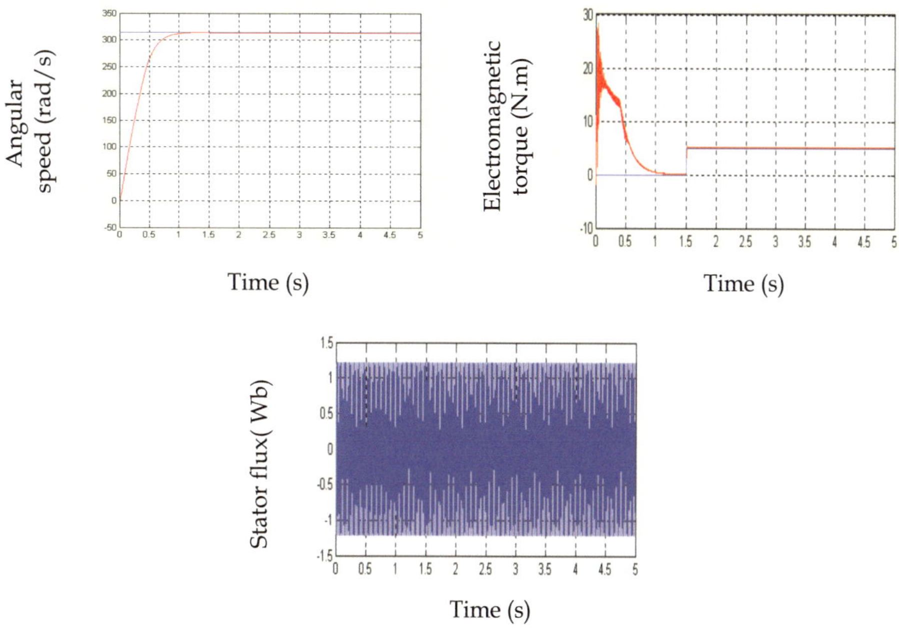

Fig. 9. Simulation results obtained with an PI regulator

10. Control of the regulator

a) Speed variation

Figure.10 shows the simulation results obtained for a speed variation for the values: (Ω_{ref} = 314 and 260 rad/s), with the load of 5 N.m applied at t=1.5s.
These results shows that the variation lead to the variation in flux and the torque. The response of the system is positive, the speed follow its reference value while the torque return to its reference value with a little error.

b) Robust control for load variation

Figure.11 shows the simulation results obtained for a load variation (C_r = 5 N.m, 2.5 N.m). As can be seen the speed, the torque, the flux and current are influenced by this variation. The torque and the speed follow their reference values.
We can see that the control is robust from the point of view load variation.

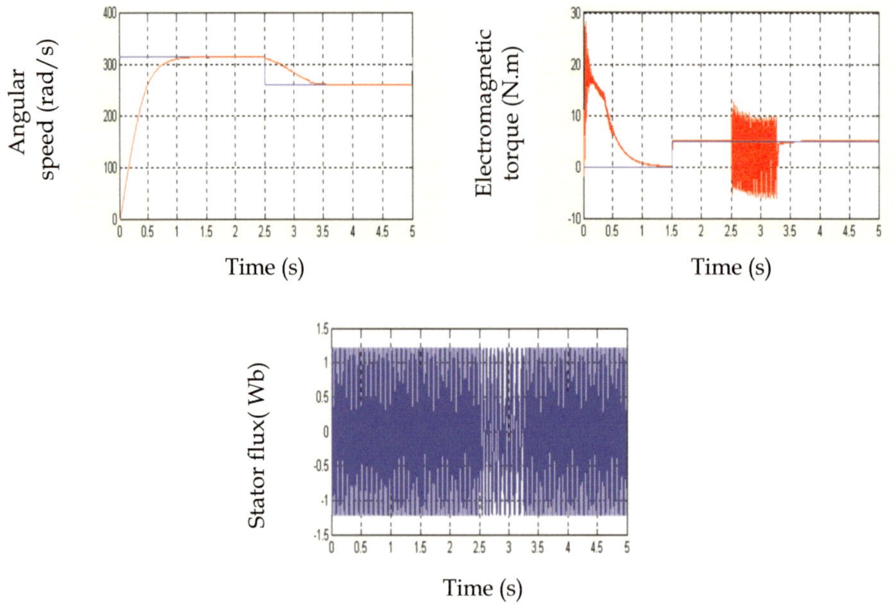

Fig. 10. Robust control for a speed variation

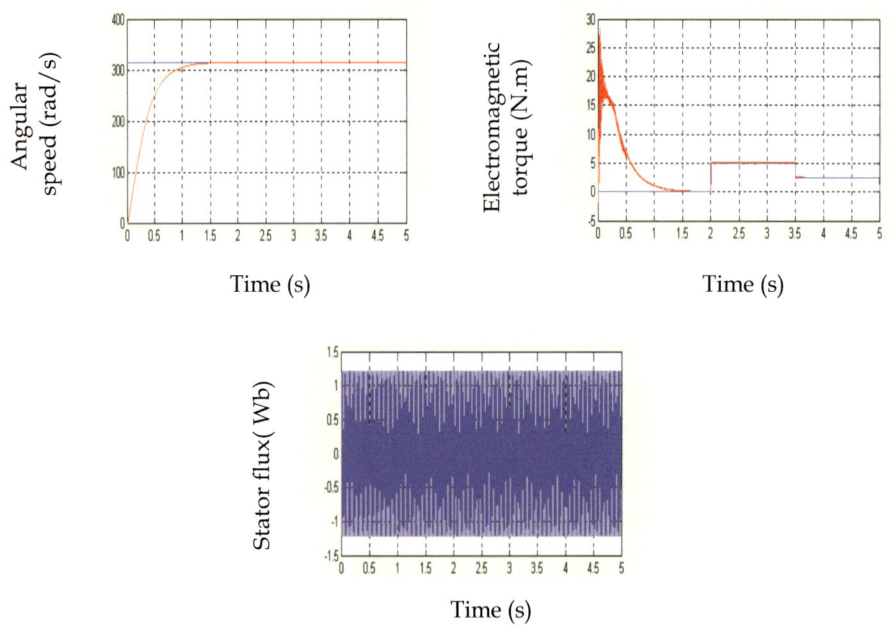

Fig. 11. Robust control under load variation

c) Robust control of the regulator under star resistance variation

In order to verified the robustess of the regulator under motor parameters variations we carried out a test for a variation of 50% in the value of star resistance at time t= 1.5s. The speed is fixed at 314 rad/s and a resistant torque of 5N.m is applied at t= 1s. Figure 6 shows in order the torque response, the current, the stator flux and the speed. The results indicate that the regulator is very sensitive to the resistance change which results in the influence on the torque and the stator flux

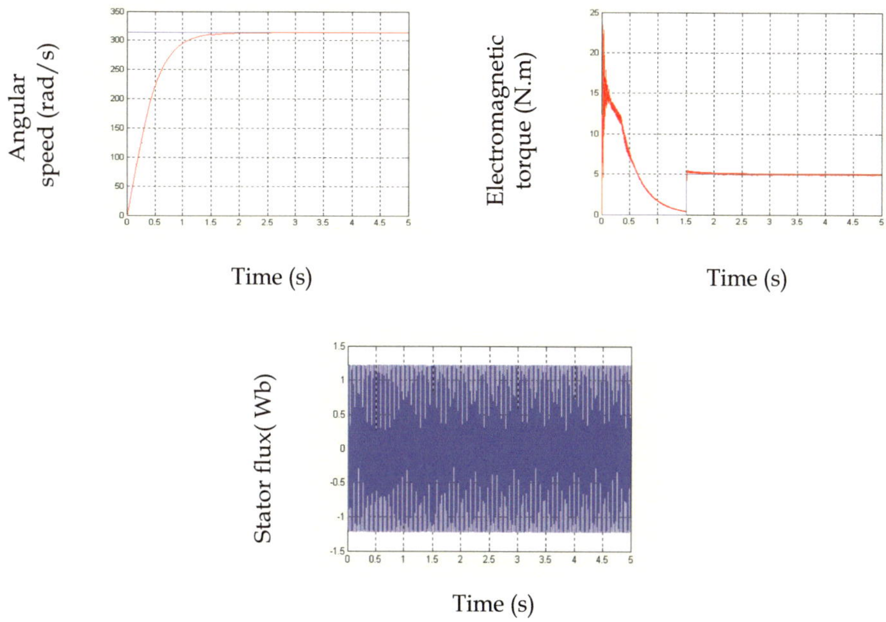

Fig. 12. Robust control under stator resistance variation

11. Conclusion

This chapter presents a control strategy for a double feed induction machine and double star induction machine based on the direct control torque (DTC) using a PI regulator. The simulation results show that the DTC is an excellent solution for general-purpose induction drives in a wide range of power.

The main features of DTC compared to the classical flux oriented control FOC can be summarized as follows:

- DTC has a simple and a robust control structure.
- DTC operates with closed torque and flux loops but without current controllers.
- DTC needs stator flux and torque estimation and, therefore, is not sensitive to rotor parameters.
- DTC is inherently a motion sensor less control method.

The simulation results show that the DTC is an excellent solution for general-purpose induction drives in a very wide power range. The short sampling time required by the DTC scheme makes it suited to very fast torque and flux controlled drives beside the simplicity of the control algorithm.

However the DTC presents two major problems:

- The absence of the harmonic of the couple restraint (electromagnetic compatibility, audible noise, variation of the acoustic quality).
- The excitation of some mechanical resonant modes which lead to a serious ageing of the system.

The DTC applies the same control effort to regulate flux as it does for the torque. Finally, we believe that the DTC principle will continue to play a strategic role in the development of high performance motion sensor less AC drives.

12. References

Carlos Ortega, Antoni Arias, Xavier del Toro. *(2005).* Novel direct torque control for induction motors using schort voltage vectors of matrix converters. *IEEE Trans.ind.Appl, pp 1353- 1358, 2005.*

Casdei.D, Serra.G, Tani.A,.*(* 2001). The use of matrix converters in direct torque control of induction machines. *IEEE Trans.on Industrial Electronics,* Vol 48,N 6,.

Hadiouche. D, H.Razik, A.Rezzoug.(2000). Stady and simulation of space vector PWM control of Double-Star Induction Motors · *IEEE-CIEP,* Acapulco, Mexico, pp 42-47.

Hadiouche.D, H.Razik, A.Rezzoug.(2000). Modelling of a double-star induction motor with an arbitrary shift angle between its three phase windings. *EPE-PEMC,* Kosice.

Kalantari.A, M. Mirsalim, H.Rastegar.(2002). Adjustable speed drive based on fuzzy logic for a dual three-phase induction machine. *Electrimacs,* pp 18-21.

Kouang-kyun La, Myoung-Ho Schin, Dong-Seok Hyun.(2000). Direct torque control of induction motor with reduction of torque ripple. *IEEE Trans.ind.Appl,* pp 1087 -1092

Petersson.AA, (2003). Analysis, modeling and control of doubly fed induction machine for wind turbines," *tutorials thesis, chalmers university of technology,* Goteborg, Sweden

Presada.S , E. Chekhet, I Shapoval. (1998). Asymptotic control of torque and unity stator side power factor of the doubly fed induction machine. *Proceedings Intern. Conf" Problems of Electrical Drives.* Alushta, pp 81-86.

Prescott.JC et Alii.(1958).The Inherent instability of Induction Motor under conditions of Doubly –Supply. *IEEE Proceedings ,UK,* vol.105,pp 319-330.

Radwan,.(2005). Perfect speed tracking of directe torque controlled induction motor drive using Fuzzy logic. *IEEE Trans.ind.Appl,* pp 38-43, 2005.

ROYS et S.COURTINE. *(1995).* Commande directe du couple d'une machine asynchrone par le contrôle direct de son flux statorique. *Journal de Physique III,*pp.863-880.France.

Xu. L, W. Cheng. (1995). Torque and reactive power control of a doubly fed induction machine by position sensorless scheme. *IEEE Transactions on Industry Applications,* vol 31,no. 3, pp 636-642.

Part 2

Oriented Approach of Recent Developments Relating to the Control of the Permanent Magnet Synchronous Motors

6

Direct Torque Control of Permanent Magnet Synchronous Motors

Selin Ozcira and Nur Bekiroglu
Yildiz Technical University
Turkey

1. Introduction

Modern electrical drive systems consist of; power electronics components, transformers, analog/digital controllers and sensors or observers. The improvements in the semiconductor power electronic components have enabled advanced control techniques with high switching frequency and the high efficiency. Complex control algorithms have been widely used and got simplified in drivers due to the developments in software technology. DC, asynchronous and synchronous motors are frequently used motor types with these driver systems. New kinds of motors are developed like linear motors, step motors, switching reluctance motors, and permanent magnet synchronous motors. Permanent magnet synchronous motors are used where in general high demands are made with regard to speed stability and the synchronous operation of several interconnected motors. They are suitable for applications where load-independent speeds or synchronous operation are required under strict observance of defined speed relations within a large frequency range.

As the technology gets improved, studies on PMSM such as direct torque control method have been improved as well. DTC has many advantages such as faster torque control, high torque at low speeds, and high speed sensitivity. The main idea in DTC is to use the motor flux and torque as basic control variables, identical to the DC drives. In order to emulate the magnetic operating conditions of a DC motor, the information regarding the rotor status is required to perform the field orientation process of the flux-vector drive. This information should be obtained by feeding the rotor speed and angular position back by using a pulse encoder. Encoders are costly and they add more complexity to the overall system.

Several methods have been developed to obtain the rotor position and angular speed from the electrical measurements and computations in order to eliminate the need for sensors. The idea behind such methods is to manipulate the motor equations in order to express motor position and speed as functions of the terminal quantities. In (Adreescu & Rabinovici, 2004) self tuning speed controller and Luenberger observer was proposed, however it is still necessary to use an encoder for rotor position detection. Most of the methods; however, work only when the rotor is anisotropic and if the dependence of the inductance on the rotor position is accurately known; moreover, the speed is not estimated, thus the drive is not completely without mechanical sensors (Yan et al, 2005). On the other hand, it can be reported that the method in (Bolognani et al, 1999) can a viable solution for the online determination of the rotor position and speed of a PMSM. In (Ichikawa et al, 2003) extended electromotive force model is used to estimate the rotor position. However, this method is based on expert knowledge. Some researchers have proposed a combination of current and flux linkage estimation (Rahman & Toliyat, 1996), (Toliyat et al, 2002) or methods including Kalman filters, fuzzy logic, and neural network observers to obtain the rotor position angle (Dan et al, 2004), (Grzesiak & Kazmierkowski, 2007). Recently, researchers have tried to reduce the torque pulses and harmonics in PMSM. In addition, (Sozer et al, 2000) presented an inverter output filter system for PWM derives to reduce the harmonics of surface mounted permanent magnet synchronous motor (SPMSM), it shows some effectiveness in reducing switching harmonics, but very large circulating current between inverter output and filter elements is required to reshape the motor terminal voltage which may violate the inverter's current limit.

Efficiency is important due to the energy scarcity of the world and higher performance is needed for modern motion control applications. PMSM offer efficiency advantages over induction machines when employed in variable speed drives. Since much of the excitation in the PMSM is provided by the magnets, the PMSM will have smaller losses associated with the magnetizing component of the stator current. The stator current may be almost purely torque producing in a PMSM drive while in an induction machine drive there is always a large magnetization current present. Due to the synchronous operation of the PMSM, rotor losses are greatly reduced. The application of the PMSMs for low speed operation in direct drives is an economic alternative for the induction motors with gearboxes. Since the speed of the direct drive PMSMs is lower than the speed of the induction motors with the gearboxes, the risk of torque harmonics appearing at the mechanical resonances is increased in the speed range of normal operation. Permanent magnet motors have been used for decades in low power applications such as servo drives and domestic appliances. Recently, the PMSM drives have been developed further and are used in industrial applications requiring high torque at low speed. PMSM drives are replacing standard induction motors with gearboxes in, for example, paper and textile industry, special applications for marine (Laurila, 2004). However, PM synchronous motors can not be fed directly from the mains supply and need to be driven by the AC motor drives. Similar to the induction machine, vector control method is employed for the PM synchronous motors to obtain high bandwidth torque control performance. For vector control, the rotor flux angle needs to be known by the AC motor drive. Therefore, sensors (e.g. an incremental encoder) on the motor shaft are utilized to sense the rotor flux angle and AC motor drives use this angle information for vector coordinate transformations. With vector coordinate transformations, the AC motor in the control coordinates is converted to a DC motor where torque control is a simple issue of current control to be achieved by the current regulator. Thus, the vector control method enables high bandwidth torque control of an AC machine, which brings high bandwidth of speed and position control (Geyer et al, 2010).

Although machine drives with modern control techniques have brought high performance and robustness to the motion control area, research has continued for developments in the AC motors and drives technology. The motivation is to improve the technology for high efficiency and for high performance. The literature study regarding to the direct torque control of the permanent magnet synchronous machines is presented in this chapter.

The theory of the direct torque control was developed by (Depenbrock, 1985) for the first time for asynchronous motors.

(Takahashi and Noguchi, 1986) developed the direct torque control only using the torque routines for asynchronous motors.

(Depenbrock, 1988) presented the direct self control theory. According to this theory, the variations in operation frequency are obtained by the algebraic calculations from the Heyland-Ossanna circular diagram based on the torque demand feed-back.
(Pillay and Krishnan 1989) modeled the permanent magnet synchronous motor by state-space variables and accomplished the torque analysis. This study was a very significant step for the permanent magnet synchronous machine studies.
(Adnanes, 1990) developed the torque analysis of the permanent magnet synchronous machine in per unit mode and obtained the detailed the mathematical relationship between the flux and torque.
(Raymond and Lang, 1991) presented the real-time adaptive control of the PMSM by using the Motorola 68020 microprocessor. The motor model was linearized for the proposed controller and the nonlinear effects of the inverter and system dynamics that can not be modeled were omitted. The mechanical parameters were estimated so that they could be taken into account for the controller.
(Pelczewski et al., 1991) performed the optimal model tracking control of the PMSM. In order to have the controller doing the computations, motor model and its linearization were required.
(Matsui and Ohashi, 1992) proposed a DSP based adaptive controller for PMSM. Therefore, they proved that the DSPs can be implemented in motor controls as well.
(Chern and Wu, 1993) presented the position control of the PMSM by using the variable oriented controller. The controller instantaneously does the calculations depending on the unknown load and motor parameters. The system model is required and the computations require very long periods of time.
(Ogasawara and Akagi, 1996) realized the position estimation of the PM motors for zero and low speed conditions according to the saliency.
(Zhong et al., 1997) completed one of the first academic studies in the field of direct torque control of the permanent magnet synchronous machines.
(Zhong et al., 1999) proposed the direct torque control of the permanent magnet synchronous motor using two level torque hysteresis controller.
(Rahman et al., 1999) achieved the direct torque control by using a method based on obtaining the d and q-axis voltages using certain coefficients.
(Luukko, 2000) developed the switching table for the direct torque control by adding the zero vectors to the vector selection algorithm.
(Vaez Zadeh, 2001) experimentally achieved the constant torque control on a vector controller by using the TMS320C31 DSP. In this study, the torque did not respond well in terms of the desired torque value and response time, since the DSP technology was not sufficient to implement the dynamic behavior of the motor.
(Tan et al., 2001) and Martins et al. (2002) proposed to reduce the torque ripples and fix the switching frequency in AC drive systems by using a multi-level inverter. These methods result in better waveforms, reduce the distortions, and are capable of operating in lower switching frequency. However, on the other hand, they require more number of switching devices. Moreover, the control strategies of these methods are very complicated.
(Dariusz et al., 2002) implemented the space vector modulation by using a DSP and achieved the direct torque control.
(Balazovic, 2003) published a technical guide that describes the torque control and vector control of the permanent magnet synchronous motor. This publication has been referred in many similar studies of the industry and academia.

(Tang et al., 2004) used the space vector modulation in order to reduce the torque ripples and they got good results. Developed control algorithm required two PI controllers and the estimation of the reference voltage and the switching sequence of the selected vectors.
(Zhao et al., 2004) achieved the control of the very high speed (200000 rpm) permanent magnet synchronous machine by using a DSP.
(Popescu et al., 2006) investigated the torque behavior of a single phase permanent magnet synchronous motor.
(Jolly et al., 2006) performed the control of the permanent magnet synchronous machine in constant power region.
(Luukko et al., 2007) presented the different rotor and load angle estimation methods for the direct torque control. They directly calculated the load angle from the PMSM equations. In these calculations, they used the tangent function. When the results of the DSP controlled inverter and the motor test setup are investigated, it is seen that the rotor angle estimation has oscillations. The error between the actual and estimated rotor angle variations get larger in the periods when the oscillations are larger. It is revealed that PID coefficients should be kept very high in order to compensate this error.
(Chen et al., 2007) designed an output filter for a direct torque controlled inverter. The filter is composed of an RLC filter and an isolation transformer. This study is interesting since it includes both the transformer design and the soft switching techniques in power electronics. Furthermore, it is true that the transformer and RLC based filter will add significant cost to the system instead as compared to the developments in the generation of inverter switching signals in terms of controls.
(Noriega et al., 2007) designed a fuzzy logic controller for the DTC. They used the torque error and the stator current for the fuzzy logic membership functions. In addition to the simulation studies, they used an AC motor drive setup called Platform III by implementing fuzzy logic functions to the software of this setup. Both simulation and the experimental results show that the stator current is not in a wave form and it has some uncertain and random shapes. As compared to the PI controlled DTC method, stator current is much more distorted.
(Wang et al., 2007) developed the reference flux vector calculation in space vector modulation for DTC. They extracted the voltage as a trigonometric function of the period and using the frame transformations, they calculated the usage periods of the zero vectors depending on the angular frequency of the current. However, this complicated control structure has been implemented in simulations but not experimentally completed. In torque graphs, there are long delay periods between the actual and calculated values.
(Zhao et al., 2007) developed a fourth order sliding mode observer for the surface mounted permanent magnet synchronous motor. According to the motor parameters, it is seen that the motor was a high power low speed motor. In simulations, the model available in Matlab&Simulink library was implemented. The authors stated that fuzzy logic can be employed to address the chattering issues commonly occur in sliding mode controls.
(Swierczynski et al., 2008) applied the DTC method onto a high power PMSM by using an inverter that is driven by the space vector modulation method. In this study, they used DSPACE 1103 control unit in which the Matlab Simulink simulation models can be directly applied. In this study, it was not required to design a speed controller. This is mainly due to the fact that the vehicle operator can control the speed according to the different driving conditions.

(Yutao et al., 2008) used the radial basis neural network functions to exploit the reference torque in a rectangular step functions. They attempted to reduce the torque vibrations and make it in a rectangular shape which is not in fact.
(Cui et al., 2008) researched on a high performance DTC system based on a DSP. They performed simulation and experimental studies. However, in their studies, the controls execution time completed in 130 μs.
(Li et al., 2008) used zero vectors in space vector modulation for DTC. Zero vectors are theoretically used in asynchronous motor's direct torque control. They tried to increase the application duration of the vectors that are used to enlarge the torque angles in low speed operation of the PMSM applications. However, in low speeds, using zero vectors for long period of time causes the fast changes in stator flux and the limit values are enforced. Moreover, switching losses of this implementation will be higher since 8 vectors are used instead of 6.
(Jilong et al., 2008) proposed an improved Kalman filter in order to sensor-less estimation of the rotor's initial position in DTC. For this reason, they employed the high frequency signal injection method. Since the high frequency signal is weak, it would not help the rotor's motion. Therefore, rotor speed is assumed to be zero. The computation intensity is very high due to the fact that the voltage and current quantities are obtained through differential inequalities depending on the speed.
(Guo et al., 2009) applied the space-vector modulation in a matrix converter for DTC application in naval vehicles. In this study, the signals for the matrix converter is generated by a DSP where the dual space modulation method was used. However, it is seen that the current drawn by the matrix converter has very high total harmonic distortion.
(Sanchez et al., 2009) achieved the direct torque control without using a speed sensor but using only current and voltage senor in order to determine the stator voltage vectors. In their results that they used a closed loop controller, they indicated that the calculated speed data oscillates too much.
(Siahbalaee et al., 2009) studied the copper losses by flux optimization for their direct torque controlled PMSP in order to reduce the torque and flux oscillations.
(Liu et al., 2009) tried to use the predictive control method in direct torque control. There are limited number of studies in the literature using predictive control in direct torque control. Experimental results were obtained by using a DSP. However, in the experimental results, they implemented complicated trigonometric functions. In the experimental results, much more flux drop can be observed as compared to their simulation results. They could apply the proposed application by reducing the flux reference.
(Inoue et al., 2010) linearized the torque control system of a direct torque controlled buried magnet synchronous motor and they acquired the torque response depending on a constant gain coefficient. They calculated the PI coefficients that are updated according to the estimated torque values by a new gain scheduling method.
(Geyer et al., 2010) achieved the direct torque control of the PMSM by implementing a model predictive control algorithm that reduces the switching frequency and hence the switching losses. The proposed algorithm could reduce the switching losses by 50% and the THD by 25%.

2. Permanent magnet synchronous motor technology

Permanent magnet synchronous motors are different from the wound field synchronous motors. However the stator structure of a permanent magnet synchronous motor is similar

to the wound field synchronous motors, the difference is only between the rotor structures. In the wound field synchronous motors, field is created on the rotor by separate excitation through the brushes (slip rings), where the field of the permanent magnet synchronous motors is created by the permanent magnets placed on the rotor. Therefore, permanent magnet synchronous motors are brushless motors. Since they are brushless, they are more robust than the DC motors; and since the field is created by permanent magnets and there are not any rotor currents, they are more efficient than the induction motors, where the rotor field currents cause rotor copper losses. However costs of PMSM are higher than DC and induction motors because of the high permanent magnet and production costs. Also, the reliability of these motors is questionable under certain circumstances such as magnetic property loss due to high working temperatures etc. Nevertheless, such properties as high efficiency, high torque, high power, small volume, and accurate speed control make permanent magnet synchronous motors preferred for chemical fiber industry (spinning pumps, godets, drive rollers), texturing plants (draw godets), rolling mills (roller table motors), transport systems (conveyor belts), glass industry (transport belts), paper machines, robotic automation, electrical household appliances, ship engines and escalators.
Permanent magnet synchronous motors are classified mainly into two groups with respect to their rotor structures as; Surface Mount Permanent Magnet (SMPM) Synchronous Motors and Interior Permanent Magnet (IPM) Synchronous Motors. SMPM motors have the permanent magnets mounted on the outer surface the rotor, and IPM motors have the permanent magnets buried in the rotor core. SMPM motors are also classified into two types with respect to the stator winding as; concentrated winding and distributed winding. Concentrated winding SMPM motor's back-emf waveform is trapezoidal; distributed winding SMPM motor's back-emf waveform is sinusoidal. Concentrated winding SMPM motors are called as Brushless DC (BLDC) motors and driven with trapezoidal signals. The distributed winding SMPM motors are called as Permanent Magnet AC (PMAC) motors and driven with sinusoidal signals. PMAC motors are also designated as servo motors or brushless AC motors. PMAC motors are generally built with strong magnetic material Samarium Cobalt ($SmCo_5$, Sm_2Co_{17}) and Neodymium Iron Boron (NdFeB). They have high dynamic performance, high efficiency, robustness, high torque density and significantly better short-time overload capability than induction motors (400% to 150%). PMAC motors are mostly employed in high performance servo (robotics, machining, etc.) applications. Some PMAC motors are built with low cost permanent magnet materials (Ferrites) to be used in low cost (fan) applications. IPM motors are newly developed motors with high torque density, high efficiency characteristics as the SMPM motors and additionally provide field weakening operation, which is impossible with the SMPM motors. IPM motors are preferred in the industrial applications such as adjustable speed drives as a replacement for the squirrel cage induction motors, to improve the efficiency and the performance. In contrast to the induction motors, IPM motors also have the advantage of providing position control loop with accuracy, without a shaft encoder (Omer & Hava, 2010). Standard induction motors, designed to run at 750-3000 rpm, are not particularly well suited for low speed operation. Normally gearboxes are used to reduce the speed from, for example, 1500 rpm to 600 rpm. A gearless PMSM drive of 600 rpm replaces as a solution, in contrast to conventional using of 1500 rpm induction motor. Because a gearbox takes up space and needs maintenance as well as considerable quantities of oil. Eliminating the gearbox saves space, installation costs and improved efficiency of the drive.

2.1 Permanent magnet synchronous motor design types

The location of the magnets on the rotor and their specifications determine the performance of the motor therefore various designs are possible. The simple representations of the frequently used designs are given below. Other designs are derived from these two.

1. Placing the magnets on the rotor surface (SMPM)
2. Placing the magnets inside the rotor (IPM)

2.1.1 Placing the magnets on the rotor surface (Surface mounted magnet rotor)

Magnets are mounted on the rotor in forms of strips or arcs. The rotor configuration shown in Fig.1 adopts surface-mounted magnets which are often glued onto rotor surface. Since the relative permeability of the magnets is almost the same as for air, such design possesses very small rotor saliency and these are sometimes referred to as "nonsalient" designs. By filling the gaps between the magnets partially with iron, a significant rotor saliency can be achieved which offers the possibility to utilize the reluctance torque. The greatest drawback of this common design is the low endurance of the magnets to the centrifugal forces. Therefore these motors are preferred in low-speed applications to avoid detachment of the magnets. These motors are commonly known as surface permanent magnet motors (SMPM). A simple representation is shown in Fig.1.

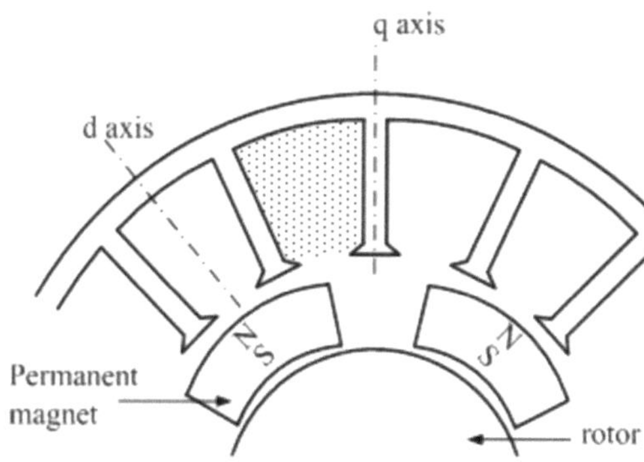

Fig. 1. Magnets placed on the rotor surface

2.1.2 Placing the magnets inside the rotor

Air gap induction of the previous design is limited and the magnets being exposed to high centrifugal forced under high speeds have lead to different designs. Here, the magnets are placed in the cavities bored in the rotor. The magnets are surrounded by magnetic materials instead of air.

The magnets now have a better resistance to centrifugal forces therefore they are more suitable for high speed applications. The efficiency values of these motors are also higher than other magnet motors. The main disadvantage is their high costs. Placement of the magnets in the rotor is a high tech process that requires fine labour. These motors are commonly known as interior permanent magnet synchronous motor (IPMSM). These magnet motor designs are mainly in two types.

2.1.2.1 Radially placed interior magnet structure

As seen in Fig.2 magnets are placed around the rotor axis buried and radially magnetized. These motors have small air gap, and low armature reaction. Flux density in the air gap can be higher than inside the magnet, thus the low-cost Ferrit magnets can be utilized for high torque density. The surface where the magnets come in contact with the rotor is coated with a non magnetic material to avoid magnetic short-circuit. However these materials have high costs.

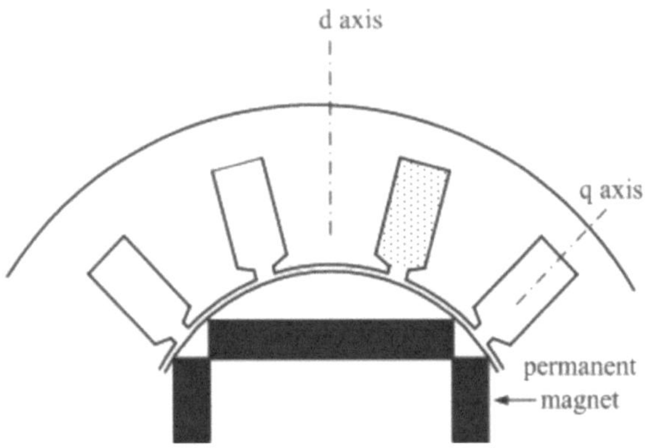

Fig. 2. Radially placed interior magnet structure

2.1.2.2 Symmetrically buried magnet structure

Permanent magnets are again buried in the rotor but are placed pointing the main axis. The most important feature of this design is to constitute induction at the poles independent from the working point of magnets. Through to this feature, air gap induction can be increased to high levels. Since the magnets are buried in the rotor, they have a great resistance to the centrifugal forces. Symmetrically buried magnet structure is shown in Fig.3.

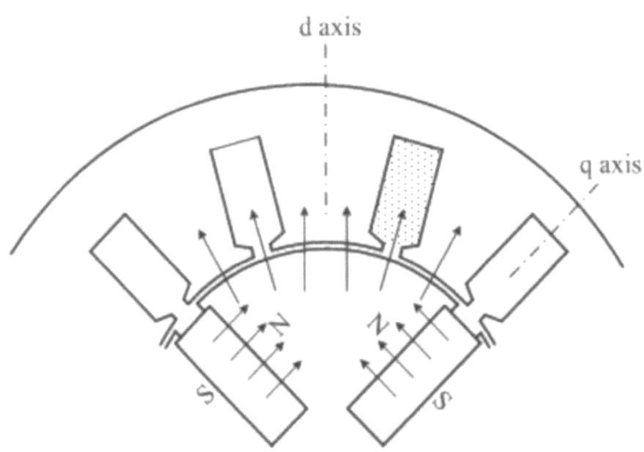

Fig. 3. Symmetrically buried magnet structure

When buried magnets are used, $L_{sd} \neq L_{sq}$ and the electromagnetic torque also contains a reluctance torque. The fact that three-phase symmetrical sinusoidal quantities are transformed into two DC components through the well known Park transformation has made modelling of PMSM in the rotor-fixed dq reference frame used almost exclusively for control purposes.

2.2 Model of PMSM on the rotor reference frame and motor equation

For high dynamic performance, the current control is applied on rotor flux (dq) reference system that is rotated at the synchronous speed. Stator magnetic flux vector ψ_s and rotor magnetic flux vector ψ_M can be represented on rotor flux (dq), stator flux (xy) reference system as shown in Fig.4. The angle between the stator and rotor magnetic flux (δ), is the load angle that is constant for a certain load torque. In that case, both stator and rotor fluxes rotate at synchronous speed. However under different loads, δ angle varies. Here, by controlling the stator current variation or the δ angle variation, the increase of the torque can be controlled.

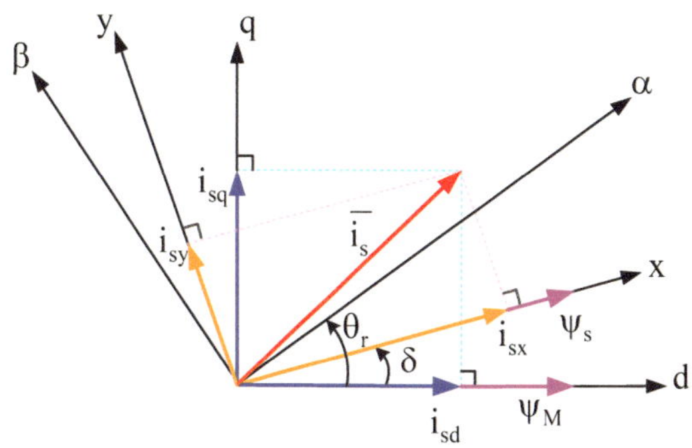

Fig. 4. Stator and rotor magnetic fluxes in different reference systems

The mathematical equations are given below (Vas, 1998). Stator current vector can be represented on rotor flux (dq) reference system as (i_{sd}) (i_{sq}) and the electromagnetic torque is related with these vectors.

$$\psi_{sd} = L_{sd} i_{sd} + \psi_M \tag{1}$$

$$\psi_{sq} = L_{sq} i_{sq} \tag{2}$$

$$u_{sd} = R_s i_{sd} + \frac{d}{dt}\psi_{sd} - \omega_r \psi_{sq} \tag{3}$$

$$u_{sq} = R_s i_{sq} + \frac{d}{dt}\psi_{sq} + \omega_r \psi_{sd} \tag{4}$$

$$\frac{d}{dt} i_{sd} = \frac{1}{L_{sd}} u_{sd} - \frac{R_s}{L_{sd}} i_{sd} + \frac{L_{sq}}{L_{sd}} \omega_r i_{sq} \tag{5}$$

$$\frac{d}{dt} i_{sq} = \frac{1}{L_{sq}} u_{sq} - \frac{R_s}{L_{sq}} i_{sq} + \frac{L_{sd}}{L_{sq}} \omega_r i_{sd} - \frac{\psi_M \omega_r}{L_{sq}} \tag{6}$$

$$T_e = \frac{3}{2} p\left(\psi_{sd} i_{sq} - \psi_{sq} i_{sd}\right) \tag{7}$$

$$T_e = \frac{3}{2} p\left[\psi_M i_{sq} - \left(L_{sq} - L_{sd}\right) i_{sd} i_{sq}\right] \tag{8}$$

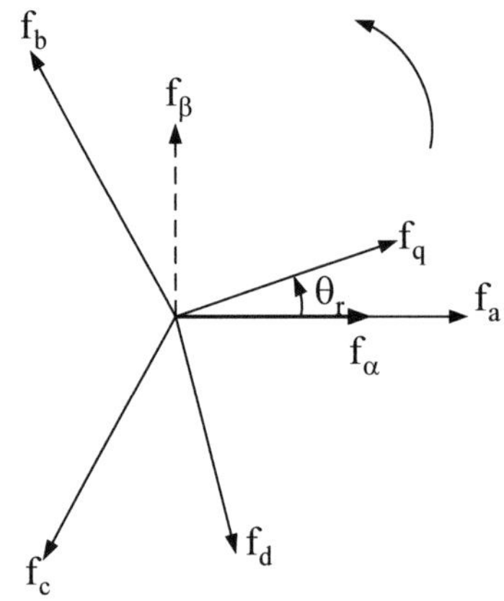

Fig. 5. Rotating reference frames

To simplify the modelling of the DTC drive, it is common practice to transform all variables from the three-phase system (abc) to an orthogonal (dq) reference frame with a direct (d) and quadrature (q) axis, where f represents the voltage, current or magnetic flux and θ_r is the rotor angle. $abc \rightarrow \alpha\beta$, Clarke Transformation (9) and $\alpha\beta \rightarrow dq$, Park Transformation (10) can be applied regarding reference frame theory. Here, equation (11) presents these both transformations in one matrix, $abc \rightarrow dq$.

$$\begin{bmatrix} \alpha \\ \beta \end{bmatrix} = \frac{2}{3} \begin{bmatrix} 1 & \cos(2\pi/3) & \cos(4\pi/3) \\ 0 & \sin(2\pi/3) & \sin(4\pi/3) \end{bmatrix} \begin{bmatrix} a \\ b \\ c \end{bmatrix} \tag{9}$$

$$\begin{bmatrix} \alpha \\ \beta \end{bmatrix} = \begin{bmatrix} \cos(\theta_r) & -\sin(\theta_r) \\ \sin(\theta_r) & \cos(\theta_r) \end{bmatrix} \begin{bmatrix} d \\ q \end{bmatrix} \tag{10}$$

$$\begin{bmatrix} d \\ q \end{bmatrix} = \frac{2}{3} \begin{bmatrix} \cos(\theta_r) & \cos(\theta_r\text{-}2\pi/3) & \cos(\theta_r\text{-}4\pi/3) \\ \text{-}\sin(\theta_r) & -\sin(\theta_r\text{-}2\pi/3) & -\sin(\theta_r\text{-}4\pi/3) \end{bmatrix} \begin{bmatrix} a \\ b \\ c \end{bmatrix} \tag{11}$$

In Eq. (12), by using the vector representation shown in Fig.5, Eq. (13) is obtained and using the Park transformation, Eq. (14) is obtained.

$$\begin{bmatrix} f_d \\ f_q \end{bmatrix} = \begin{bmatrix} \cos\delta & -\sin\delta \\ \sin\delta & \cos\delta \end{bmatrix} \begin{bmatrix} f_x \\ f_y \end{bmatrix} \tag{12}$$

$$\sin\delta = \frac{\psi_{sq}}{|\psi_s|}$$

$$\cos\delta = \frac{\psi_{sd}}{|\psi_s|} \tag{13}$$

$$T_e=\frac{3}{2}p\left[\psi_{sd}\left(i_{sx}\sin\delta+i_{sy}\cos\delta\right)-\psi_{sq}\left(i_{sx}\cos\delta-i_{sy}\sin\delta\right)\right]$$

$$T_e=\frac{3}{2}p\left[i_{sx}\frac{\psi_{sd}\psi_{sq}}{|\psi_s|}+i_{sy}\frac{\psi_{sd}^2}{|\psi_s|}-i_{sx}\frac{\psi_{sd}\psi_{sq}}{|\psi_s|}+i_{sy}\frac{\psi_{sq}^2}{|\psi_s|}\right]$$

$$T_e=\frac{3}{2}p|\psi_s|i_{sy} \tag{14}$$

It is clear that the electromagnetic torque is directly proportional to the y-axis component of the stator current (Zhong et al, 1997). Dependency on less number of parameters is the main advantage of the stator current control. It is possible to say that in a practical application, the estimation technique shown in equation (8) requires knowledge of inductances. The estimated instantaneous electric torque is easily compared with a reference value to achieve a fast torque control. At the same time, the stator flux linkage is compared with the reference value to ensure sufficient magnetization of the motor. The torque of the PMSM is controlled by monitoring and controlling the armature current since electromagnetic torque is proportional to the current.

3. Direct torque control of permanent magnet synchronous motors

In general, there are two control methods used for the PMSM; field oriented control and direct torque control. The AC drives in which flux oriented control (FOC) is used field control leads to flux control. Here, rotor flux space vector is calculated and controlled by using the angular velocity which is derived from the speed feedback and the stator current vector. The greatest drawback of the flux vector control is the need for a tachogenerator or an encoder for high accuracy. This need definitely increases the costs of the system.

The basic principle of DTC is to directly select the stator voltage vectors according to the errors between the reference and actual values of the torque and stator flux. Torque and flux are resolved and directly controlled using nonlinear transformations on hysteresis controllers, without performing coordinate transformations. A double layer hysteresis band controller is utilized for stator flux control and a three-layer hysteresis band controller is used for torque control. DTC is an alternative to field oriented control method in high performance applications due to the advantages of reduced computations (Swierczynski et al, 2008) Since the torque and flux estimators in DTC requires and relies on the parameters identification and accuracy of the estimations, the estimation of the electromagnetic torque is essential for the entire system performance.

In classical PWM and flux vector controlled drives, voltage and frequency are used as basic control variables and that are modulated and then applied to the motor. This modulator layer needs an additional signal processing time and restricts the torque and speed response. The key notion behind DTC is to directly steer the stator flux vector by applying the appropriate voltage vector to the stator windings. This is done by using a pre-designed switching table to directly update the inverter's discrete switch positions whenever the variables to be controlled, the electromagnetic torque and the stator flux, exceed the hysteresis bounds around their references. The switching table is derived on the basis of the desired performance specifications on the controlled variables also include the balancing of the inverter's neutral point potential around zero.

3.1 Two - level inverter nonlinear continuous-time model

An equivalent representation of a three-phase two-level inverter driving a PMSM is shown in Fig.6. At each phase, the inverter can produce two different voltages $-u_{dc}/2$, $u_{dc}/2$ where u_{dc} denotes the voltage of the dc-link. The switch positions of the inverter can therefore be fully described using the three integer variables $v_a, v_b, v_c \in \{-1, 1\}$ where each variable corresponds to one phase of the inverter, and the values -1, 1 correspond to the phase potentials $-u_{dc}/2$, $u_{dc}/2$ respectively.

There are $2^3 = 8$ different vectors of the form $v_{abc} = [v_a, v_b, v_c]^T$

Using Eq. (11) these vectors can be transformed into the dq frame resulting in vectors of the form $v_{dq} = [v_d, v_q]^T$ as shown in Fig.7 where they are mapped into the two-dimensional dq plane. Even though they are commonly referred to as voltage vectors, this term describes the switch positions rather than the actual voltages applied to the machine terminals. The voltage vectors can be divided in two groups: six long vectors forming the outer hexagon and two zero vectors. The zero vectors correspond to the switch combinations (+1,+1,+1) and (-1,-1,-1), and short-circuit the machine terminals.

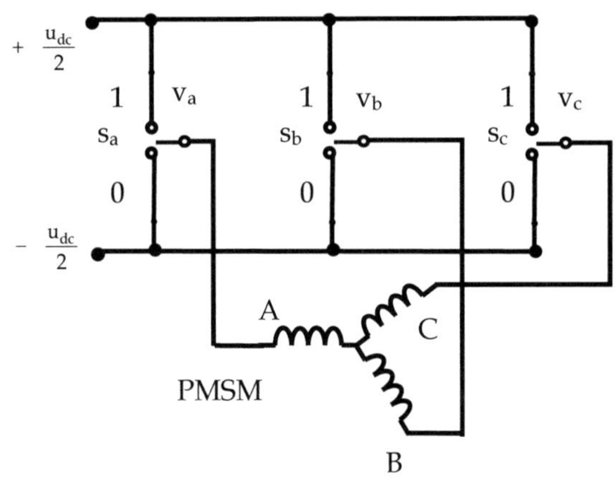

Fig. 6. Three-phase two-level inverter

3.2 Determination of the voltage space vector

The main principle of DTC is determination of correct voltage vectors using the appropriate switching table. The determination process is based on the torque and stator magnetic flux hysteresis control. Stator magnetic flux can be calculated using equation (15).

$$\overline{\psi}_s = \int_t^{t+\Delta t} \left(\overline{u}_s - R_s \overline{i}_s\right) dt \tag{15}$$

Eq. (15) shows that the stator magnetic flux and the voltage space vector are in the same direction. Therefore, amplitude and direction control of the stator magnetic flux is possible by selecting the suitable voltage space vectors. Voltage vector plane is divided into six parts as shown in Fig.7. Two adjacent vectors that yield the lowest switching frequency are selected in order to increase or decrease the amplitude respectively.

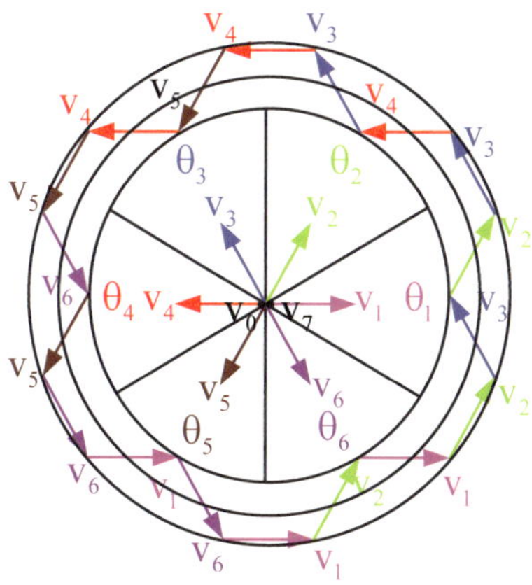

Fig. 7. Vectors of space vector modulation

Here, when the stator magnetic flux is moved clockwise in section 1, voltage space vector v_2 is selected in order to increase the stator magnetic flux amplitude and voltage space vector v_3 is selected in order to decrease the amplitude. When the stator magnetic flux moves clockwise, if still in section 1, v_6 is used to increase the amplitude and v_5 is used to decrease the amplitude. The torque of the permanent magnet synchronous motor can be controlled using DTC by means of controlling the stator magnetic flux rotation speed in cases where the stator magnetic flux amplitude is kept constant (Ozcira et al, 2008). Since the magnets on the rotor are continuously rotating, stator magnetic flux does not change when v_0 and v_7 zero vectors are used (Rahman et al, 1998). Zero vectors can be used to estimate the initial position of the rotor by a fixed active voltage vector while limiting current by applying zero vectors. In this position data is not used; therefore, zero vectors are not used within DTC for PMSM. Table 1. shows the suggested switching sequences. For these different states, the hysteresis controllers can be used as flux and torque hysteresis controllers.

ψ	T	θ					
		θ(1)	θ(2)	θ(3)	θ(4)	θ(5)	θ(6)
1	1	V_2 (110)	V_3 (100)	V_4 (101)	V_5 (001)	V_6 (011)	V_1 (010)
	0	V_7 (111)	V_0 (000)	V_7 (111)	V_0 (000)	V_7 (111)	V_0 (000)
	-1	V_6 (101)	V_1 (001)	V_2 (011)	V_3 (010)	V_4 (110)	V_5 (100)
0	1	V_3 (010)	V_4 (110)	V_5 (100)	V_6 (101)	V_1 (001)	V_2 (011)
	0	V_0 (000)	V_7 (111)	V_0 (000)	V_7 (111)	V_0 (000)	V_7 (111)
	-1	V_5 (001)	V_6 (011)	V_1 (010)	V_2 (110)	V_3 (100)	V_4 (101)

Table 1. Switching vectors

In Table 1. ψ denotes stator magnetic flux hysteresis controller output, T denotes torque hysteresis controller output, and θ represents magnetic flux sector. These vectors are selected in order to provide the stator flux error within $2\Delta\psi_s$ bandwidth and the actual torque error in the $2\Delta t_e$ bandwidth at each switching period. The flux hysteresis controller output is $d\psi_s$. If an increase is needed for flux, it is assumed that $d\psi_s = 1$ and when a decrease is needed, it is assumed that $d\psi_s = 0$. Two level hysteresis controller is determined by Eq. (16).

$$d\psi_s = \begin{cases} 1, & |\psi_s| \le \psi_{sref} - \Delta\psi_s \\ 0, & |\psi_s| \ge \psi_{sref} + \Delta\psi_s \end{cases} \tag{16}$$

The torque hysteresis controller output is dt_e. If an increase is needed at torque, it is assumed that $dt_e = 1$ and when a decrease is needed, it is assumed that $dt_e = -1$. If there is no change at actual torque, $dt_e = 0$. Three level hysteresis controller is determined by (17) for clockwise rotation and counter clockwise rotation by (18).

$$dt_e = \begin{cases} 1, & |t_e| \le |t_{eref}| - \Delta t_e \\ 0, & |t_{eref}| \ge |t_{eref}| \end{cases} \tag{17}$$

$$dt_e = \begin{cases} -1, & |t_e| \le |t_{eref}| + \Delta t_e \\ 0, & |t_{eref}| \ge |t_{eref}| \end{cases} \tag{18}$$

In direct torque method, three level torque comparator is used to select whether the inverter output voltage vector should be a torque-increasing vector or a torque-reducing vector. The appropriate vector is then applied for the duration of the sampling period. At low speed the torque increasing vectors are very effective at increasing the torque, whereas the torque reducing vectors are less effective. In contrast, at high rotor speeds, the torque-increasing vectors are less effective, whereas the torque reducing vectors are more effective. The result of this is that, at low speed, the torque tends to make a considerable excursion above the maximum torque hysteresis limit.

3.3 Determination of the stator flux space vector's sector

Stator flux space vector's sector (θ) should be known in order to select the appropriate switching vector. Space vector's angle determines the sector. Equation (19) helps to find the sector. This equation can be used when the switching signals will be generated.

$$\theta = \arctan \frac{\psi_{sd}}{\psi_{sq}} \tag{19}$$

In practice this mathematical operation is too complex for a DSP. Therefore, another alternative way can be used to determine θ. Equation (20) is calculated and due to result of this operation a Table 2 is constituted.

$$\sqrt{3}\,|\psi_{sq}| - |\psi_{sd}| \tag{20}$$

ψ_{sd}	ψ_{sq}	$\sqrt{3}\lvert\psi_{sq}\rvert-\lvert\psi_{sd}\rvert$	Sector(θ)
+	+/-	-	1.
+	+	+	2.
-	+	+	3.
-	+/-	-	4.
-	-	+	5.
+	-	+	6.

Table 2. Stator flux space vector's sector

3.4 Stator flux control by using LP filter

In order to eliminate the errors of a pure integration in voltage model, a low pass (LP) filter should be used. Equation (15) defines stator magnetic flux in general form.

$$\bar{\psi}_s=\left(\bar{V}_s-R_s\bar{I}_s\right)/j\omega_e \tag{21}$$

In equation (21) this general form is given in sinusoidal form where ω_e denotes stator flux angular frequency.

$$\bar{\psi}_s'=\left(\bar{V}_s-R_s\bar{I}_s\right)/\left(j\omega_e+\omega_c\right) \tag{22}$$

ω_c denotes the low pass filter's cut off frequency. In equation (22) ψ_s' represents estimated stator flux but estimated flux isn't equal to stator flux $\bar{\psi}_s$ which used in equation (21). By using equation (21) and equation (22) stator flux can be obtained.

$$\bar{\psi}_s=\bar{\psi}_s'-j\left(\omega_c/\omega_e\right)\bar{\psi}_s' \tag{23}$$

For this equation in the case of $\omega_e \gg \omega_c$ LP filter converges to the pure integrator. However maintain the clarity equation (24) can be derived.

$$\frac{\psi_s'}{\psi_s}\angle\varphi=\frac{\omega_e}{\sqrt{\omega_e^2+\omega_c^2}}\angle\varphi \tag{24}$$

$$\varphi=\frac{\pi}{2}-\arctan\left(\omega_e/\omega_c\right) \tag{25}$$

By choosing a low cut off frequency, amplitude and phase errors decrease; however, as LP filter efficiency decreases, DC shift in current and voltage cannot be eliminated.

3.5 Observerless scheme for sensorless speed control based on DTC

The study includes low pass filter - flux estimator which utilizes voltage model on a direct torque controlled permanent magnet synchronous motor. Flux estimation techniques used for the high performance motor drivers are based on using the flux model, voltage model or both of these models together. Flux estimation with current model is used at low frequencies and requires the stator current and the rotor speed data. The drawbacks of this method are its susceptibility to the changes of the rotor parameters at high speeds and its need of a speed sensor. For sensorless flux estimation, the voltage model is preferred. This model has a very high accuracy at high speeds. However, at low speeds due to the very low stator voltage, decrease of the ohmic voltage and component voltage diminish the error of the integration process increases. Accuracy of the flux estimation depends on the accuracy of the measured current, voltage and real parameters. In practice, the structure of the flux sensors and the error due to noise or a small DC shift of the transitional region, results in error accumulation at the integrator. In order to avoid that effect, an integrator with a LP filter is used instead of a pure integrator. DTC performance is highly sensible to determined voltage vector and is highly dependant to stator flux space vector being accurately estimated.

Current model isn't affected by the variations of the stator resistance however is sensible to changes of the rotor parameters at high speeds. Estimation process of the voltage model does not require rotor speed data. Therefore voltage model is preferred in sensorless applications and operation at moderate or high speeds. In voltage model, errors of current and voltage measurements affect the accuracy of the integral function. The errors occur because of the phase delay of the sensors, the errors of the transformer gain, shift of the measurement system and fault of the quantal errors of the digital system. Variation of the stator resistance with temperature also causes errors.

In order to avoid the shift of a pure integration in voltage model, a low pass (LP) filter integrator is used. LP filter eliminates the shifting but causes phase and amplitude errors. Therefore, the driver performance decreases especially at frequencies close to the cut-off frequency of the filter. Further study is available to improve the flux estimation using LP filter. The method utilizes an adaptive control system that depends on the force and the stator flux being orthogonal. Using an adaptive control complicates the simple DTC system. In this study a low pass filter which utilizes voltage model, is presented. The effects of the LP filter to the flux estimation performance and the practical aspects are analyzed in detail. Simulation results show that a low pass filter which utilizes voltage model on a direct torque controlled PMSM driver, can achieve a robust control.

The torque of the permanent magnet synchronous motor is controlled by monitoring and controlling the armature current since electromagnetic torque is proportional to the current. In DTC, torque and flux can be controlled independently since the stator flux is controlled directly with the stator voltage using Eqs. (26) and (27). Stator flux can be quickly controlled and the motor performance can be kept high. Moreover, torque is estimated by using current data provided by Eq. (28).

$$\psi_\beta = \int_t^{t+\Delta t} \left(u_\beta - R_s i_\beta\right) dt \tag{26}$$

$$\psi_\alpha = \int_t^{t+\Delta t} \left[\left(u_\alpha - R_s i_\alpha\right) + \psi_M\right] dt \tag{27}$$

$$T_{est} = p\left(\psi_\alpha i_\beta - \psi_\beta i_\alpha\right) \tag{28}$$

Here, current and voltage data on the α-β axis of the PMSM are the input variables. Position and speed data resulted from the estimation process are conveyed to the proper evaluations within the DTC control system. Estimation process of the voltage model is independent from the rotor speed data.

$$\omega_{est} = \frac{\psi_\alpha \frac{d\psi_\beta}{dt} - \psi_\beta \frac{d\psi_\alpha}{dt}}{\psi_\alpha^2 + \psi_\beta^2} \tag{29}$$

In this speed sensorless algorithm, only stator phase currents and inverter output voltages are measured. Torque and speed are estimated in an open-loop without using an observer, as given in the Eqs. (28), (29). A low-pass filter is accompanied to the results obtained from (29) in order to monitor the speed estimation. On the other hand, this method gives only the synchronous speed, not the mechanical speed for induction motor due to the slip speed. This implementation is applicable for general purpose PMSM drives.

Voltage vector for the DTC is determined by comparison of the estimated stator flux and torque values with their reference values. For high performance driver systems of permanent magnet synchronous motors, it is basically desired that the drive system should not be affected by parameter variations. The other criterion is that the speed must reach the reference value as quickly as possible when speed reference changes. Traditional PI and PID controllers are widely used as speed controllers in driver systems. Here, the speed controller consists of both a PI controller and an acceleration compensator.

Observerless scheme for sensorless DTC method is highly sensible to determined voltage vector and is highly dependent to the stator flux space vector. Therefore, stator flux space vector should be accurately estimated. The measured input values to the DTC control are only motor current and voltage. The voltage is defined from the DC-bus voltage and inverter switch positions. Since the inverter operates at high switching frequencies, undesired harmonic components can be produced.

4. Low-Pass Filter for Harmonics

It is desirable that the voltages and currents provided to the motor terminals do not include harmonic components. These harmonics may cause many unwanted effects such as electromagnetic interference (EMI) noise and high dv/dt which affect the motor control system as well (Chen et al, 2007). In such hysteresis controlled inverters, when the error per sampling time is large, the voltage vectors can no longer remain within the small hysteresis band. The LP filter can be used to affect the inverter switching frequency in order to decrease the stress on the inverter, so that PMSM is not affected by the variations and noises in entire system. Fig.8 shows the RLC filter, which plays the main role in reducing the high dv/dt of line to line voltages at motor terminals.

The transfer function of the filter is given by

$$\frac{V_o}{V_i} = \frac{R_1 C_1 s + 1}{C_1 L_1 s^2 + (r_1 + R_1) C_1 s + 1} \tag{30}$$

To obtain over damping response, the filter resistances is selected to be,

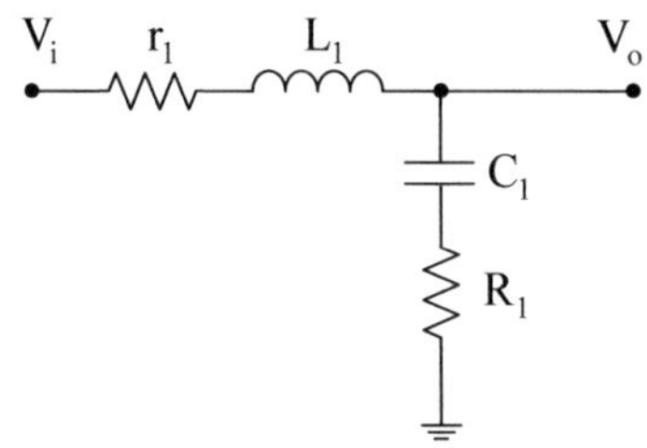

Fig. 8. RLC Low pass filter for harmonics

$$R_1 + r_1 > \sqrt{\frac{4L_1}{C_1}} \quad (31)$$

Cut off frequency ω_c is given by Eq. (32).

$$\omega_c = 1/\sqrt{L_1 C_1} \quad (32)$$

Here, the cut off frequency is approximately calculated to be 300 Hz and LP filter parameters are given in Table 3.

L_1	22 mH
C_1	500 μF
R_1	200 Ω
r_1	0.2 Ω

Table 3. Filter parameters

By using (30) and the filter parameters, (33) is obtained. Here, the LP block is constructed with (33). In order to reduce ohmic losses, the series resistance r_1 is normally of small value, as shunt resistance R_1 is selected high enough to limit the currents drawn by the filter (Ozcira et al, 2008).

$$\frac{V_o}{V_i} = \frac{1}{11.10^{-5}s + 1} \quad (33)$$

The filter system is characterized by affecting inverter switching frequency in such a way to decrease switching stresses. The filter system uses dissipative elements to reshape motor voltage waveform to provide voltage to the motor windings.

5. Model verification of direct torque control algorithm

Direct torque control is used as the control method in simulations thus the operating principle of the DTC is described briefly. In principle the DTC is a hysteresis control - of the stator flux linkage and the torque that directly selects one of the six non-zero and two zero discrete voltage vectors of the inverter. The principal operation of the DTC is shown as a block diagram in Fig.9. The system includes a flux estimator which utilizes a voltage model on a direct torque controlled PMSM, and speed can also be estimated by using calculations with aid of the current and voltage data. The voltage and current signals are inputs to an accurate motor model which produces the exact value of the stator flux and torque. Motor torque and flux comparators compare the actual values to the reference values that are produced by torque and flux reference controllers. The outputs from these controllers are updated every 10μs and they indicate whether the torque or flux has to be varied. Depending on outputs, the switching logic directly determines the optimum inverter switching signals (Bekiroglu & Ozcira, 2010). DTC method was applied and both flux and torque were controlled by hysteresis controllers; thus, the delays related to the PWM modulator were eliminated. Besides, a low-pass filter was implemented to reduce the harmonics and noise in the PMSM.

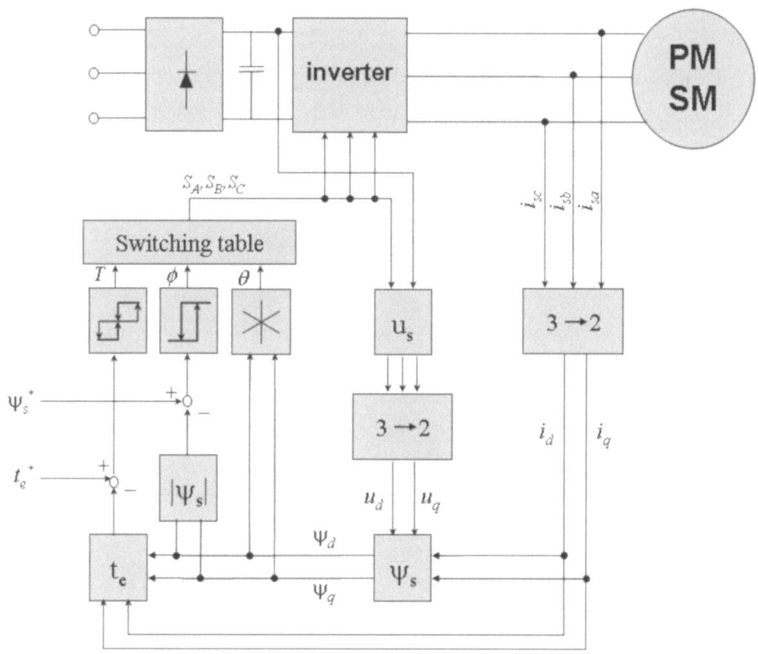

Fig. 9. Block diagram of the direct torque control (Luukko, 2000)

The control logic is implemented by a MATLAB® Function block that includes the software developed for this study. This function block determines the optimal switching vectors according to their sectors. The inverter is operated based on the different IGBT states given in Table 1 and Table 2. Since the inverter operates at high switching frequencies, undesired harmonic components can be produced. Simulink® model of low voltage - high power surface mounted PMSM is developed according to the dq model. The inverter DC bus voltage is 510V. During $t=[0,\ 0.4]$ interval, $\omega = 13\ \text{rad/s}$ and at $t = 0.4\text{s}$ $\omega = -13\ \text{rad/s}$ is applied as the reference speed. According to observe adaptation of the actual speed at $t = 0.2\text{s}$ 60 Nm is applied to load torque value.

Motor parameters are; $P_n = 18\text{kW}$, $I_n = 50\text{A}$, $R_s = 0.43\Omega$, $L_d = 25\text{mH}$, $L_q = 25\text{mH}$, $\psi_M = 1.58\text{Wb}$, $J = 2.16\text{kgm}^2$, $B_m = 0.005538$, $p = 10$

Fig.10 shows the speed control Simulink® diagram of the direct torque control for permanent magnet synchronous motor.

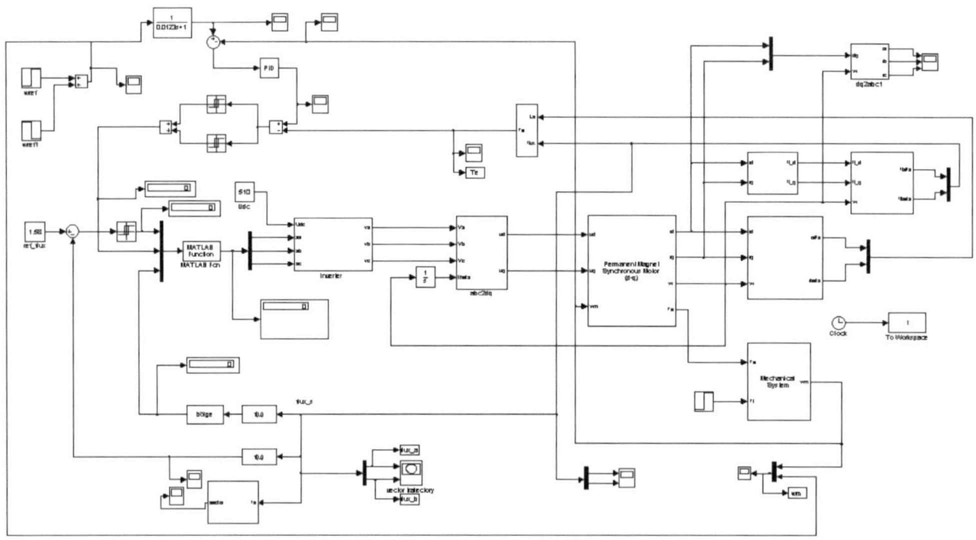

Fig. 10. Closed loop speed control system Simulink diagram of the PMSM with DTC.

6. Simulation results

The system dynamic responses are shown below with a sampling time 10μs.

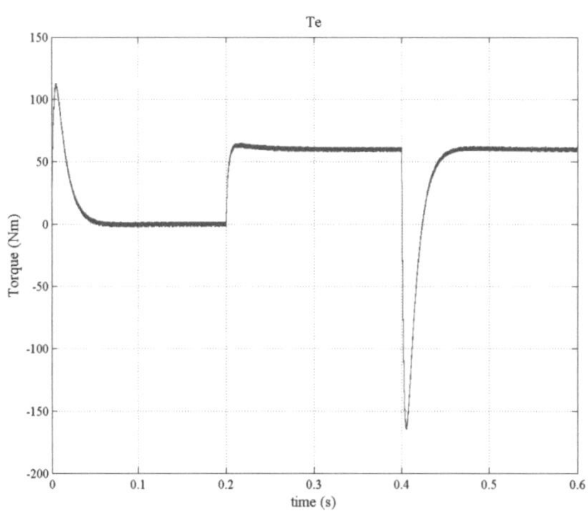

Fig. 11. Torque dynamic response

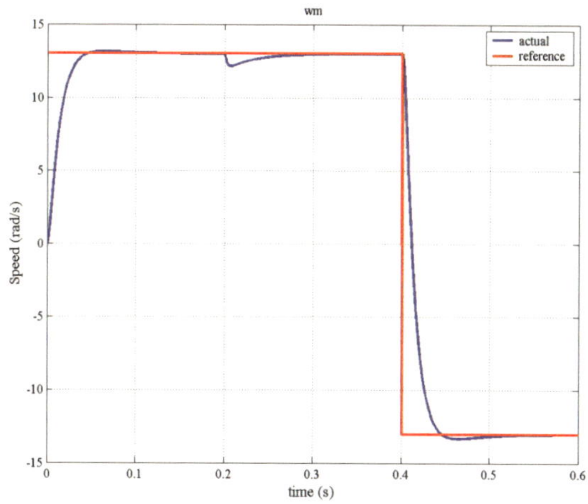

Fig. 12. Torque dynamic response

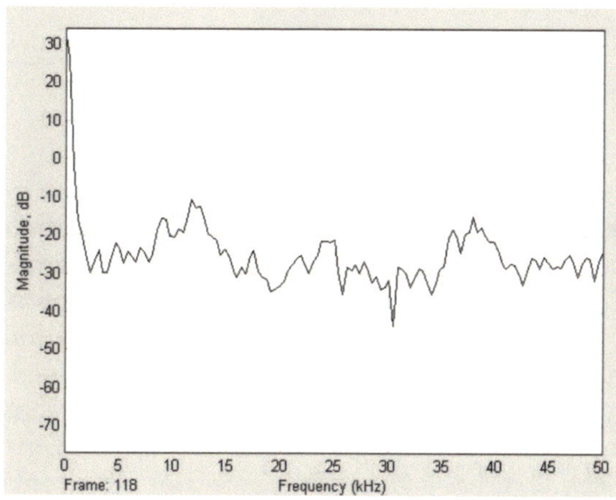

Fig. 13. Current harmonics without LP filter

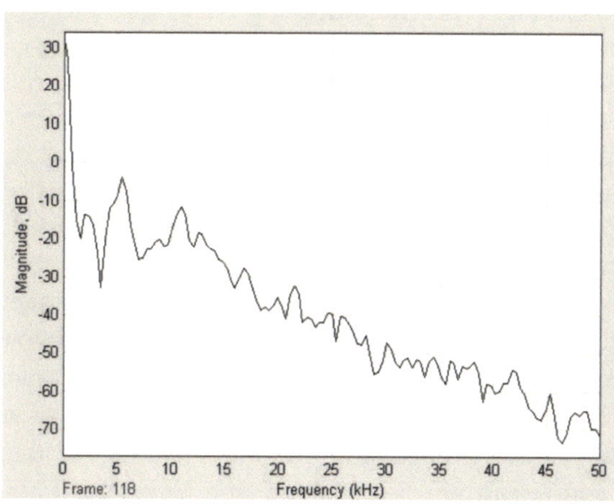

Fig. 14. Current harmonics without LP filter

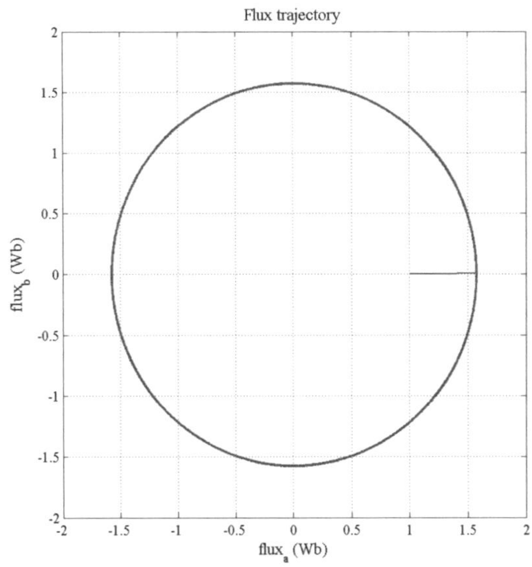

Fig. 15. Stator magnetic flux vector trajectory

7. Conclusion

DTC is intended for an efficient control of the torque and flux without changing the motor parameters and load. Also the flux and torque can be directly controlled with the inverter voltage vector in DTC. Two independent hysteresis controllers are used in order to satisfy the limits of the flux and torque. These are the stator flux and torque controllers. DTC process of the permanent magnet synchronous motor is explained and a simulation is constituted. It is concluded that DTC can be applied for the permanent magnet synchronous motor and is reliable in a wide speed range. Especially in applications where high dynamic performance is demanded DTC has a great advantage over other control methods due to its property of fast torque response. In order to increase the performance, control period should be selected as short as possible. When the sampling interval is selected smaller, it is possible to keep the bandwidth smaller and to control the stator magnetic flux more accurately. Also it is important for the sensitivity to keep the DC voltage in certain limits.

As an improvement approach, a LP filter can be added to the simulation in order to eliminate the harmonics. In simulation, certain stator flux and torque references are compared to the values calculated in the driver and errors are sent to the hysteresis comparators. The outputs of the flux and torque comparators are used in order to determine the appropriate voltage vector and stator flux space vector.

When results with and without filters are compared, improvement with the filters is remarkable, which will effect the voltage in a positive manner. Choosing cut off frequency close to operational frequency decreases DC shift in the stator voltage. However, this leads to phase and amplitude errors. Phase error in voltage leads to loss of control. Amplitude error, on the other hand, causes voltage and torque to have higher values than the reference values and field weakening can not be obtained due to voltage saturation. Hence, cutoff frequency of LP filter must be chosen in accordance to operational frequency.

8. References

Adnanes, A. K. (1990). Torque Analysis of Permanent Magnet Synchronous Motors, *Proceedings of PESC*, pp. 695-701, San Antonio, June 1990, Texas

Adreescu, G. D. & Rabinovici, R. (2004). Torque-speed adaptive observer and inertia identification without current transducers for control of electrical drives. *Proc. of the Romanian Academy Series A*, Vol. 5, No. 1, pp. 89-95, ISSN 1454-9069

Bekiroglu, N. & Ozcira, S. (2010). Observerless Scheme for Sensorless Speed Control of PMSM Using Direct Torque Control Method with LP Filter. *Advances in Electrical and Computer Engineering*, Vol. 10, No. 3, pp. 78-83, ISSN 1582-7445

Balazovic, P. (2003). 3-Phase PM Synchronous Motor Torque Vector Control. *Application Note DRM018*, Motorola Czech System Laboratories Roznov pod Radhostem

Bolognani, S.; Oboe, R. & Zigliotto, M. (1999). Sensorless full-digital PMSM drive with EFK estimation of speed and rotor position. *IEEE Trans. on Ind. Electronics*, Vol. 46, No. 1, pp. 184-191, ISSN 0278-0046

Chen, X.; Xu, D.; Liu, F. & Zhang, J. (2007) A Novel Inverter-Output Passive Filter for Reducing Both Differential and Common-Mode dv/dt at the Motor Terminals in PWM Drive Systems. *IEEE Transactions on Industrial Electronics*, Vol. 54, No. 1, pp. 419-426, ISSN 0278-0046

Chern, T. L. & Wu, Y.C. (1993). Design of brushless DC position servo systems using integral variable structure approach. *Electric Power Applications IEE Proceedings B* Vol.140, No.1, pp. 27-34, ISSN 1350-2352

Cui, J.; Wang, H.; Wang, C.; Wan, J. & Mu, G. (2008). Research on high performance direct torque control system based on DSP, *Proceedings of 7th World Congress on Intelligent Control and Automation WCICA*, pp. 1494-1497, June 2008, IEEE, Chongqing

Dan, S.; Yikang, H. & Zhu, J. G. (2004). Sensorless direct torque control for permanent magnet synchronous motor based on fuzzy logic, in *Proceedings of 4th International Power Electronics and Motion Control Conference IPEMC*, pp. 1286 - 1291, Aug. 2004, Xi'an

Dariusz, S.; Martin, P. K. & Frede, B. (2002). DSP Based Direct Torque Control of Permanent Magnet Synchronous Motor Using Space Vector Modulation, *Proceeding of the IEEE International Symposium on Industrial Electronics ISIE*, pp. 723-727, Nov 2002, L'Aquila

Depenbrock, M. (1985). DTC leans on the theory of field-oriented control of induction machines and the theory of direct self-control *Etz-Archive* BD.7, H7, pp. 211-218

Depenbrock M. (1988). Direct Self-control of inverter-fed machine. *IEEE Transactions on Power Electronics*, Vol.3, No.4, pp. 420-429, ISSN 0885-8993

Geyer, T.; Beccuti, G. A.; Papafotiou, G. & Morari, M. (2010). Model Predictive Direct Torque Control of Permanent Magnet Synchronous Motors, *Proceedings of IEEE Energy Conversion Congress and Exposition ECCE*, pp. 1-8, Sept 2010, Atlanta

Grzesiak L. M. & Kazmierkowski, M. P. (2007). Improving flux and speed estimators for sensorless AC drives. *IEEE Industrial Electronics Magazine*, Vol. 1, No. 3, pp. 9-19, ISSN 1932-4529

Guo, Y.; Xianbing, C. & Chen, C. L. P. (2009). DTC-SVM-based Matrix Converter for a PMSM in a Vessel Electric Propulsion System, *Proc. of the IEEE Control and Decision Conference and Chinese Control Conference*, pp. 3397–3401, Dec 2009, Shanghai

Ichikawa, S.; Chen, Z.; Tomita, M.; Doki, S. & Okuma, S. (2003). Sensorless controls of salient-pole permanent magnet synchronous motors using extended electromotive force models. *Electrical Eng. in Japan*, Vol. 146, No. 3, pp. 55-64, ISSN 1520-6416

Inoue, Y.; Morimoto, S. & Sanada, M. (2010). Examination and Linearization of Torque Control System for Direct Torque Controlled IPMSM. *IEEE Transactions on Industry Applications*, Vol. 46, No.1, pp. 159-166, ISSN 0093-9994

Jilong, Q.; Yantao, T.; Yimin, G. & Cheng, Z. (2008). A sensorless initial rotor position estimation scheme and an Extended Kalman Filter observer for the direct torque controlled Permanent Magnet Synchronous Motor Drive, *Proceedings of the International Conference on Electrical Machines and Systems ICEMS*, pp. 3945 - 3950, Oct 2008, IEEE, Wuhan

Jolly, L.; Jabbar, M.A. & Liu, Q. (2006). Optimization of The Constant Power Speed Range of A Saturated Permanent-Magnet Synchronous Motor. *IEEE Transactions on Industry Applications*, Vol. 42, No. 4, pp.1024-1030, ISSN 0093-9994

Laurila, L. (2004). Analysis of Torque and Speed Ripple Producing Non-idealities of Frequency Converters in Electric Drives. *Disseration Lappeenranta University of Technology*, Stockholm

Li, Y.; Gerling, D. & Liu, W. (2008). A Novel Switching Table Using Zero Voltage Vectors for Direct Torque Control in Permanent Magnet Synchronous Motor *18th International Conference on Electrical Machines ICEM*, pp. 1-6, Vilamoura, Sept 2008, IEEE, Algarve

Liu, G.; Zhang, Y.; Chen, Z. & Jia, H. (2009). PMSM DTC Predictive Control System Using SVPWM Based on the Subdivision of Space Voltage Vectors, *Proceedings of the IEEE 6th International Power Electronics and Motion Control Conference IPEMC*, pp. 1818-1821, May 2009, Wuhan

Luukko, J. (2000). Direct Torque Control of Permanent Magnet Synchronous Machines-Analysis and Implementation. *Disseration Lappeenranta University of Technology*, Stockholm

Luukko, J.; Niemela, M. & Pyrhonen, J. (2007). Estimation of Rotor and Load Angle of Direct Torque Controlled Permanent Magnet Synchronous Machine Drive. *IET Electronics Power Applications*, Vol.1, No. 3, pp. 209-306, ISSN 1751-8679

Martins, C.; Roboam, X.; Meynard, T. A. & Carylho, A. S. (2002). Switching Frequency Imposition and Ripple Reduction in DTC Drives by A Multilevel Converter. *IEEE Transactions on Power Electronics*, Vol.17, pp. 286-297, ISSN 0885-8993

Matsui, N. & Ohashi, H. (1992). DSP-based adaptive control of a brushless motor. *IEEE Transactions on Industry Applications*, Vol.28, No.2, pp. 448–454, ISSN 0093-9994

Noriega, G.; Restrepo, J.; Guzman, V. ; Gimenez, M. & Aller, J. (2007). Direct Torque Control of PMSM using Fuzzy Logic with PWM, *Proceedings of 42nd International Universities Power Engineering Conference UPEC*, pp. 203-209, Sept 2007, IEEE, Brighton

Ogasawara, S. & Akagi, H. (1996). An approach to real-time position estimation at zero and low speed for a PM motor based on saliency, *Proceedings on Industry Applications Conference*, pp. 29-35, San Diego, Oct 1996, IEEE, California

Omer, G. & Hava, A. (2010). Experimental investigation of shaft transducerless speed and position control of AC induction and interior permanent magnet motors. *Turkish Journal of Electrical Engineering & Computer Sciences*, Vol.18, No.5, pp. 1-18, ISSN 1303-6203

Ozcira, S.; Bekiroglu, N. & Aycicek, E. (2008). Speed Control of Permanent Magnet Synchronous Motor Based on Direct Torque Control Method, *Proceedings on International Symposium on Power Electronics, Electrical Drives, Automation and Motion SPEEDAM*, pp. 268-272, Ischia, June 2008, IEEE, Napoli

Ozcira, S.; Bekiroglu, N. & Aycicek, E. (2008). Direct Torque Control of Permanent Magnet Synchronous Motor using LP Filter, *18th International Conference on Electrical Machines ICEM*, pp. 1-5, Vilamoura, Sept 2008, IEEE, Algarve

Pelczewski, P. M.; Oberschelp, W. & Kunz, U. H. (1991). Optimal Model-Following Control of a Positioning Drive System with a Permanent-Magnet Synchronous Motor, *IEE Proceedings on Control Theory and Applications* Vol.138, No.3, pp. 267 - 273, ISSN 1350-2379

Pillay, P. & Krishnan, R. (1989). Modeling, Simulation, and Analysis of Permanent-Magnet Motor Drives Part II: The brushless DC Motor Drive. *IEEE Transactions on Industry App*, Vol.25, No.2, pp. 274-279, ISSN 0093-9994

Popescu, M.; Miller, T.J.E.; McGilp, M. I.; Strappazzon, G.; Trivillin, N. & Santarossa, R. (2006). Torque Behavior of One-Phase Permanent-Magnet AC Motor. *IEEE Transactions on Energy Conversion*, Vol. 21, No. 1, pp. 19-26, ISSN 0885-8969

Rahman K. M. & Toliyat, H. A. (1996). Sensorless operation of permanent magnet AC (PMAC) motors with modified stator windings, *Procs in Conference Records IEEE-IAS 31st Annual Meeting*, pp. 326-333, San Diego, Oct. 1996, IEEE, California

Rahman, M. F.; Zhong, L. & Lim, K. W. (1998). A direct torque-controlled interior permanent magnet synchronous motor drive incorporating field weakening. *IEEE Transactions on Industry Applications*, Vol.34, No.6, pp. 1246-1253, ISSN 0093-9994

Rahman, M. F.; Zhong, L. & Haque, E. (1999). Selection of Voltage Switching Tables for DTC Controlled Interior Permanent Magnet Motor. *School of Electrical Engineering and Telecommunications Lect.* The University of New South Wales, Sydney

Raymond, B. & Lang, J. H. (1991). Real-time Adaptive Control of the Permanent-Magnet Synchronous Motor. *IEEE Transactions on Industy Applicaitons*, Vol. 27, No. 4, pp. 704-716, ISSN 0093-9994

Sanchez, E.; Al-rifai, F. & Schofield, N. (2009). Direct Torque Control of Permanent Magnet Motors using a Single Current Sensor, *Proceedings of the IEEE International Electric Machines and Drives Conference IEMDC*, pp. 89-94, May 2009, Miami

Siahbalaee, J.; Vaez-Zadeh, S. & Tahami, F. (2009). A New Loss Minimization Approach With Flux And Torque Ripples Reduction of Direct Torque Controlled Permanent Magnet Synchronous Motors, *Proceedings of 13th European Conference on Power Electronics and Applications EPE*, pp. 1-8, Sept 2009, IEEE, Barselona

Sozer, Y.; Torrey, D. A. & Reva, S. (2000). New inverter output filter topology for PWM motor drives, *IEEE Transactions on Power Electronics*, Vol. 15, No. 6, pp. 1007-1017, ISSN 0885-8993

Swierczynski, D.; P. Wojcik, P.; Kazmierkowski, M. P. & Janaszek, M. (2008). Direct Torque Controlled PWM Inverter Fed PMSM Drive for Public Transport, *Proceedings on IEEE International Workshop on Advanced Motion Control AMC*, pp. 716-720, March 2008, Trento

Takahashi, I. & Noguchi, T. (1986). A New Quick-Response and High-Efficiency Control Strategy of an Induction Motor. *IEEE Transactions on Industry Applications*, Vol. IA-22, No.5, pp. 820-827, ISSN 0093-9994

Tan, Z.; Li, Y. & Li, M. (2001). A Direct Torque Control of Induction Motor Based on Three Level Inverter, *Proceedings of PESC*, pp. 1435–1439, June 2001, IEEE, Vancouver

Tang, L.; Zhong, L.; Rahman, M. F. & Hu, Y. (2004). A Novel Direct Torque Controlled Interior Permanent Magnet Synchronous Machines Drive with Low Ripple in Flux and Torque and Fixed Switching Frequency. *IEEE Transactions on Power Electronics* Vol. 19, No. 2, pp.346-354, ISSN 0885-8993

Toliyat, H. A.; Hao, L.; Shet, D. S. & Nondahl, T. A. (2002). Position-sensorless control of surface-mount permanent-magnet AC (PMAC) motors at low speeds. *IEEE Trans. on Industrial Electronics*, Vol. 49, No. 1, pp. 157-164, ISSN 0278-0046

Vaez-Zadeh, S. (2001). Variable Flux Control of Permanent Magnet Synchronous Motor Drives for Constant Torque Operation. *IEEE Transactions on Power Electronics*, Vol. 16, No.4, pp. 527-534, ISSN 0885-8993

Vas, P. (1998). *Sensorless Vector and Direct Torque Control*, Oxford University Press, ISBN 0-19-856465-1, New York.

Wang, L. & Gao, Y. (2007). A Novel Strategy of Direct Torque Control for PMSM Drive Reducing Ripple in Torque and Flux, *Proceedings of IEEE International Electric Machines & Drives Conference IEMDC*, pp. 403-406, May 2007, Antalya

Yan, Y.; Zhu, J. & Lu, H. (2005). Direct torque control of surface-mounted permanent magnet synchronous motor based on accurate modeling, *Proceedings of Australasian Universities Power Engineering Conference AUPEC*, Sept. 2005, Tasmania

Yutao, L.; Fanzhen, M.; Xingfeng, F. & Peng, K. (2008). The Rectangular Fluctuation Control of Improved Direct Torque Controlled Permanent Magnet Synchronous Machines Based on RBF Network, *Proceedings of 27th Chinese Control Conference, CCC*, pp. 628-631, July 2008, Kunming

Zhao, L.; Ham, C. H.; Wu, T. X.; Zheng L.; Sundaram, K.B.; Kapat, J. & Chow, L. (2004). DSP-Based Super High-Speed PMSM Controller Development and Optimization. *Digital Signal Processing Workshop and the 3rd IEEE Signal Processing Education Workshop*, pp. 187- 190, August 2004, New Mexico

Zhao, S. & Peng, S. (2007). A Modified Direct Torque Control Using Space Vector Modulation (DTC-SVM) for Surface Permanent Magnet Synchronous Machine (PMSM) with Modified 4-order Sliding Mode Observer, *Proc. of IEEE International Conference on Mechatronics and Automation ICMA*, pp. 1207-1212, August 2007, Heilongjiang

Zhong, L.; Rahman, M. F.; Hu, W. Y. & Lim, K. W. (1997). Analysis of Direct Torque Control in Permanent Magnet Synchronous Motor Drives. *IEEE Trans. on Power Electronics*, Vol.12, No.3, pp. 528-536, ISSN 0885-8993

Zhong, L.; Rahman, M. F.; Hu, W. Y.; Lim, K. W. & Rahman, M. A. (1999). A Direct Torque Controller for Permanent Magnet Synchronous Motor Drives. *IEEE Transactions on Energy Conversion*, Vol.14, No.3 pp. 637-642, ISSN 0885-8969

7

Torque Control of PMSM and Associated Harmonic Ripples

Ali Ahmed Adam[1], and Kayhan Gulez[2]
[1]*Fatih University, Engineering Faculty, Electrical-Electronics Eng. Dept., 34500 Buyukcekmece-Istanbul,*
[2]*Yildiz Technical University, Electrical-Electronics Eng. Faculty, Control and Automation Engineering Dept., 34349 Besiktas- Istanbul, Turkey*

1. Introduction

Vector control techniques have made possible the application of PMSM motors for high performance applications where traditionally only dc drives were applied. The vector control scheme enables the control of the PMSM in the same way as a separately excited DC motor operated with a current-regulated armature supply where then the torque is proportional to the product of armature current and the excitation flux. Similarly, torque control of the PMSM is achieved by controlling the torque current component and flux current component independently.
Torque Control uses PMSM model to predict the voltage required to achieve a desired output torque or speed. So by using only current and voltage measurements (and rotor position in sensor controled machine), it is possible to estimate the instantaneous rotor or stator flux and output torque demanded values within a fixed sampling time. The calculated voltage is then evaluated to produce switching set to drive the inverter supplying the motor. PMSM torque control has traditionally been achieved using Field Oriented Control (FOC). This involves the transformation of the stator currents into a synchronously rotating d-q reference frame that is typically aligned to the rotor flux. In the d-q reference frame, the torque and flux producing components of the stator current can separately be controlled. Typically a PI controller is normally used to regulate the output voltage to achieve the required torque.
Direct Torque Control (DTC), which was initially proposed for induction machines in the middle of 1980's (Depenbrock, 1984 and 1988; Takahashi, 1986), was applied to PMSM in the late 1990's (French, 1996; Zhong, 1997). In the Direct Torque Control of the PMSM, the control of torque is exercised through control of the amplitude and angular position of the stator flux vector relative to the rotor flux vector. Many methods have been proposed for direct torque control of PMSM among which Hysteresis based direct torque control (HDTC) and Space Vector Modulation direct torque control (SVMDTC).
In 2009 Adam and Gulez, introduced new DTC algortim for IPMSM to improve the performance of hysteresis direct torque control. The algorithm uses the output of two hysteresis controllers used in the traditional HDTC to determine two adjacent active vectors. The algorithm also uses the magnitude of the torque error and the stator flux linkage position to select the switching time required for the two selected vectors. The selection of

the switching time utilizes suggested table structure which, reduce the complexity of calculation. The simulation and experimental results of the proposed algorithm show adequate dynamic torque performance and considerable torque ripple reduction as well as lower flux ripple, lower harmonic current and lower EMI noise reduction as compared to HDTC. Only two hysteresis controllers, current sensors and built-in counters microcontroller are required to achieve torque control.

Torque ripple and harmonic noise in PMSM are due to many factors such as structural imperfectness associated with motor design, harmonics in control system associated with measurement noises, switching harmonics and harmonic voltages supplied by the power inverter which constitute the major source of unavoidable harmonics in PMSM. These harmonics cause many undesired phenomena such as electromagnetic interference "EMI" and torque ripples with consequences of speed oscillations, mechanical vibration and acoustic noise which, deteriorate the performance of the drive in demanding applications (Holtz and Springob 1996). These drawbacks are especially high when the sampling period is greater than 40μs (Zhong, et al. 1997).

Recently many research efforts have been carried out to reduce the torque ripples and harmonics in PMSM due to inverter switching with different degree of success. Yilmaz (Yilmaz, et al. 2000) presented an inverter output passive filter topology for PWM motor drives to reduce harmonics of PMSM, the scheme shows some effectiveness in reducing switching harmonics, but however, very large circulating current between inverter output and filter elements is required to reshape the motor terminal voltage which violate current limitation of the inverter. Many researchers (Hideaki et al, 2000; Darwin et al., 2003; Dirk et al , 2001) have addressed active filter design to reduce or compensate harmonics in supply side by injecting harmonics into the line current which have no effect on the current supplying the load. Satomi (Satomi, et al. 2001) and Jeong-seong (Jeong-seong, et al. 2002) have proposed a suppression control method to suppress the harmonic contents in the d-q control signals by repetitive control and Fourier transform but, however, their work have nothing to do with switching harmonics and voltage harmonics provided by the PWM inverter supplying the motor. Se- Kyo, et al. (1998), Dariusz et al. (2002), and Tang et, al. (2004) have used space vector modulation to reduce torque ripples with good results; however, their control algorithm depends on sophisticated mathematical calculations and two PI controllers to estimate the required reference voltage and to estimate the switching times of the selected vectors. Holtz and Springob (1996, 1998) presented a concept for the compensation of torque ripple by a self- commissioning and adaptive control system.

In this chapter, two different methods to improve torque ripple reduction and harmonic noises in PMSM will be presented. The first method is based on passive filter topology (Gulez et al., 2007). It comprises the effects of reducing high frequency harmonic noises as well as attenuating low and average frequencies. The second method is based on active series filter topology cascaded with two LC filters (Gulez et al., 2008).

Modern PMSM control algorthims

2. Algorithm 1: Rotor Field Oriented Control "FOC"

The control method of the rotor field-oriented PMSM is achieved by fixing the excitation flux to the direct axis of the rotor and thus, it is position can be obtained from the rotor shaft by measuring the rotor angle θ_r and/or the rotor speed ω_r.

Consider the PMSM equations in rotor reference frame are given as:

$$\begin{bmatrix} v_{sd} \\ v_{sq} \end{bmatrix} = \begin{bmatrix} R + pL_{sd} & -P\omega_r L_{sq} \\ P\omega_r L_{sd} & R + pL_{sq} \end{bmatrix} \begin{bmatrix} i_{sd} \\ i_{sq} \end{bmatrix} + \begin{bmatrix} 0 \\ P\omega_r \psi_F \end{bmatrix} \quad T_e = \frac{3}{2} P(\psi_F i_{sq} + (L_{sd} - L_{sq}) i_{sd} i_{sq})) \tag{1}$$

Where,
v_{sd}, v_{sq}: d-axis and q-axis stator voltages;
i_{sd}, i_{sq}: d-axis and q-axis stator currents;
R: stator winding resistance;
L_{sd}, L_{sq}: d-axis and q-axis stator inductances;
$p=d/dt$: differential operator;
P: number of pole pairs of the motor;
ω_r: rotor speed;
Ψ_F: rotor permanent magnetic flux;
Te: generated electromagnetic torque;
To produce the largest torque for a given stator current, the stator space current is controlled to contain only i_{sq}.
And since for PMSM $L_d \leq L_q$, the second torque component in Eq.(1) is negative with positive values of i_{sd} and zero for SPMSM. Thus, to ensure maximum torque, the control algorithm should be such that i_{sd} is always zero, which result in simple torque expression as:

$$T_e = 3/2\ P\psi_F\, i_{sq} = 3/2\ \psi_F\, |\, i_s|\ sin(\alpha - \theta_r) \tag{2}$$

The stator windings currents are supplied from PWM inverter, using hysteresis current controller. The actual stator currents contain harmonics, which, produce pulsating torques, but these may be filtered out by external passive and active filters, or using small hysteresis bands for the controllers.

2.1 Implementation of rotor field oriented control

The block diagram of rotor-field oriented control of PMSM in polar co-ordinate is shown in Fig.1 (Vas, 1996). The stator currents are fed from current controlled inverter. The measured stator currents are transformed to stationary D-Q axis. The D and Q current components are then transformed to polar co-ordinate to obtain the modulus $|i_s|$ and the phase angle α_s of the stator-current space phasor expressed in the stationary reference frame.

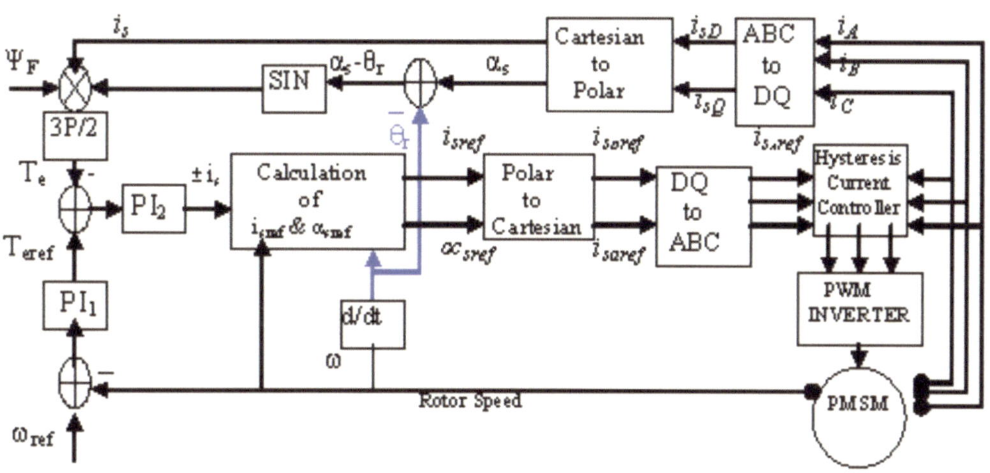

Fig. 1. Rotor Field Oriented Control of PMSM

The rotor speed ω_r and rotor angle θ_r are measured; and the position of the stator current in the rotor reference frame is obtained. Then, the instantaneous electromagnetic torque *Te* can be obtained as stated in Eq. (2).
The necessary current references to the PWM inverter are obtained through two cascaded PI controllers. The measured rotor speed ω_r is compared with the given reference speed ω_{ref} and the error is controlled to obtained the reference torque T_{eref}. The calculated torque is subtracted from the reference torque and the difference is controlled to obtain the modulus of i_{sref}. The reference angle α_{sref} is set equal to $\pi/2$, and the actual rotor angle is added to $(\alpha_{sref} - \theta_r)$ to obtain the angle α_{sref} of the stator current in the stationary D-Q frame. Theses values are then transformed to the three-reference stator currents i_{sAref}, i_{sBref} and i_{sCref} and used to drive the current controller.
The functions of the PI controllers (other controllers such as Fuzzy Logic, Adaptive, Slide mode or combinations of such controller may be used) are to control both the speed and torque to achieve predetermined setting values such as:

1. Zero study state error and minimum oscillation,
2. Wide range of regulated speed,
3. Short settling time,
4. Minimum torque ripples,
5. Limited starting current.

Based on the above description a FOC model was built in MatLab/Simulink as shown in Fig. 2. The model responses for the data setting in Table 1 of SPMSM with ideal inverter were displayed in Fig.3 to Fig.7. The PI controllers setting and reference values are:

Ts=1 μs, ω_{ref} =300, T_L =5Nm, PI_2: Kp=10, Ki=0.1 PI_1: Kp=7, Ki=0.1.

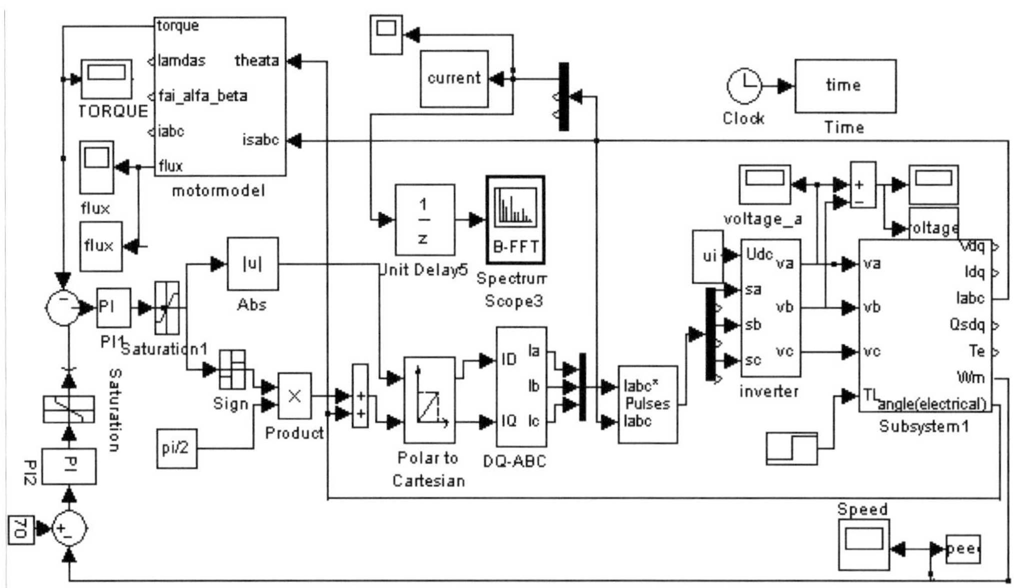

Fig. 2. FOC model in Matlab/Simulink

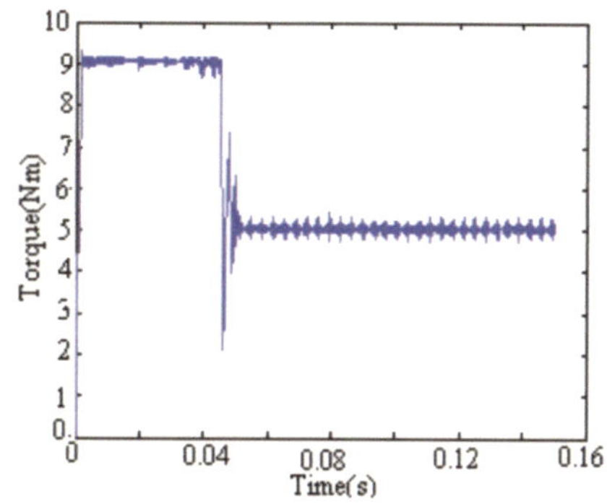

Fig. 3. Torque response

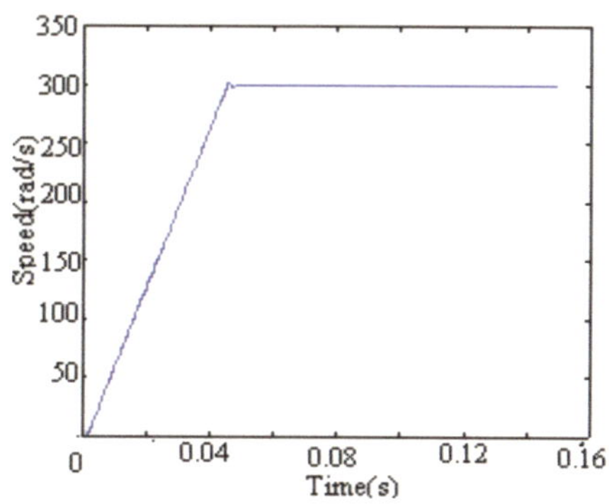

Fig. 4. Speed response

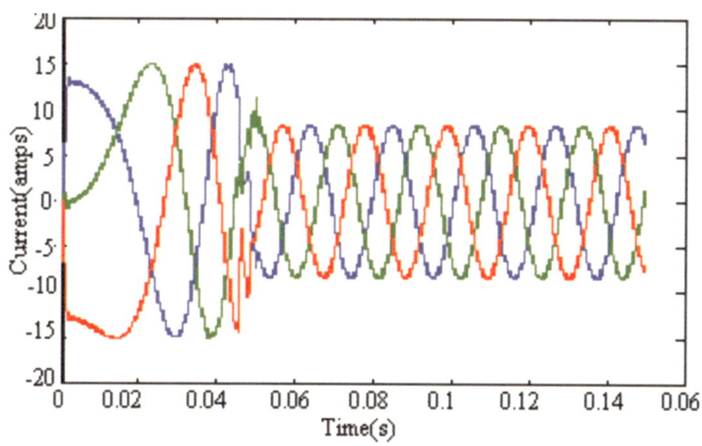

Fig. 5. Line current response

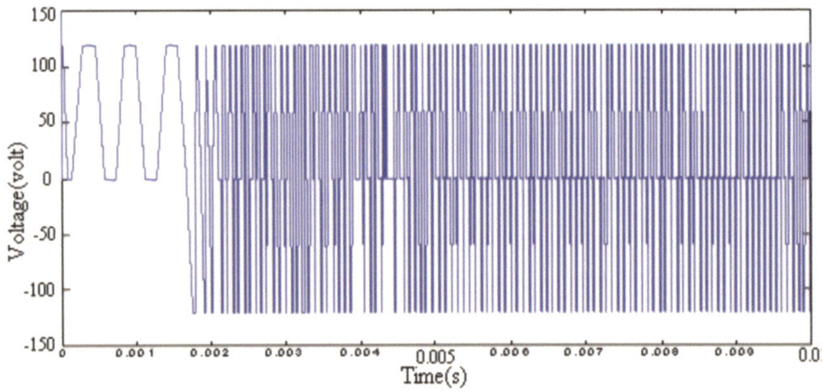

Fig. 6. V_{ab} switching pattern

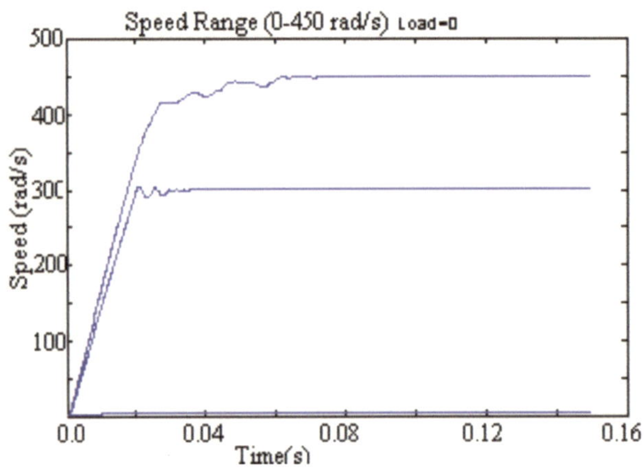

Fig. 7. Regulated Speed range (0-450) rad/s

Vdc	120V
Ψ_F	0.1546 web.
Rs	1.4 ohm
Ld	0.0066 H
Lq	0.0066 H
J	0.00176 kGm^2
B	0.000388 N/rad/s

Table 1. Motor parameters

The above figures show acceptable characteristics however, the torque pulsation cannot be avoided and the line currents are almost sinusoidal with some harmonic values. The speed can be regulated up to the rated value (300rad/s) with acceptable response. Bearing in mind that sensors, analog/digital converters, switching elements of the inverter and algorithm processing in DSP are time consuming, it is practically difficult to achieve such system with small sampling period. Thus, in practice convenient sampling periods, such as 100 μs (or larger) is normally selected for processing. In the following, simulation practical values will be adopted to obtain reasonable results for comparison. So, PMSM with parameters shown in Table 2 was simulated in the same model with the following setting values:

Ts=100 μs, T_L =2Nm, ω_{ref} =70 rad/s PI$_2$: Kp=10, Ki=0.1 and PI$_1$: Kp=7, Ki=0.1

Number of pole pairs	P	2
Stator leakageresistance	R_s	5.8 Ohm
d-axis inductance	L_{sq}	102.7 mH
q-axis inductance	L_{sd}	44.8 mH
Permanent magnet flux	Ψ_F	533 mWb
Inertia constant	J	0.000329Nms^2
Friction constant	B	0.0
Reference speed	ω	70 rad/s
Load torque	T_L	2 Nm

Table 2. IPMSM parameters

The simulation responses were shown below:

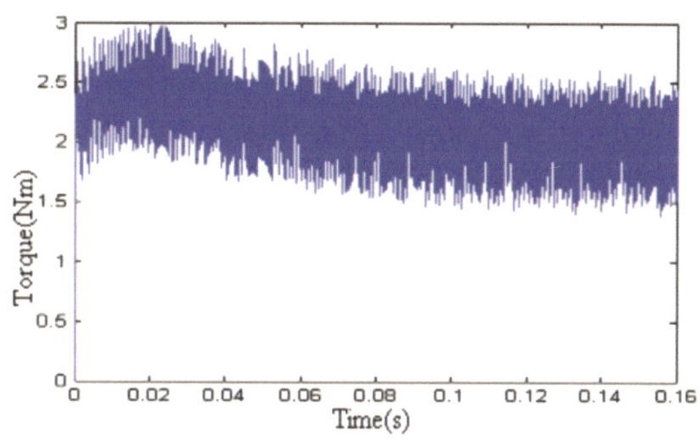

Fig. 8. FOC Torque response

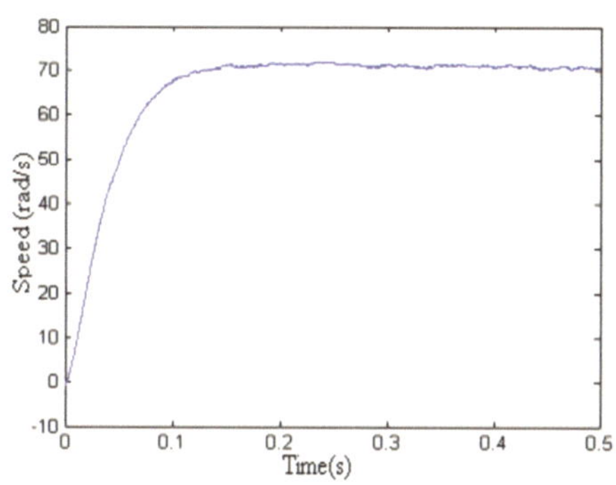

Fig. 9. FOC Speed response

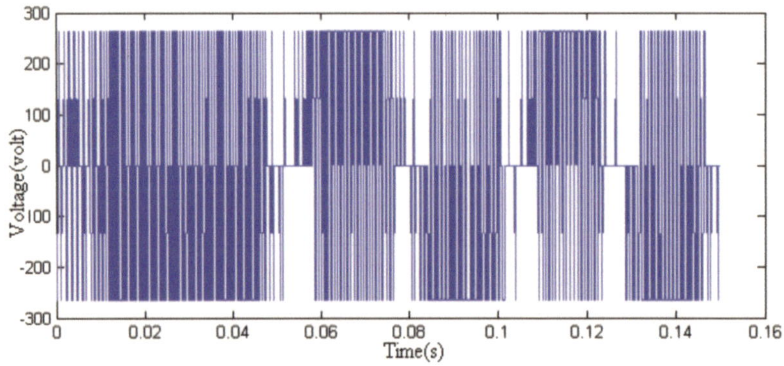

Fig. 10. FOC Line Voltage Switching

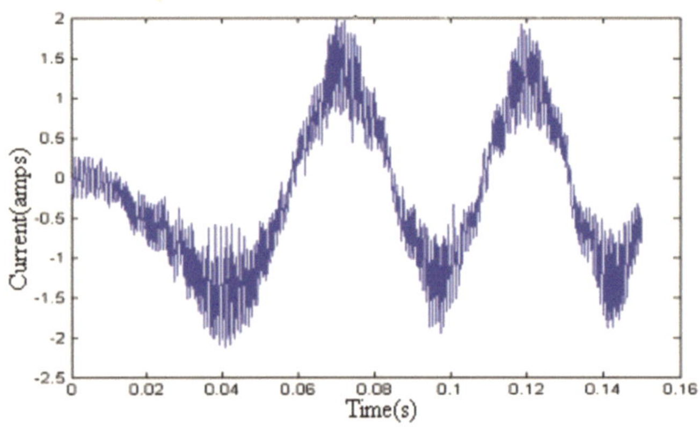

Fig. 11. FOC Current response

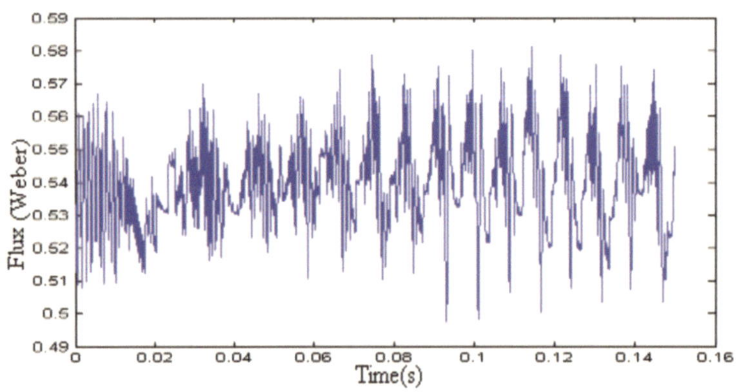

Fig. 12. Flux response

The responses showed that the torque pulsation is very high and line currents are full of harmonic components which give rise to EMI noises, in addition flux and speed are not free of ripples which result in unwanted phenomena such as machine vibration and acoustic noise.

3. Algorithm 2: Hysteresis Direct Torque Control (HDTC)

This method which is also called Basic DTC can be explained by referring to Fig.13. In this figure, the angle between the stator and rotor flux linkages δ is the load angle when the stator resistance is neglected. In the study, state δ is constant corresponds to a load torque, where stator and rotor flux rotate at synchronous speed. In transient operation, δ varies and the stator and rotor flux rotate at different speeds. Since the electrical time constant is normally much smaller than the mechanical time constant, the rotating speed of stator flux with respect to rotor flux, can easily be changed also that the increase of torque can be controlled by controlling the change of δ or the rotating speed of the stator flux (Zhong, 1997) as will be explained in the following analysis.

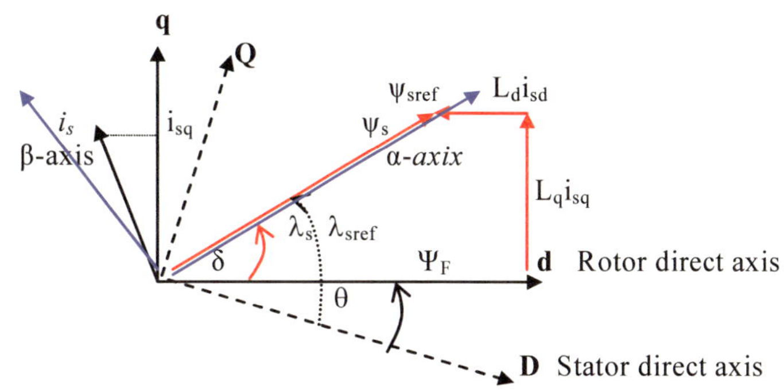

Fig. 13. Stator and rotor flux space phasors

3.1 Flux and torque criteria

Referring to Fig. 13 the flux equations in rotor dq axis frame can be rewritten as:

$$\psi_{sd} = L_{sd} i_{sd} + \psi_F = |\psi_s| \cos\delta \tag{3}$$

$$\psi_{sq} = L_{sq} i_{sq} = |\psi_s| \sin\delta \tag{4}$$

Where, $|\psi_s|$ represent the amplitude of the stator flux linkage calculated as:

$$|\overline{\psi}_s|^2 = \left(L_{sd} i_{sd} + \psi_F\right)^2 + \left(L_{sq} i_{sd}\right)^2 \tag{5}$$

In the general α-β reference frame the torque equation can be written as (Zhong, 1997):

$$\mathrm{T_e} = \frac{3}{2} P |\psi_s| i_{s\beta} \tag{6}$$

Where; i_β is the component of the stator phasor space current perpendicular to the stator flux axis α.

Equation (6) suggests that the torque is directly proportional to the β-axis component of the stator current if the amplitude of the stator flux linkage is kept constant.

Now using Eq.(3) and Eq.(4) to rewrite the torque equation as:

$$T_e = \frac{3P|\psi_s|}{4L_{sd}L_{sq}}\left[2\psi_F L_{sq}\sin\delta - |\psi_s|\left(L_{sq}-L_{sd}\right)\sin 2\delta\right] \tag{7}$$

For SPMSM $L_{sd} = L_{sq} = L_s$ and this expression is reduced to

$$T_e = \frac{3P|\psi_s|}{2L_s}\psi_F \sin\delta = \frac{3P|\psi_s|}{2L_s}\psi_F \sin\delta^{\bullet}t \tag{8}$$

Where $\delta^{\bullet}$ is the angular speed of the stator flux linkage relative to the permanent magnet axis.
At constant flux values, Eq. (8) shows that T_e-δ has sinusoidal relationship and the derivative of this equation suggest that the increase of torque is proportional to the increase of δ in the range of $-\pi/2$ to $\pi/2$. So the stator flux linkage should be kept constant and the rotational speed $\delta^{\bullet}$ is controlled as fast as possible to obtain the maximum change in actual torque.
For IPMSM, the torque expression contains in addition to the excitation torque, reluctance torque and for each stator flux level value, there exist different T_e-δ curve and different maximum torque. Fig. 14 (Zhong, 1997) shows these relationship for different values of $|\psi_s|$. Observe the crossing of curve $|\psi_s|=2\psi_F$ where, the derivative of torque near zero crossing has negative value, which implies that DTC can not be applied in this case.

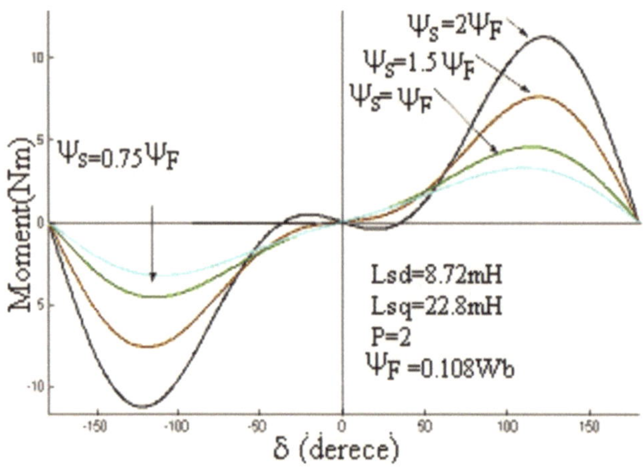

Fig. 14. Different T_e-δ curves for different stator flux values

Analytically this condition can be obtained from derivative of Eq. (7) as follows:

$$\frac{dT_e}{dt} = \frac{3P|\psi_s|}{2L_{sd}L_{sq}}\left[\psi_F L_{sq}\delta^{\bullet} - |\psi_s|\left(L_{sq}-L_{sd}\right)\delta^{*}\right] \tag{9}$$

And thus for positive torque derivative under positive $\delta^{\bullet}$, $|\psi_s|$ should be selected in such a way that (Tang et al., 2002; Zhong et al. 1997):

$$|\psi_s| \langle \frac{L_{sq}}{L_{sq}-L_{sd}}\psi_F \tag{10}$$

That if fast dynamic response is required. Also that (Tang et al., 2002) for stable torque control the following criteria should be satisfied.

$$\delta < \cos^{-1}\left(\frac{a/\psi_s - \sqrt{(a/\psi_s)^2 + 8}}{4}\right) \tag{11}$$

Where, $a = \dfrac{L_{sq}}{L_{sq} - L_{sd}} \psi_F$

3.2 Control of stator flux strategy

The stator flux linkage of a PMSM in the stationary reference frame can be expressed as:

$$\psi_s = \int (V_s - Ri_s)dt = V_s t - R\int i_s dt + \psi_s\big|_{t=0} \tag{12}$$

During switching interval each voltage vector is constant, so if stator resistance is neglected then, this equation implies that the stator flux will move in the direction of the applied voltage vector.

To select the voltage vectors or controlling the amplitude of the stator flux linkage, the voltage vector plane is divided into six sectors (FS_1 to FS_6) as shown in Fig. 15. In each region two adjacent voltage vectors are selected to increase or decrease the amplitude respectively of the flux within a hysteresis band. For example, the vectors V_2 and V_3 are used to increase and decrease the flux amplitude when ψ_s is in region one and rotating in a counter clockwise direction. If rotating in clockwise direction then V5 and V6 are used for the same reason.

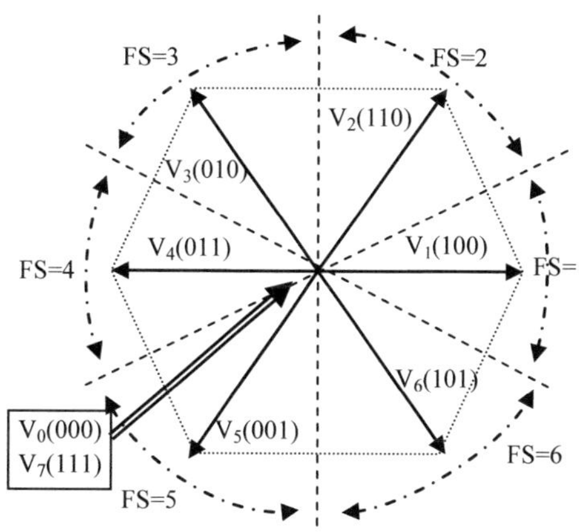

Fig. 15. Applied vectors position and flux sectors.

3.3 Implementation of Hysteresis DTC

The block diagram of a PMSM drive with HDTC may be as shown in Fig. 16, where the measured current phase values and dc voltage are transferred to D-Q stationary axis values, and the flux linkage components ψ_{sD} and ψ_{sQ} at the mth sampling instance are calculated from the stator voltages as follows:

$$\psi_{sD}(m) = \psi_{sD}(m-1) + (V_D(m-1) - Ri_{sD})Ts \tag{13}$$

$$\psi_{sQ}(m)= \psi_{sQ}(m-1) +(V_Q(m-1)-Ri_{sQ})Ts \qquad (14)$$

Where Ts is the sampling period and i_{sD} and i_{sQ} are calculated as average values of $i_s(m-1)$ and $i_s(m)$ and thus, amplitude and flux angle position with respect to stationary D-Q axis can be calculated as:

$$|\psi_s| = \sqrt{\psi_D^2(m)+\psi_Q^2(m)}$$
$$\lambda_s = \tan^{-1}\frac{\psi_Q(m)}{\psi_D(m)} \qquad (15)$$

The torque can be rewritten in the stationary reference frame as (Zhong et al., 1997):

$$T_e(m) = \frac{3}{2}P\left(\psi_{sD}(m)i_{sQ}(m) - \psi_{sQ}(m)i_{sD}(m)\right) \qquad (16)$$

However if the phase currents and the rotor speed and/or rotor position are monitored then Eq. (3) and Eq. (4) can be used to calculate torque and flux values, where then the transformation D-Q ↔ d-q is necessary to achieve the required values.

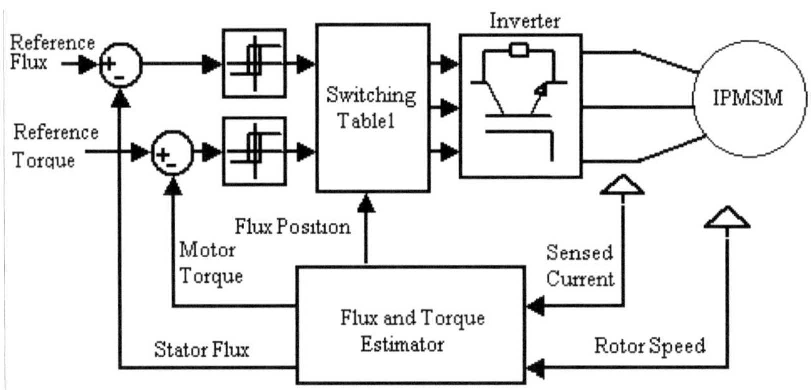

Fig. 16. HDTC of PMSM

The calculated Torque and Flux magnitude values are compared with their respective reference values and the produced errors are inputs to their respective hysteresis comparators. The flux linkage comparator is a two level comparator $\boldsymbol{\phi}\ \varepsilon$ {1, 0} and the torque comparator is a three level comparator $\boldsymbol{\tau}\ \varepsilon${1, 0, -1}. The outputs of these comparators together with stator position λ_s (or sector number) are inputs to optimum voltage switching lookup table as the one shown in Table 3 (Luukko, 2000). The output of this table is switching vector to the inverter driving the motor.

Based on the above description a HDTC of PMSM model was built in Matlab Simulink as shown in Fig.17.

The torque and flux estimator is based on monitoring of phase currents and rotor angle. The model responses for the Table 2 and controllers setting values as:

PI speed controller: Kp=0.04 and Ki=2,

Hysteresis logic: Flux band = ± 0.01; Torque Band = ±0.01; Sampling time: Ts= 0.0001s; has been simulated with results displayed in Fig.18-Fig.22

Φ	τ	FS					
		1 $-30\leq \lambda_s <30$	2 $30\leq \lambda_s <90$	3 $90\leq \lambda_s <150$	4 $150\leq \lambda_s <210$	5 $210\leq \lambda_s <270$	6 $270\leq \lambda_s <330$
1	1	$V_2(110)$	$V_3(010)$	$V_4(011)$	$V_5(001)$	$V_6(101)$	$V_1(100)$
	0	$V_7(111)$	$V_0(000)$	$V_7(111)$	$V_0(000)$	$V_7(111)$	$V_0(000)$
	-1	$V_6(101)$	$V_1(100)$	$V_2(110)$	$V_3(010)$	$V_4(011)$	$V_5(001)$
0	1	$V_3(010)$	$V_4(011)$	$V_5(001)$	$V_6(101)$	$V_1(100)$	$V_2(110)$
	0	$V_0(000)$	$V_7(111)$	$V_0(000)$	$V_7(111)$	$V_0(000)$	$V_7(111)$
	-1	$V_5(001)$	$V_6(101)$	$V_1(100)$	$V_2(110)$	$V_3(010)$	$V_4(011)$

Table 3. Optimum switching lookup table for HDTC inverter. Φ is the output of flux hysteresis controller, τ is the output of the torque hysteresis controller, the entries $V_i(\ldots)$ is the switching logic to the inverter and FS (Flux Sector) define the stator flux position sector

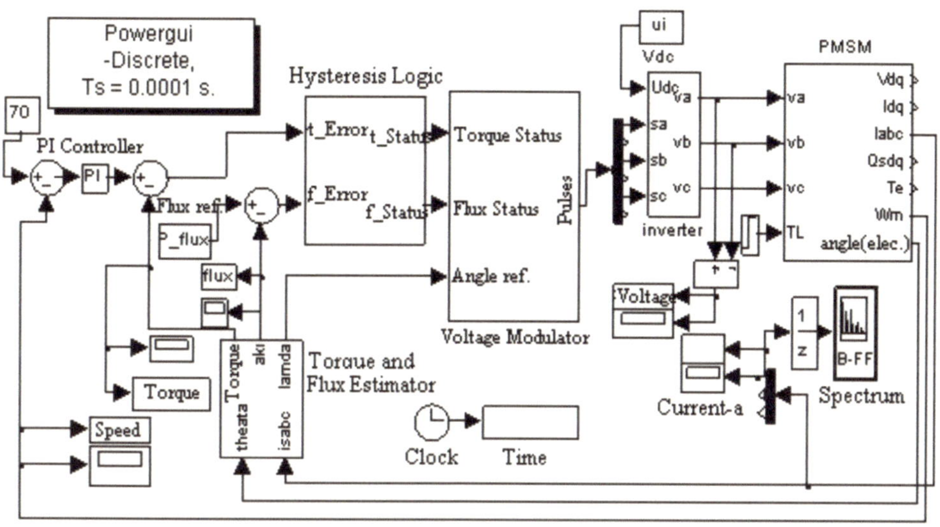

Fig. 17. HDTC of PMSM in Matlab/Simulink

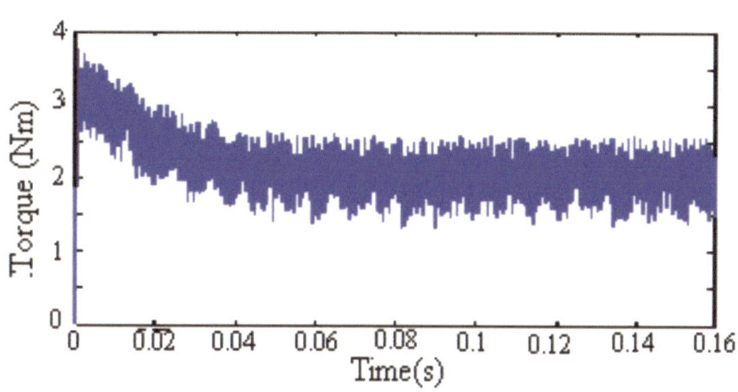

Fig. 18. Torque Response

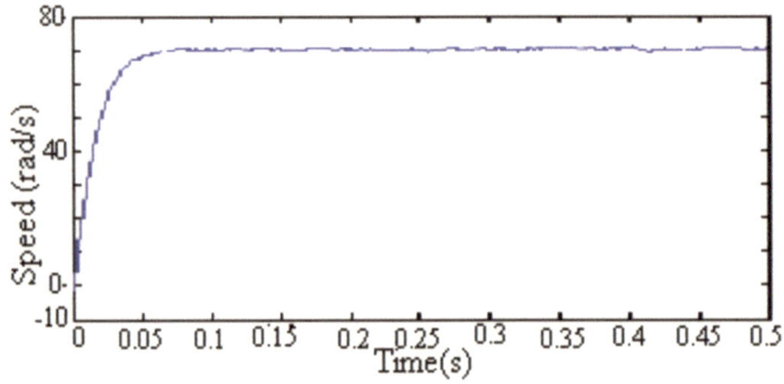

Fig. 19. Speed Response

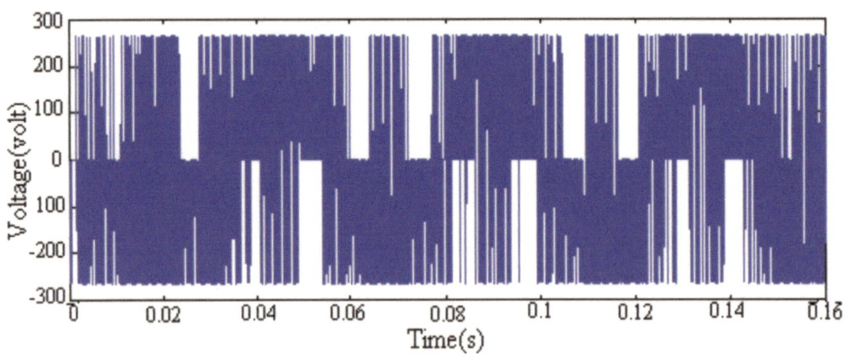

Fig. 20. Voltage switching of line a-b

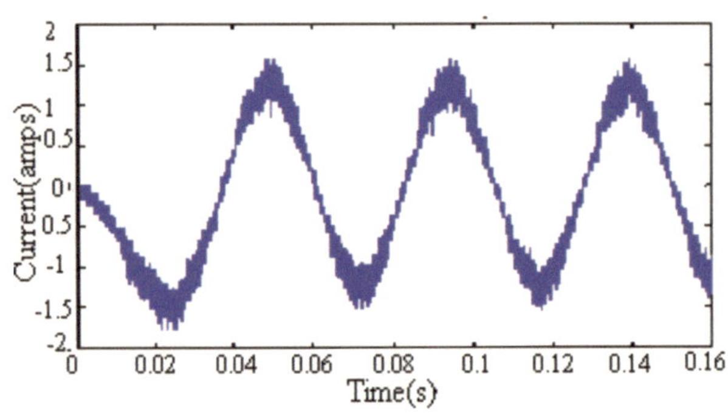

Fig. 21. HDTC Line current of phase-a

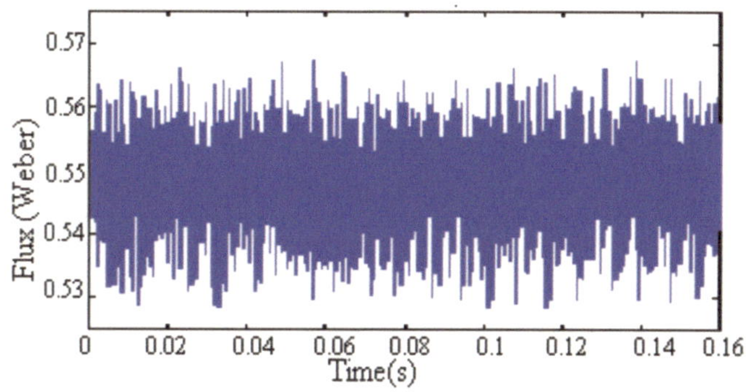

Fig. 22. HDTC Flux response

The responses showed that the torque pulsation is also high and line currents are full of switching harmonics as compared to the FOC algorithm. In addition the flux and speed are also not free of ripple which result in machine vibration and acoustic noise.

4. Algorithm 3: Space Vector Modulation Direct Torque Control (SVMDTC)

In this method, a mathematical model of PMSM and space vector modulation of inverter are used to carry out system algorithm. Thus, instead of switching table and hysteresis controller a space voltage modulation vectors depending on the flux positions are used to compensate for errors in flux and torque (Dariusz et al, 2002; Tang et al., 2004). One of the SVM-DTC block diagrams is shown in Fig. 23 (Dariusz et al., 2002).

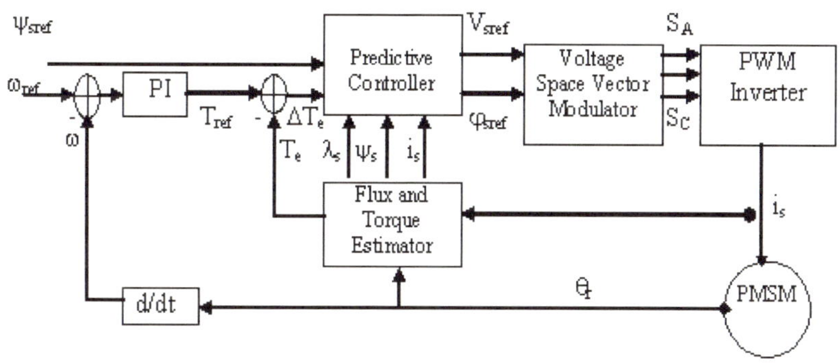

Fig. 23. Direct Torque Control SV-Modulation

In Fig. 23, the torque error signal ΔT_e and reference amplitude of the stator flux Ψ_{sref} are delivered to predictive controller, which also uses information about the amplitude and position of the actual stator flux vector and measured stator current vector. The predictive controller determines the stator voltage command vector in polar co-ordinates $V_{sref} = |V_{sref}| \angle\varphi_{sref}$ for space vector modulator (SVM) which finally generates the pulses S_A, S_B and S_C to control the PWM inverter.

Referring to Eq. (7), the electromagnetic torque produced by the motor is given by:

$$T_e = \frac{3}{2} P \frac{|\psi_s|}{L_d L_q}\left[\psi_F L_q \sin\delta + \tfrac{1}{2}|\psi_s|(L_{sd} - L_{sq})\sin 2\delta\right] \tag{17}$$

From this equation, it can be seen that for constant stator flux amplitude and flux produced by the permanent magnet, the electromagnetic torque can be changed by control of the torque angle. The torque angle δ can be changed by changing position of the stator flux vector with respect to the PM vector using the actual voltage vector supplied by the PWM inverter (Dariusz, 2002). The flux and torque values can be calculated as in Section 3.1 or may be estimated as in Section 3.3. The internal flux calculator is shown in Fig. 24.

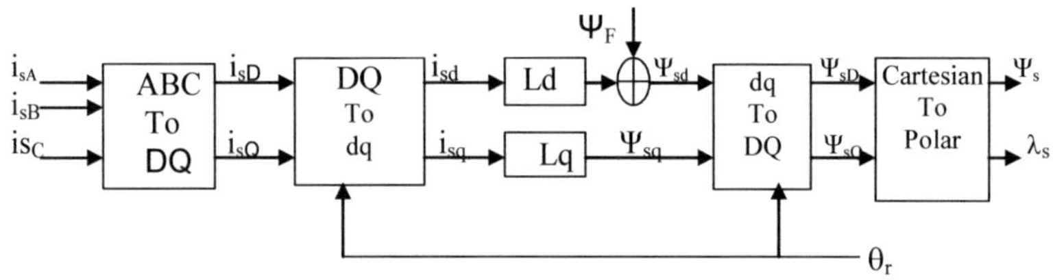

Fig. 24. Flux Estimator Block Diagram

The internal structure of the predictive controller is in Fig. 25.

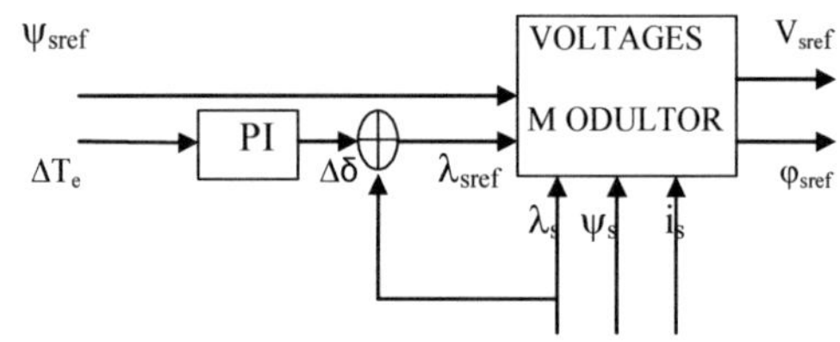

Fig. 25. Predictive Controller

Sampled torque error ΔT_e and reference stator flux amplitude Ψ_{sref} are delivered to the predictive controller. The error in the torque is passed to PI controller to generate the increment in the load angle Δδ required to minimize the instantaneous error between reference torque and actual torque value. The reference values of the stator voltage vector are calculated as:

$$V_{sref} = \sqrt{V_{sD\_ref}{}^2 + V_{sQ\_ref}{}^2} \quad and \quad \varphi_{sref} = \tan^{-1}\frac{V_{sQ\_ref}}{V_{sD\_ref}} \tag{18}$$

Where:

$$V_{sD\_ref} = \frac{\psi_{sref}\cos(\lambda_s + \Delta\delta) - \psi_s \cos\lambda_s}{T_s} + R_s i_{sD}\,. \tag{19}$$

$$V_{sQ\_ref} = \frac{\psi_{sref}\sin(\lambda_s + \Delta\delta) - \psi_s \sin\lambda_s}{T_s} + R_s i_{sQ}\,. \tag{20}$$

Where, T_s is the sampling period.

For constant flux operation region, the reference value of stator flux amplitude is equal to the flux amplitude produced by the permanent magnet. So, normally the reference value of the stator flux is considered to be equal to the permanent magnet flux.

4.1 Implementation of SVMDTC

The described system in Fig. 23 has been implemented in Matlab/Simulink, with the same data and loading condition as in HDTC with PI controllers setting as:

Predictive Controller: Ki=0.03, Kp=1 Speed Controller: Ki= 1 Kp=0.04.

The simulation results are shown in Fig. 26 to Fig. 29. As evidence from the figures, the SVM-DTC guarantee lower current pulsation, smooth speed as well as lower torque pulsation. This is mainly due to the fact that the inverter switching in SVM-DTC is uni-polar compared to that of FOC & HDTC (see Fig. 10, Fig. 20 and Fig. 28), in addition the application of SVM reduces switching stress by avoiding direct transition from +Vdc to -Vdc and thus avoiding instantaneous current reversal in dc link. However, the dynamic response in Fig. 9, Fig. 19, and Fig. 27 show that HDTC has faster response compared to the SVM-DTC and FOC.

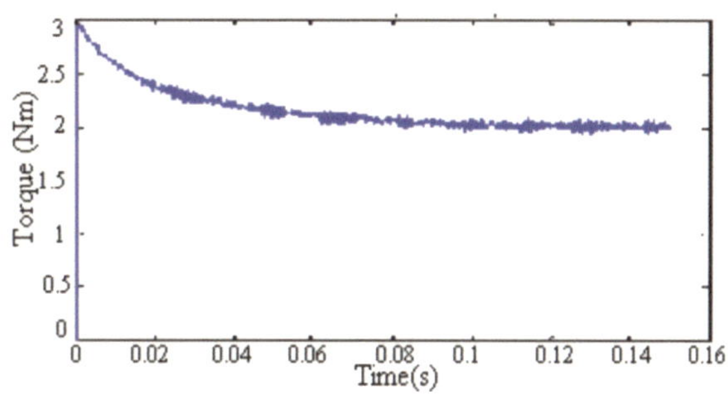

Fig. 26. SVMDTC torque response

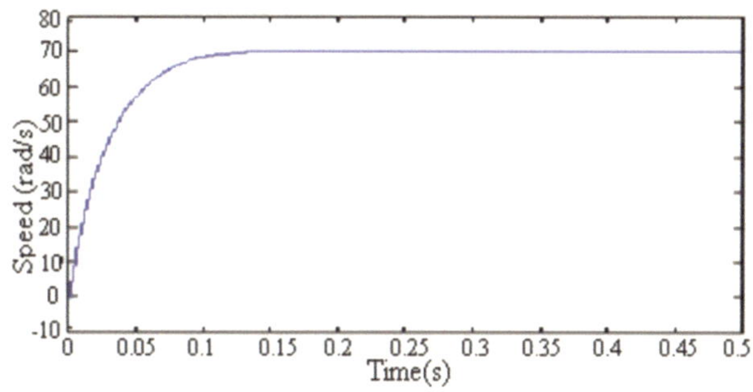

Fig. 27. SVMDTC rotor speed response

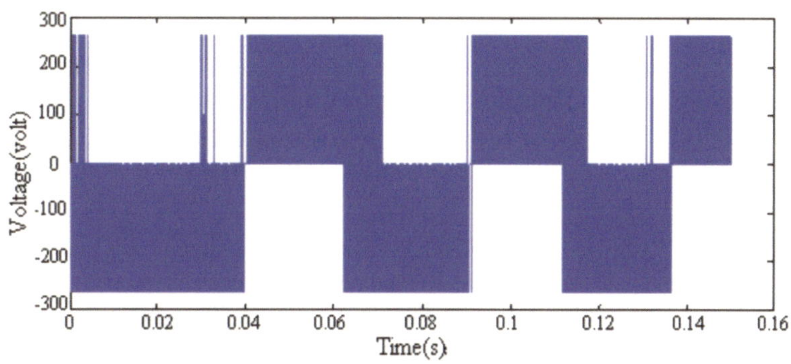

Fig. 28. SVMDTC Line voltage (V_{ab}) waveform

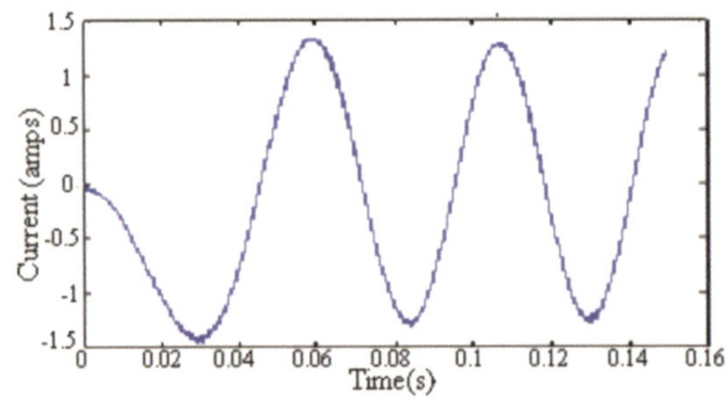

Fig. 29. SVMDTC Line current response of phase a

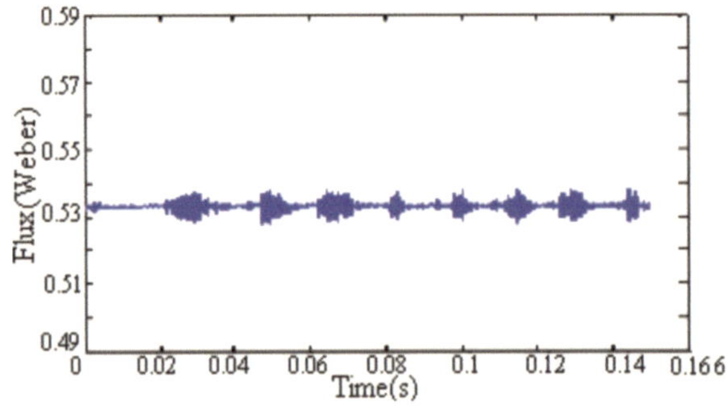

Fig. 30. Stator flux response.

5. High Performance Direct Torque Control Algorithm (HP-DTC)

In this section, a new direct torque algorithm for IPMSM to improve the performance of hysteresis direct torque control is described. The algorithm uses the output of two hysteresis controllers used in the traditional HDTC to determine two adjacent active vectors. The algorithm also uses the magnitude of the torque error and the stator flux linkage position to select the switching time required for the two selected vectors. The selection of the switching time utilizes suggested table structure which, reduce the complexity of calculation. Two Matlab/Simulink models, one for the HDTC, and the other for the proposed model are programmed to test the performance of the proposed algorithm. The simulation results of the proposed algorithm show adequate dynamic torque performance and considerable torque ripples reduction as well as lower flux ripples, lower harmonic current and lower EMI noise reduction as compared to HDTC. Only one PI controller, two hysteresis controllers, current sensors and speed sensor as well as initial rotor position and built-in counters microcontroller are required to achieve this algorithm (Adam & Gulez, 2009).

5.1 Flux and torque bands limitations

In HDTC the motor torque control is achieved through two hysteresis controllers, one for stator flux magnitude error control and the other for torque error control. The selection of one active switching vector depends on the sign of these two errors without inspections of their magnitude values with respect to the sampling time and level of the applied stator voltage. In this section, short analysis concerning this issue will be discussed.

5.1.1 Flux band

Consider the motor stator voltage equation in space vector frame below:.

$$V_s = R_s i_s + \frac{d\Psi_s}{dt} \tag{21}$$

Equation (21) can be written as:

$$dt = \frac{d\Psi_s}{V_s - R_s i_s} \tag{22}$$

For small given flux band $\Delta\Psi_s^o$, the required fractional time to reach the limit of this value from some reference flux Ψ^* is given by:

$$\Delta t = \frac{\left|\Delta\Psi_s^0\right|}{\left|V_s - R_s i_s\right|} \tag{23}$$

And if the voltage drop in stator resistance is ignored, then the maximum time for the stator flux to remain within the selected band starting from the reference value is given as:

$$\Delta t_{\max} = \frac{\left|\Delta\Psi_s^0\right|}{\left|V_s\right|} = \frac{\left|\Delta\Psi_s^0\right|}{\left|2/3\,V_{dc}\right|} \tag{24}$$

Thus if the selected sampling time Ts is large than Δt_{max}, then the stator flux linkage no longer remains within the selected band causing higher flux and torque ripples.

According to (24) if the average voltage supplying the motor is reduced to follow the magnitude of the flux linkage error, the problem can be solved, i.e. the required voltage level to remain within the selected band is:

$$V_{level} = \frac{\Delta t_{\max}}{T_s} V_{kk} \tag{25}$$

Where V_{kk} is the applied active vectors

Thus, by controlling the level of the applied voltage, the control of the flux error to remain within the selected band can be achieved. For transient states, $\Delta\Psi_s$ is most properly large which, requires large voltage level to be applied in order to bring the machine into steady state as quickly as possible.

5.1.2 Torque band

The maximum time Δt_{torque} for the torque ripples to remain within selected hysteresis band can be estimated as:

$$\Delta t_{torque} = \frac{\left|\Delta T^0\right|}{\left|Te_{ref}\right|} * t^0 \tag{26}$$

Where, ΔT^0; is the selected torque band

Te_{ref} ; is the reference electromagnetic torque
t^0; is the time required to accelerate the motor from standstill to some reference torque Te_{ref}.
The minimum of the values given in (24) and (26) can be considered as the maximum switching time to achieve both flux and torque bands requirement. However, when the torque ripples is the only matter of concern, as considered in this work, may be enough to consider the maximum time as suggested by (26).
Now due to flux change by $\Delta\Psi_s$, the load angle δ will change by Δδ as shown in Fig. 31. Under dynamic state, this change is normally small and can be approximated as:

$$\Delta\delta \approx \sin^{-1}\frac{\Delta\Psi_S}{|\Psi_S|} \approx \frac{\Delta\Psi_S}{|\Psi_S|} \tag{27}$$

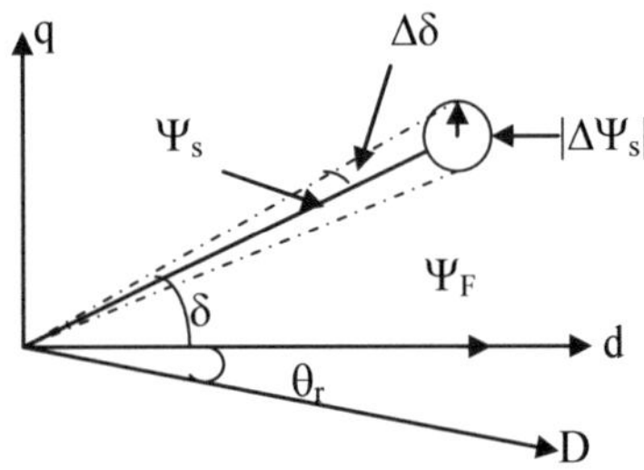

Fig. 31. Stator flux linkage variation under dynamic state

The corresponding change in torque due to change $\Delta\Psi_s$ can be obtained by differentiation of torque equation with respect to δ. Torque equation can be rewritten as:

$$T_e = \frac{3}{4}P\frac{|\Psi_s|}{L_{sd}L_{sq}}\left[2\Psi_F L_{sq}\sin\delta - |\Psi_s|(L_{sq}-L_{sd})\sin 2\delta\right] \tag{28}$$

Where, then

$$\Delta T = \frac{\partial T_e}{\partial\delta}\cdot\Delta\delta \approx \frac{\partial T_e}{\partial\delta}\cdot\frac{\Delta\psi_S}{|\Psi_s|} \tag{29}$$

Substitute (24) in (29) and evaluate to obtain:

$$\Delta T = \frac{3}{2}P\frac{|V_s|\Delta t}{L_{sd}L_{sq}}\left[\Psi_F L_{sq}\cos\delta - |\Psi_s|(L_{sq}-L_{sd})\cos 2\delta\right] \tag{30}$$

Where, $\Delta t = minimum\ (\Delta t_{max}, \Delta t_{torque})$
Equation (30) suggests that ΔT can also be controlled by controlling the level of V_s. Thus both ΔT and $\Delta\Psi_s$ can be controlled to minimum when the average stator voltage level is controlled to follow the magnitude of ΔT.

5.2 The HP-DTC Algorithm

The basic structure of the proposed algorithm is shown in Fig. 32.

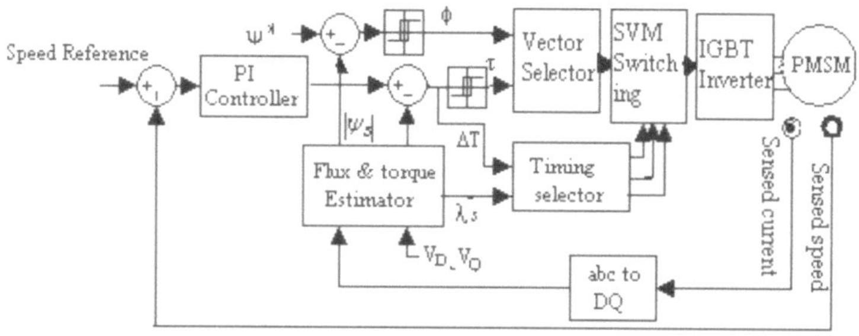

Fig. 32. The HPDTC system of PMSM

5.2.1 Vector selector

In Fig.32 the vector selector block contains algorithm to select two consecutive active vectors V_{k1}, and V_{k2} depending on the output of the hysteresis controllers of the flux error and the torque error; ϕ and τ respectively as well as flux sector number; n. The vector selection table is shown in Table 4., while vectors position and flux sectors is as shown in Fig.15

ϕ	τ	V_{k1}	V_{k2}
1	1	n+1	n+2
1	0	n-1	n-2
0	1	n+2	n+1
0	0	n-2	n-1

Table 4. Active vectors selection table

In the above table

if $V_k>6$ then $V_k=V_k-6$

if $V_k<1$ then $V_k=V_k+6$

5.2.2 Flux and torque estimator

In Fig. 32 the torque and flux estimator utilizes equation (21) to estimate flux and torque values at m sampling period as follows:

$$\psi_D(m) = \psi_D(m-1) + (V_D(m-1) - R_s i_D)T_s \tag{31}$$

$$\psi_Q(m) = \psi_Q(m-1) + (V_Q(m-1) - R_s i_Q)T_s \tag{32}$$

$$|\psi_s| = \sqrt{\psi_D^2 + \psi_Q^2} \tag{33}$$

$$\lambda_s = Tan^{-1}\frac{\psi_Q}{\psi_D} \tag{34}$$

Where; the stationary D-Q axis voltage and current components are calculated as follows:

$$V_D(m-1) = (V_{Dk1}t_{k1} + V_{Dk2}t_{k2}) / Ts \tag{35}$$

$$V_Q(m-1) = (V_{Qk1}t_{k1} + V_{Qk2}t_{k2}) / Ts \tag{36}$$

$$i_D = (i_D(m-1) + i_D(m)) / 2 \tag{37}$$

$$i_Q = (i_Q(m-1) + i_Q(m)) / 2 \tag{38}$$

The torque value can be calculated using estimated flux values as:

$$T_e = \frac{3}{2} P(\Psi_D(m) i_Q(m) - \Psi_Q(m) i_D(m)) \tag{39}$$

5.2.3 The timing selector structure

In Fig. 32 the timing selector block contains algorithm to select the timing period pairs of vectors V_{k1} and V_{k2}. The selection of timing pairs depends on two axes, one is the required voltage level and the other is the reflected flux position in the sector contained between V_{k1} and V_{k2}. The reflected flux position is given by:

$$\rho_s = \lambda_s \bmod 60 - \pi / 6 \tag{40}$$

Where λ_s; is the stator flux linkage position in D-Q stationary reference frame.
Fig. 34 shows the proposed timing table. In this figure, the angle between the two vectors V_{k1} and V_{k2} which is 60^0, is divided into 5 equal sections ρ_{-2}, ρ_{-1}, ρ_0, ρ_{+1}, and ρ_{+2}. The required voltage level is also divided into 5 levels.
The time pairs (t_{k1}, t_{k2}), expressed as points, (out of 20 points presenting the sampling period) define the timing periods of V_{k1} and V_{k2} respectively. The remaining time points, $(t_0=20-t_{k1}-t_{k2})$, is equally divided between zero vectors V_0 and V_7.

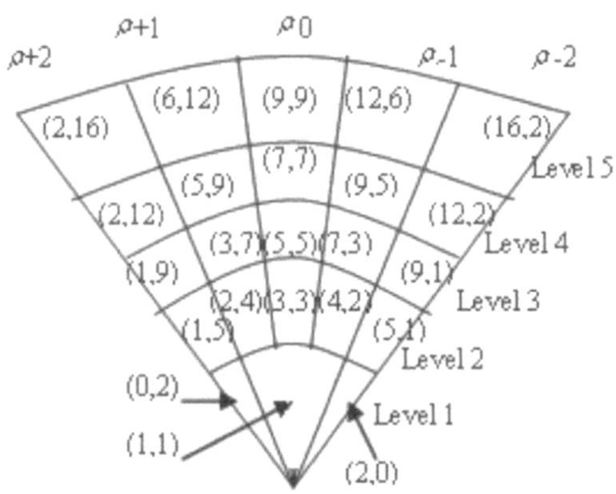

Fig. 34. Timing diagram for the suggested algorithm

The time structure shown in Fig.34 has the advantage of avoiding the complex mathematical expressions used to calculate t_{k1} and t_{k2}, as the case in space vector modulation used by (Dariusz, 2002) and (Tan, 2004). In addition, it is more convenient to be programmed and executed through the counter which controls the period t_{k1}, t_{k2} and t_0. The flow chart of the algorithm is shown in Fig. 35.

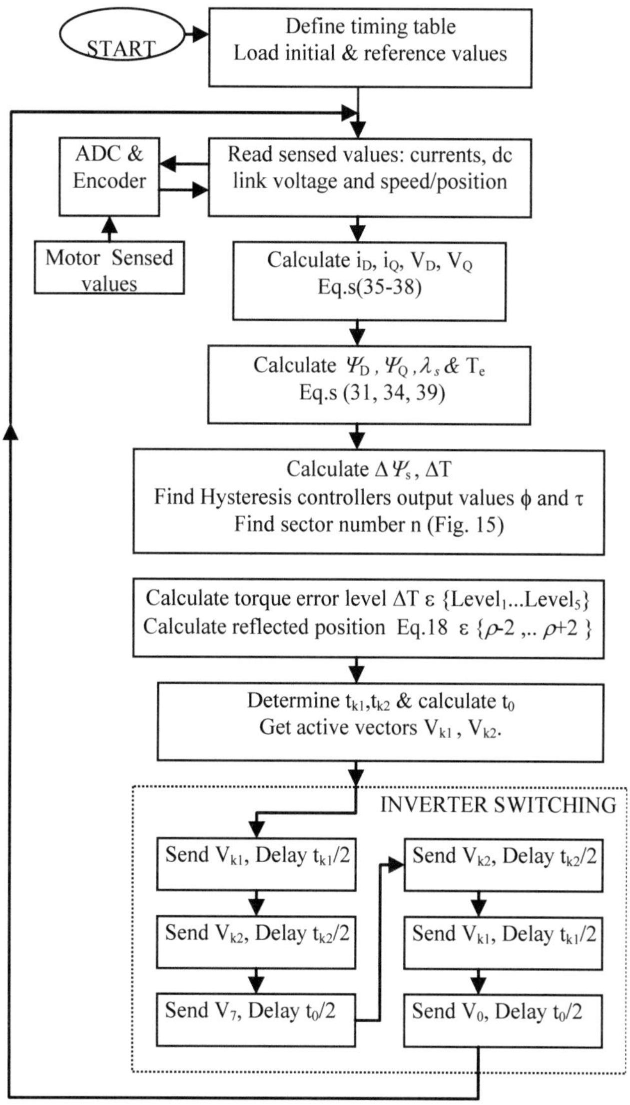

Fig. 35. A Flow chart of the proposed algorithm

5.3 Simulation and results

To examine the performance of the proposed DTC algorithm, two Matlab/Simulink models, one for HDTC and the other for the HPDTC were programmed. The motor parameters are shown in table 2. The inverter used in simulation is IGBT inverter with the following setting:

IGBT/Diode

Snubber Rs, Cs = (1e-3ohm,10e-6F)

Ron=1e-3ohm

Forward voltage (V_f Device,V_f Diode)= (0.6, 0.6)

T_f(s),T_t(s) = (1e-6, 2e-6)

DC link voltage= +132 to -132.

The simulation results with 100μs sampling time for the two algorithms under the same operating conditions are shown in Fig. 36 -to- Fig. 41. The torque dynamic response is simulated with open speed loop, while the steady state performance is simulated with closed speed loop, 70rad/s as reference speed, and 2 Nm as load torque.

5.3.1 Torque dynamic response

The torque dynamic response with HDTC and the HPDTC are shown in Fig.36-a and Fig.36-b respectively. The reference torque for both algorithms is changed from +2.0 to -2.0 and then to 3.0 Nm. As shown in the figures, the dynamic response with the proposed algorithm is adequately follows the reference torque with lower torque ripples. In the other hand, the torque response with the proposed algorithm shows fast response as the HDTC response.

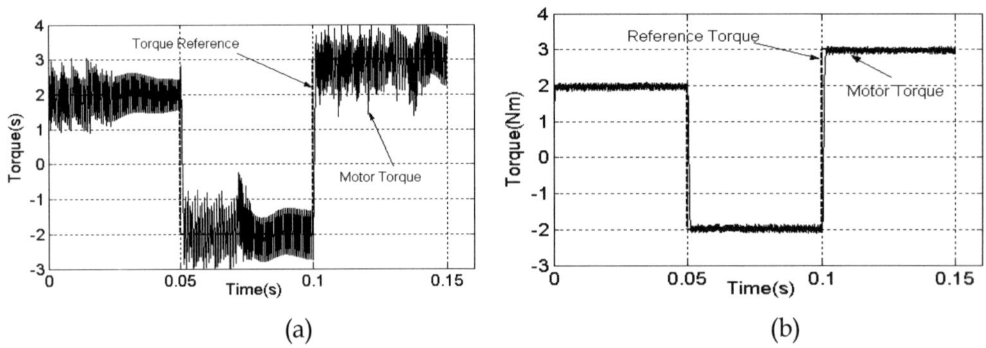

Fig. 36. Motor dynamic torque with opened speed loop: (a) HDTC (b) HP-DTC

Fig. 37 demonstrates the idea of maximum time to remain within the proposed torque band as suggested by equation (26). According to the shown simulated values, the time required to accelerate the motor to 2 Nm is ≈ 0.8ms, so if the required limit torque ripple is not to exceed 0.1 Nm, as suggested in this work, then, the maximum switching period according to Eq. (26) is ≈0.05ms which is less than the sampling period (Ts=0.1 ms).

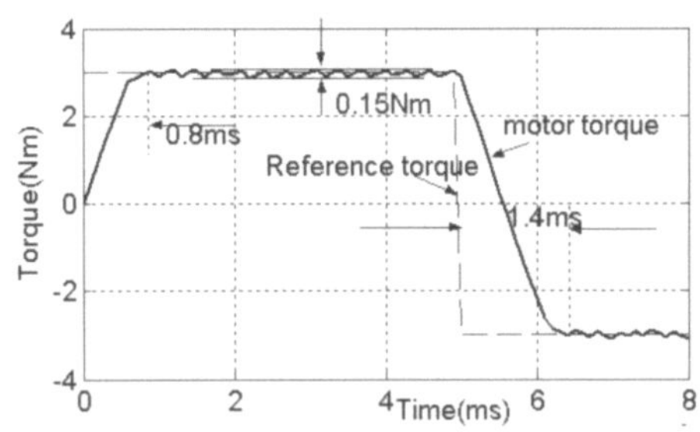

Fig. 37. Torque ripples and motor accelerating time

Although the torque ripple is brought under control, the flux ripples still high as shown in Fig. 38 which, is mainly due to control of the voltage level according to the magnitude of torque error only.

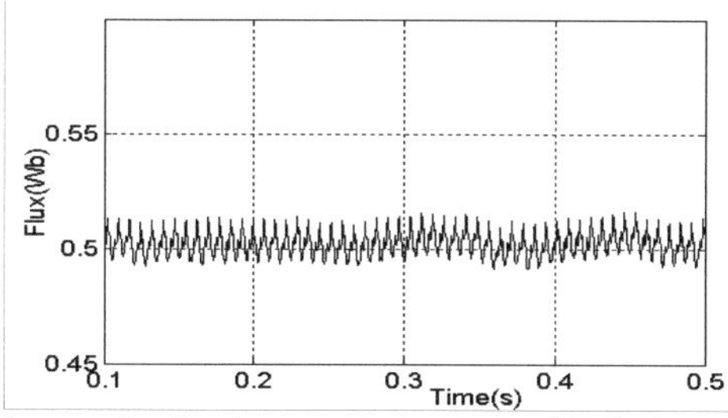

Fig. 38. Flux response when only the torque error magnitude is used to approximate the required voltage level

5.3.2 Motor steady state performance

The motor performance results under steady state are shown in Fig. 39 -to- Fig. 41. Fig. 39-a and Fig. 39-b, show the phase currents of the motor windings under HDTC and the HPDTC respectively, observe the change of the waveform under the proposed method, it is clear that the phase currents approach sinusoidal waveform with almost free of current pulses appear in Fig. 39-a. Better waveform can be obtained by increasing the partition of the timing structure, however, when smoother waveform is not necessary, suitable division as the one shown in Fig. 34 may be enough.

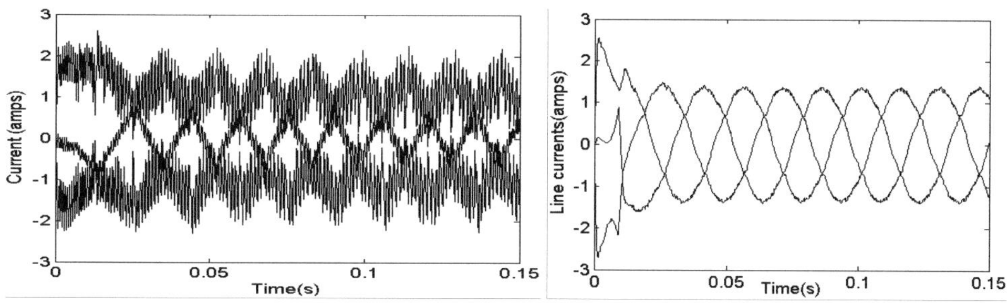

(a) (b)

Fig. 39. Motor line currents: (a) HDTC (b) HPDTC

The torque response in Fig. 40 shows considerable reduction in torque ripples from 3.2Nm (max. -to- max.) down to less than 0.15 Nm when the new method HP-DTC is used, which in turn, will result in reduced motor mechanical vibration and acoustic noise, this reduction also reflected in smoother speed response as shown in Fig. 41

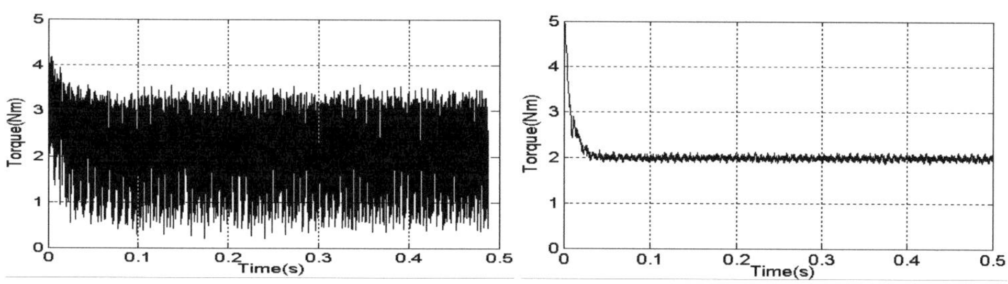

(a) (b)

Fig. 40. Motor steady state torque response: (a) HDTC (b) HPDTC

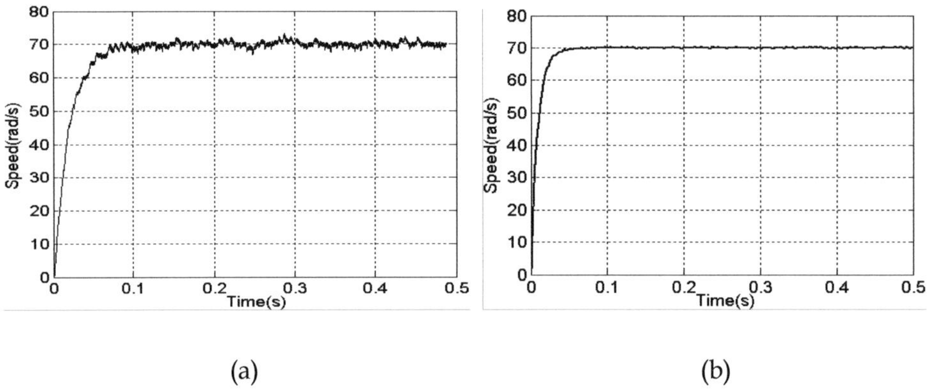

Fig. 41. Rotor speed response: (a) HDTC (b) HPDTC

6. Torque ripple and noise in PMSM algorithm

One of the major disadvantages of the PMSM drive is torque ripple that leads to mechanical vibration and acoustic noise. The sensitivity of torque ripple depends on the application. If the machine is used in a pump system, the torque ripple is of no importance. In other applications, the amount of torque ripple is critical. For example, the quality of the surface finish of a metal working machine is directly dependent on the smoothness of the delivered torque (Jahns and Soong, 1996). Also in electrical or hybrid vehicle application, torque ripple could result in vibration or noise producing source which in the worst case could affect the active parts in the vehicle.

The different sources of torque ripples, harmoinc currents and noises in permanent magnet machines can be abstracted in the following (Holtz and Springob 1996,1998):

- Distortion of the stator flux linkage distribution
- Stator slotting effects and cogging
- Stator current offsets and scaling errors
- Unbalanced magnetization
- Inverter switching and EMI noise

However switching harmonics and voltage harmonics supplied by the power inverter constitute the major source of harmonics in PMSM. In this section, the reduction of torque ripple and harmonics generated due to inverter switching in PMSM control algorithms using passive and active filter topology will be investigated.

Method1: Compound passive filter topology

6.1 The proposed passive filter topology

Fig. 42 shows a block diagram of basic structure of the proposed filter topology (Gulez et al., 2007) with PMSM drive control system. It consists of compound dissipative filter cascaded by RLC low pass filter. The compound filter has two tuning frequency points, one at inverter switching frequency and the other at some average selected frequency.

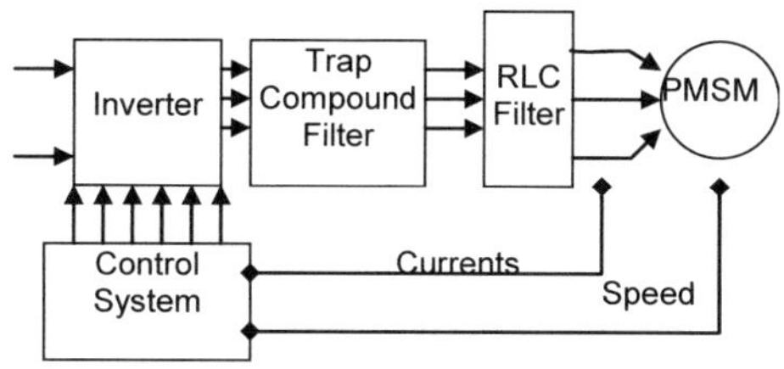

Fig. 42. Block diagram of the proposed filter topology with PMSM drive system

6.1.1 The compound trap filter

Fig. 43 shows the suggested compound trap filter. It consists of main three passes, one is low frequency pass branch through R_2 and L_2, another is the high frequency pass through C_2 and R_1 and the other is the average frequency pass through C_1, L_1 and R_1 to the earth.

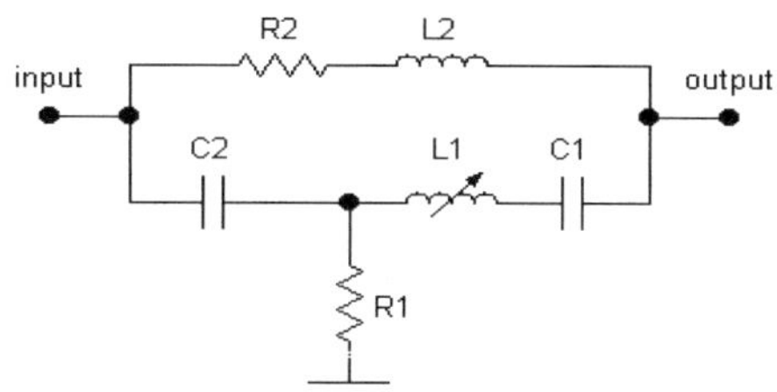

Fig. 43. The suggested compound trap filter

For some operating frequency ω_o, the component of the low pass branch constitutes low impedance path while at the same time shows high impedance for the high frequency component, which forces the high frequency to pass through C_2 to the earth. Some of the average frequency components will find their way through the low pass branch. These frequencies will be absorbed by tuning resonance of branch L_1-C_1 to some selected average frequency such that.

$$\omega_o < \omega_{av} < \omega_{sw}$$

Where

ω_o ; is the operating frequency

$\omega_{av} = 1/\sqrt{L_1 C_1}$: is the selected average frequency

ω_{sw}: is inverter switching frequency calculated as $1/(2T_s)$; T_s being the sampling period

The behavior of the Compound Trap filter can be explained by studying the behavior of the impedances constitutes the Π equivalent circuit of the Compound Trap filter shown in Fig. 44.

In Fig. 44 the impedances Z_1, Z_2 and Z_3 can be expressed as:

$$Z_1 = \frac{\omega L_1 - 1/\omega c_1}{\omega R_1 C_2} + j(\omega L_1 - \frac{1}{\omega C_1} - \frac{1}{\omega C_2}) \tag{41}$$

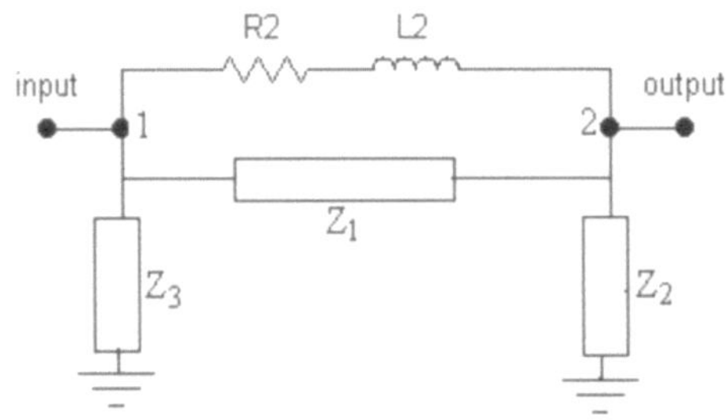

Fig. 44. Equivalent Π circuit of the compound trap filter.

$$Z_2 = R_1 - R_1 C_2 \omega(\omega L_1 - \frac{1}{\omega C_1}) + j(\omega L_1 - \frac{1}{\omega C_1}) \quad (42)$$

$$Z_3 = R_1 - \frac{R_1}{\omega^2 L_1 C_2 - C_2 / C_1} - j\frac{1}{\omega C2} \quad (43)$$

Figures 45, 46 and 47 show the frequency-magnitude characteristics of the impedances Z_2, Z_3 and Z_{12} respectively. Z_{12} is the equivalent impedance value between point 1 and 2 as shown in Fig. 44.

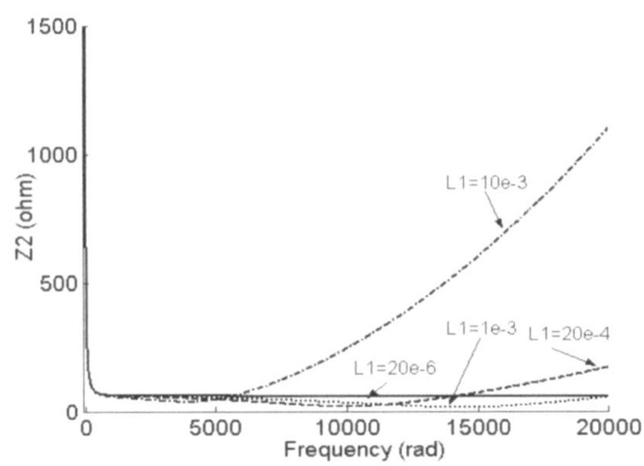

Fig. 45. Z_2 characteristics at C1=52.0e-6F and different L_1 values.

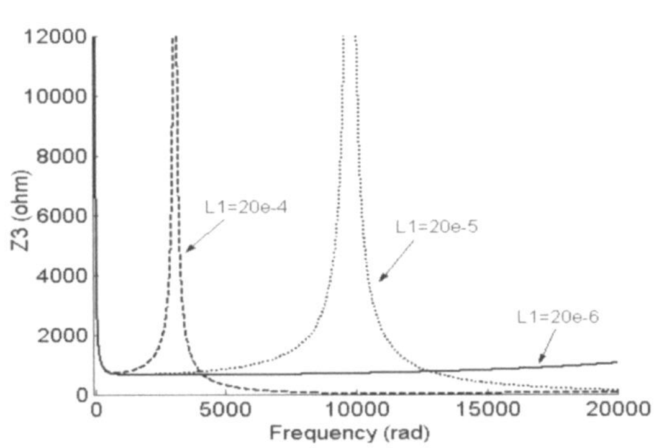

Fig. 46. Z_3 characteristics at C1=52.0e-6F and different L_1 values.

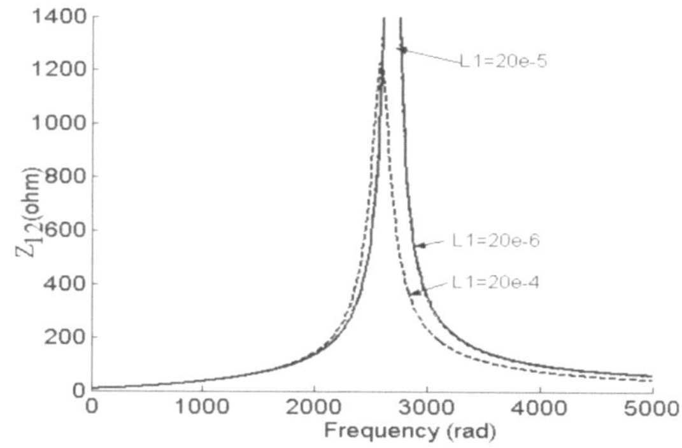

Fig. 47. Z_{12} characteristics at C1=52.0e-6F and different L_1 values.

As can be inferred from Equations (41 to 43) and Figures (Fig. 46 to Fig. 47) it is evidence that both Z_2 and Z_3 show capacitive behavior at small values of L_1. So if the value of L_1 is kept small (L1 < 1e-5 H), the tuning of high frequency current components is ensured through the compound trap filter.

In the other hand the characteristics of Z_{12} shown in Fig. 47 demonstrates that at average frequencies the impedance of Z_{12} is high while at low and high frequencies the impedance is low thus Z_{12} constitute band stop filter to the average frequency current components. The magnitude of Z_{12} can effectively be changed by changing the value of L1 at constant C1, while the range of the average frequencies can effectively be changed by changing the value of C1 at constant L1. Thus through proper tuning of L1 and C1 the desired average frequency range can be selected.

6.1.2 RLC filter

Fig. 48 shows the suggested RLC filter, which play main role in reducing the high *dv/dt* of line to line voltages at motor terminals. The transfer function of this circuit is given by:

$$\frac{V_0}{V_i} = \frac{R_3C_3s+1}{C_3L_3s^2+(r_3+R_3)C_3s+1} \tag{44}$$

To obtain over damping response, the filter resistances are selected such that:

$$(R_3+r_3) > \sqrt{\frac{4L_3}{C_3}} \tag{45}$$

With cutoff frequency ω_c is given by:

$$\omega_c = 1/\sqrt{L_3C_3} \tag{46}$$

To reduce ohmic losses the series resistance r_3 is normally of small value, while the shunt resistance R_3 is selected high enough to limit the currents drawn by the filter. This current can be expressed as:

$$i_{CR3} = \frac{z_{PMSM}}{z_{PMSM}+\sqrt{R_3^2+(1/sC_3)^2}} i_{in} \tag{47}$$

Where, Z_{PMSM} is PMSM motor input impedance.
At the selected cutoff frequency, this current should be large compared to i_{motor} drawn by the motor; while at operating frequency this current should be very small compared to i_{motor}. Another point in selection of the RLC parameters is that, the filter inductors are essentially shorted at line frequency while the capacitors are open circuit and for EMI noise frequencies, the inductors are essentially open circuit while the capacitors are essentially shorted, thus considerable amount of EMI noises will pass through the filter resistors to the earth and cause frequency dependent voltage drop across the series branch of the filter which, in turn, helps in smoothing of the voltage waveform supplying the motor.

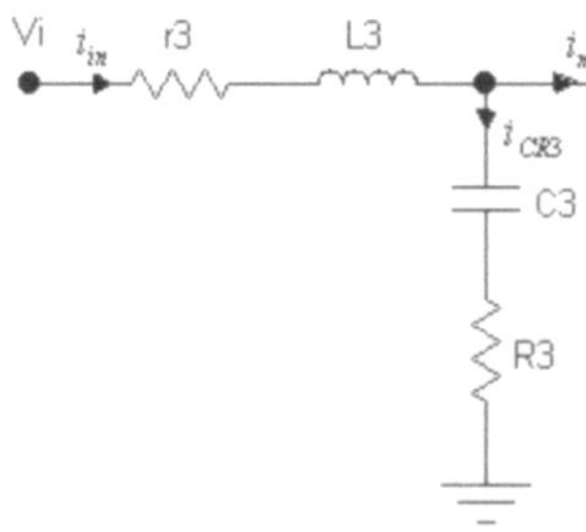

Fig. 48. RLC filter cascaded to the trap compound filter

To evaluate the performance of the suggested passive filter topology, it was applied to HDTC algorithms under MatLab simulation. The following subsections show the results of the simulations.

6.1.3 Torque ripples and noise reduction in HDTC using passive filter

Fig. 49 shows the basic structure of HDTC of PMSM with the proposed passive filter topology. The switching table in Fig.49 is the same as that shown in Table 3. In this figure, the switching of the inverter is updated only when the outputs of the hysteresis controllers change states, which result in variable switching frequency and associated large harmonic range and high current ripples.

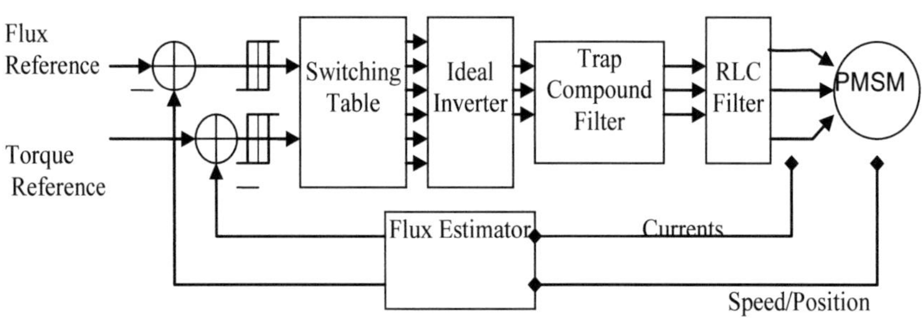

Fig. 49. The basic structure of HDTC of PMSM with the proposed filter topology

6.1.4 Simulations and results

To simulate the performance of the proposed passive filter topology under HDTC Matlab/Simulink was used.

Under base speed operation, the speed control was achieved through PI controller with K_i= *2.0 and* K_p= *0.045*. The flux reference is set equal to Ψ_F and hysteresis bands are set to 0.01 for both the torque and flux hysteresis controllers. The motor parameters are shown in Table 2 and the passive filter parameters are in Table 5.

L_1	20μH	L_2	30mH	L_3	30mH
C_1	52μF	C_2	5.1μF	C_3	12.5 μF
R_1	56kΩ	R_2	2.2Ω	R_3	128 Ω
				r_3	2Ω

Table 5. Passive Filter Parameters

The simulation results with 100μs sampling time are shown in Fig. 50 to Fig. 55. Fig. 50-a in particular, shows the motor line voltage V_{ab} without applying the proposed filter topology. When the filter topology is connected, the switching frequency is reduced as depicted in Fig. 50-b compared to the one shown in Fig. 50-a. The line voltages provided to the motor terminals approach sinusoidal waveform, observe the change of the waveform at the output of the compound filter in Fig. 50-c and at the motor terminal in Fig. 50-d. Better waveform as mentioned before can be obtained by increasing the series inductance L_3 and decreasing the resistance r_3.

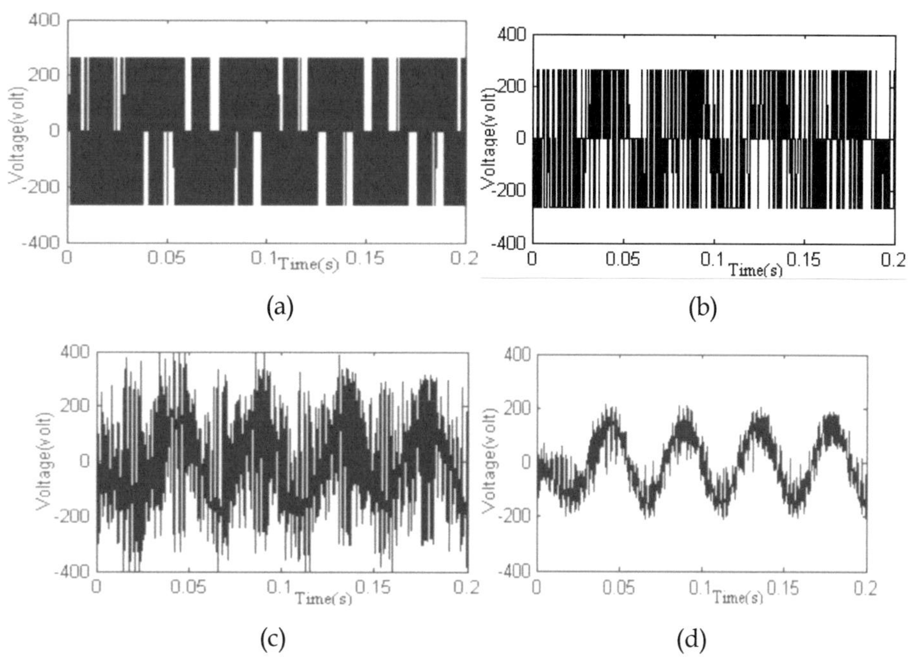

Fig. 50. Motor line voltage (a) before applying the filter topology (b) at inverter terminals after applying the filter topology (c) at the compound Filter output terminals (d) at the motor terminals as output of the RLC filter

The motor performance before and after applying the filter topology are shown in Fig. 51 to Fig. 54. In Fig. 51, the motor line currents show considerable reduction in noise and harmonic components after applying the filter which reflects in smoother current waveform.

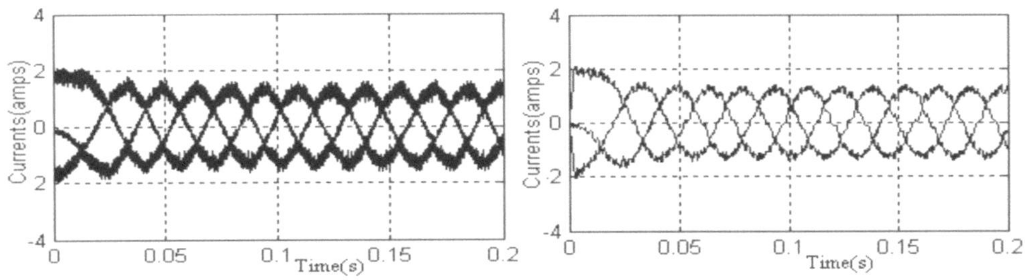

(a) (b)

Fig. 51. Motor line currents: (a) before (b) after applying the filter topology

The torque response in Fig. 52 shows considerable drop in torque ripples from 1.4Nm (ripples to ripples) down to 0.6 Nm after applying the Filter topology, which will result in reduced motor mechanical vibration and acoustic noise. The speed response in Fig. 52, shows slight smoothness after applying the passive filter topology.
The status of the line current harmonics and EMI noise before and after connecting the filter topology are shown in Fig. 53 to Fig. 54.

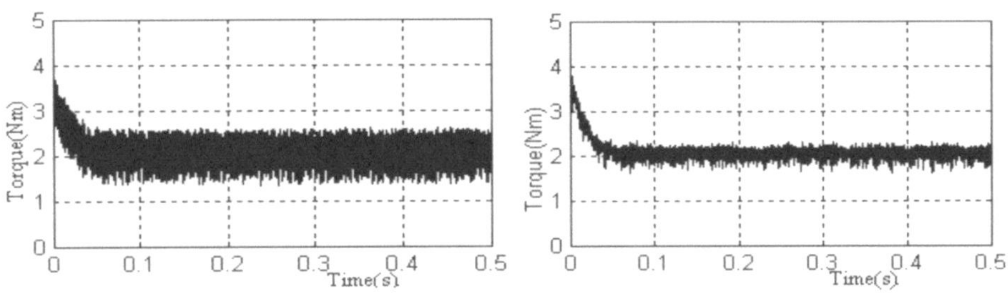

(a) (b)

Fig. 52. Motor torque: (a) before (b) after applying the filter topology (load torque is 2Nm)

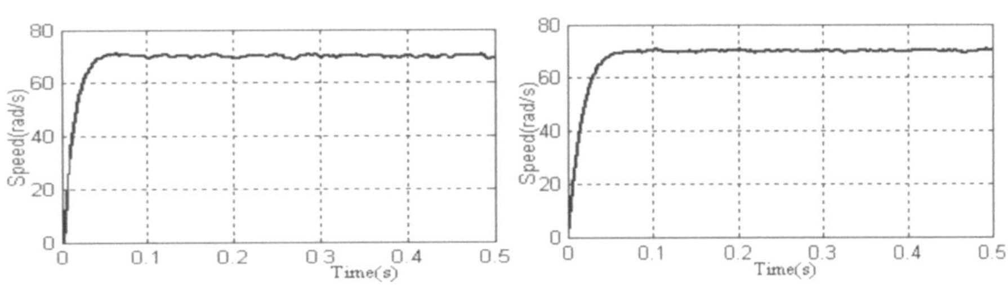

(a) (b)

Fig. 53. Rotor speed: (a) before (b) after applying the filter topology

In Fig. 54-a the spectrum of the line current without connecting the filter shows that harmonics currents with THD of ~3% have widely distributed with a dominant harmonics concentration in the range around 2 kHz.. After connecting the filter topology, the THD is effectively reduced to less than 1.7% with dominant harmonics concentration in the low frequency range (less than 0.5 kHz.), while the high frequency range is almost free of parasitic harmonics as shown in Fig. 54-b.

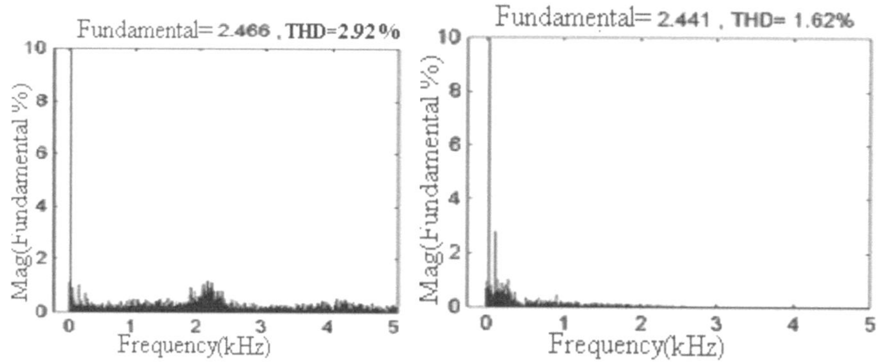

(a) (b)

Fig. 54. Phase-a current spectrum: (a) before applying the filter topology (b) after applying the filter topology

The EMI noise level near zero crossing before applying the filter topology in Fig. 55-a shows a noise level of ~ -5dB at operating frequency , ~-10dB at switching frequency (5KHz) and ~-47dB at the most high frequencies (greater than 0.2 MHz.). When the filter is connected, the EMI noise level is damped down to ~-20dB at operating frequency , ~-30dB at switching frequency and ~-67dBs at the most high frequencies as shown in Fig. 55-b.

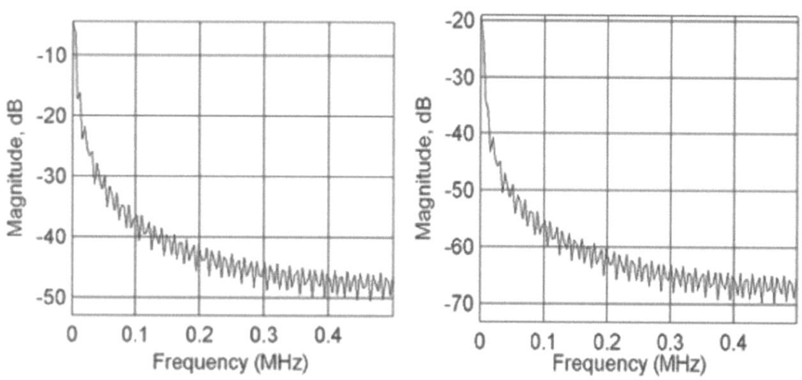

(a) (b)

Fig. 55. EMI level: (a) before applying the filter topology (b) after applying the filter topology

6.2 Method 2: active filter topology

In this section an active filter topology will be proposed to reduce torque ripples and harmonic noises in PMSM when controlled by FOC or HDTC equipped with hysteresis controllers. The filter topology consists of IGBT active filter (AF) and two RLC filters, one in the primary circuit and the other in the secondary circuit of a coupling transformer. The AF is characterized by detecting the harmonics in the motor phase voltages and uses hysteresis voltage control method to provide almost sinusoidal voltage to the motor windings.

6.2.1 The proposed active filter topology

When the PMSM is controlled by HDTC, the motor line currents and/or torque are controlled to oscillate within a predefined hysteresis band. Fig. 56, for example, shows typical current waveform and the associated inverter output voltage switching.
In the shown figure the inverter changes state at the end of a sampling period only when the actual line current increases or decrease beyond the hysteresis band which result in high ripple current full of harmonic components.

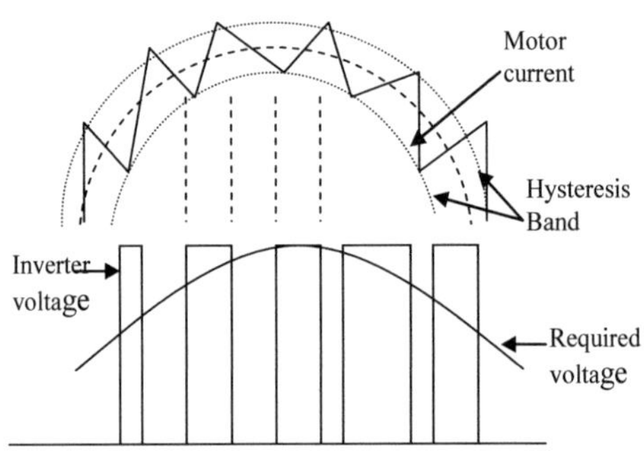

Fig. 56. Current waveform and associated inverter voltage switching equipped with hysteresis controllers

To reduce the severe of these ripples two methods can be mentioned, the first one is to reduce the sampling period which implies very fast switching elements, and the second one is to affect the voltage provided to the motor terminals in such a way to almost follow sinusoidal reference guide. The last method will be adopted here so, active filter topology is used to affect inverter voltage waveform to follow the required signal voltage.
Series active power filters were introduced by the end of the 1980s and operate mainly as a voltage regulator and as a harmonic isolator between the nonlinear load and the utility system (Hugh et al, 2003). Since series active power filter injects a voltage component in series with the supply voltage, they can be regarded as a controlled voltage source. Thus this type of filters is adopted here to compensate the harmonic voltages from the inverter supplying the motor.
Fig. 57 shows a schematic diagram of basic structure of the proposed filter topology; including the active filter, coupling transformer, RLC filters and block diagram of the active filter control circuit
In Fig. 57 V_{sig} is the desired voltage to be injected in order to obtain sinusoidal voltage at motor terminals and V_{AF} is the measured output voltage of the active filter. V_{AF} is subtracted from V_{sig} and passed to hysteresis controller in order to generate the required switching signal to the active filter. The active filter storage capacitor C_F which operates as voltage source should carefully be selected to hold up to the motor line voltage. The smoothing inductance L_F should be large enough to obtain almost sinusoidal voltage at the motor

terminals. The reference sinusoidal voltage V^* which should be in phase with the main inverter output voltage V_{inv}, is calculated using information of the motor variables.
In the following sections firstly, the operating principle of voltage reference control circuit will be explained then the two other parts will follow.

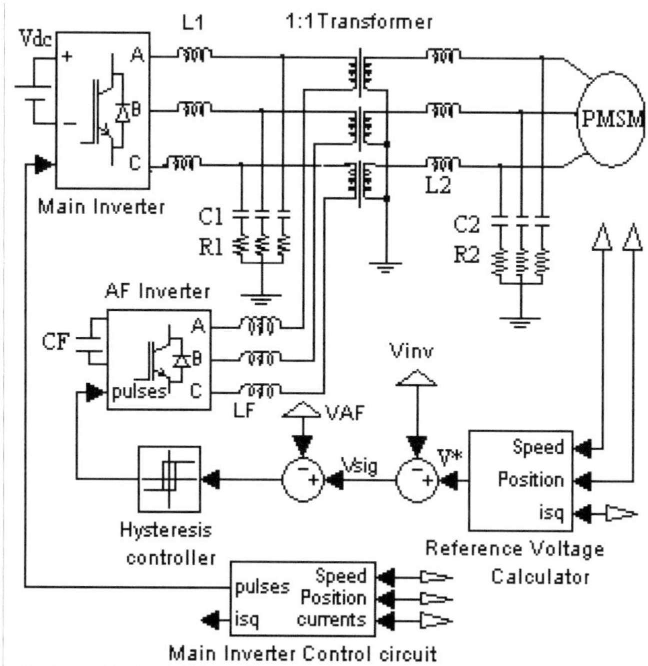

Fig. 57. Basic Structure of the Proposed Filter Topology

6.2.2 Voltage reference signal generator

The effectiveness of the active filter is mainly defined by the algorithm used to generate the reference signals required by the control system. These reference signals must allow current and voltage compensation with minimum time delay. In this study the method used to generate the voltage reference signals is related to FOC algorithm, which use motor model in rotor d-q reference frame and rotor field oriented control principles with monitored rotor position/speed and monitored phase currents. The motor model in this synchronously rotating reference frame is given by:

$$\begin{bmatrix} v_{sd} \\ v_{sq} \end{bmatrix} = \begin{bmatrix} R + pL_{sd} & -\omega_r L_{sq} \\ \omega_r L_{sd} & R + pL_{sq} \end{bmatrix} \begin{bmatrix} i_{sd} \\ i_{sq} \end{bmatrix} + \begin{bmatrix} 0 \\ \omega_r \psi_F \end{bmatrix} \tag{48}$$

$$T_e = \frac{3}{2} P(\psi_F i_{sq} + (L_{sd} - L_{sq}) i_{sd} i_{sq})) \tag{49}$$

Under base speed operation, the speed or torque control can be achieved by forcing the stator current component i_{sd} to be zero while controlling the i_{sq} component to directly proportional to the motor torque T_e as in (50):

$$T_e = \frac{3}{2} P \psi_F i_{sq} \tag{50}$$

The instantaneous q-axis current can be extracted from (50) and hence by setting i_{sd} to zero, the instantaneous d and q axis voltages can be calculated from (48) as:

$$V_{sd} = -\omega_r L_{sq} i_{sq} \tag{51}$$

$$V_{sq} = R i_{sq} + p L_{sq} i_{sq} + \omega_r \psi_F \tag{52}$$

Once the values of d-axis and q-axis voltage components are obtained, Park and Clarke transformation can be used to obtain the reference sinusoidal voltages as:

$$\begin{bmatrix} v_a^* \\ v_b^* \\ v_c^* \end{bmatrix} = K \begin{bmatrix} 1 & 0 \\ -1/2 & \sqrt{3}/2 \\ -1/2 & -\sqrt{3}/2 \end{bmatrix} \begin{bmatrix} \cos\theta & -\sin\theta \\ \sin\theta & \cos\theta \end{bmatrix} \begin{bmatrix} V_{sd} \\ V_{sq} \end{bmatrix} \tag{53}$$

Where, K is the transformation constant and θ is rotor position

6.2.3 Active filter compensation circuit

Fig. 58 shows simplified power circuit of the proposed topology (*the passive RCL filters are not shown*). In this circuit V_{dc} is the voltage of the main inverter circuit, $V_{CF}^{\pm}$ is equivalent compensated voltage source of the active filter. In order to generate the required compensation voltages that follow the voltage signal v_{sig}; *bearing in the mind that the main inverter change switching state only when the line current violates the condition of the hysteresis band and that the capacitor voltage polarity can not change abruptly,* the switches sw_1 and sw_2 are controlled within each consecutive voltage switching of the main inverter to keep the motor winding voltages with acceptable hysteresis band.

The motor line current i_m is controlled within the motor main control circuit with hysteresis current controller to provide the required load torque; therefore, two hysteresis controller systems, one for voltage and the other for current are working independently to supply the motor with almost sinusoidal voltage

In Fig. 58, when switching signal (eg.100) is send to the main inverter, i.e. phase a is active high while phase b and c are active low, then, following the path of the current i_m in Fig.58 the voltage provided to the motor terminal can be expressed as:

$$V_s = \frac{2}{3}(V_{dc} - V_{CF}^{\pm} - \frac{3}{2} L_F \frac{di_m}{dt}) \tag{54}$$

The limit values of inductor L_F and the capacitor C_F can be determined as follows:

During a sampling period T_s, the change in the capacitor voltage can be calculated as:

$$\Delta V_{CF} = \frac{1}{C_F} \int_0^{Ts} i_m dt \tag{55}$$

So if maximum capacitor voltage change is determined as Vdc, the minimum capacitor value can be calculated as:

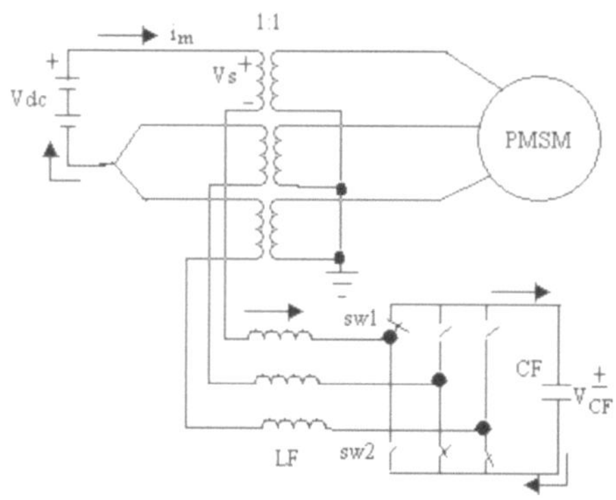

Fig. 58. Simplified power circuit of the proposed active filter topology.

$$C_F \geq \left| \frac{\int_{nTs}^{(n+1)Ts} i_m dt}{Vdc} \right| = \left| \frac{Ts \bullet i_{mav}}{V_{dc}} \right| \tag{56}$$

Where, i_{mav} is the maximum of the average current change which can be occurred per sample periods.
The limit values of the smoothing inductance L_F can be expressed as:

$$\frac{1}{(2\pi f_{sw})^2 C_F} < L_F \leq \left| \frac{V_{LF\max}}{\frac{3}{2}\max(\frac{di_m}{dt})} \right| \tag{57}$$

Where, the lower limit is determined by selecting the resonance frequency of the combination $C_F L_F$ to be less than the inverter switching frequency f_{sw} to guarantee reduced switching frequency harmonics. The upper limit is calculated by determining the maximum voltage drop across the inductors V_{LFmax}, and the maximum current change per sampling period di_m/dt.

6.2.4 The Coupling

The coupling between the main inverter circuit and the active filter circuit is achieved through 1:1 transformer, and to attenuate the higher frequency EMI noises, LCR filters are used at the transformer primary and secondary windings as suggested by Fig. 59

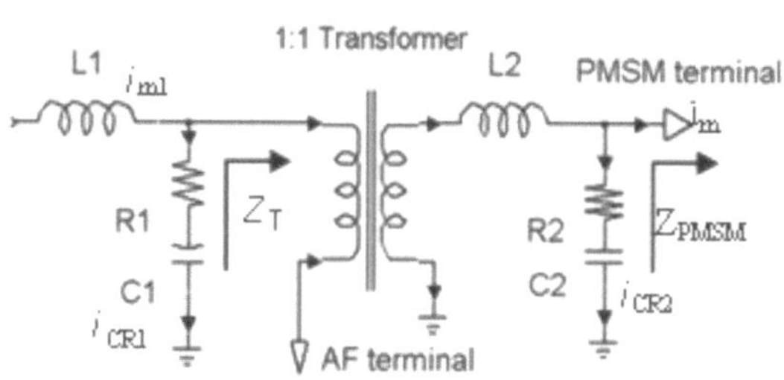

Fig. 59. Coupling between AF and main inverter from one side, and PMSM in the other side.

The important point here is that, the resonance which may arise between capacitor C_1 and transformer primary winding and between capacitor C_2 and motor inductance winding should be avoided when selecting capacitor values.
At selected cutoff frequency, the currents i_{CR1} and i_{CR2} derived by the RLC filters are given by

$$i_{CR1} = \frac{z_T}{z_T + \sqrt{R_1^2 + (1/sC_1)^2}} i_{m1} \quad and$$
$$i_{CR2} = \frac{z_{PMSM}}{z_{PMSM} + \sqrt{R_2^2 + (1/sC_2)^2}} i_{m2} \qquad (58)$$

Where, Z_T and Z_{PMSM} are as defined in Fig. 59.
Bearing in the mind the conditions required in the selection of RLC, these currents should be large compared to i_{m1}, drawn by the transformer, and/or i_m, drawn by the motor at selected cutoff frequency; while at operating frequency these currents should be very small compared to i_{m1} and i_m.

6.2.5 Simulation and results

In order to verify that the proposed filter topology does actually improve the performance of the conventional HDTC methods, the HDTC is implemented in Matlab/Simulink to compare the performance of the PMSM with and without the filter topology under the same operating and loading conditions
The motor parameters are in Table 2 and the filters parameters are in Table 6. The AF capacitor used is 200μF and its inductors are 200mH. The drive is IGBT inverter.

L_1	1μH	L_2	1.5μH
C_1	2μF	C_2	2μF
R_1	250Ω	R_2	750Ω

Table 6. Active Filter Topology parameters

The simulation results with 100μs sampling time and ±0.1 Nm hysteresis torque band are shown in Fig. 60 to Fig. 66. The torque dynamic response is simulated with open speed loop, while the steady state performance is simulated with closed speed loop at 70rad/s as reference speed, and 2 Nm as load torque.
The torque dynamic responses before and after connecting the AF are shown in Fig. 60-a and Fig. 60-b respectively. The reference torque for both figures is changed from +2.0 to -2.0 and then to 3.0 Nm. As shown in the figures, the dynamic response with the proposed filter topology is adequately follows the reference torque with lower torque ripples and settles down within ±0.1 Nm band of the reference torque; while the torque dynamic under HDTC without filter topology can not settle down within the specified torque bands due to presence of high torque ripples (± 1.0 Nm). On the other hand, the torque response time without filter topology is shorter (~1.2ms) than the torque response time with the proposed filter topology (2.5ms). This delay in the torque response with the proposed filter topology is mainly due to delay of current propagation through the L_FC_F loop of the active filter however; this is not significant if compared with the results provided by Tang et al (2004).

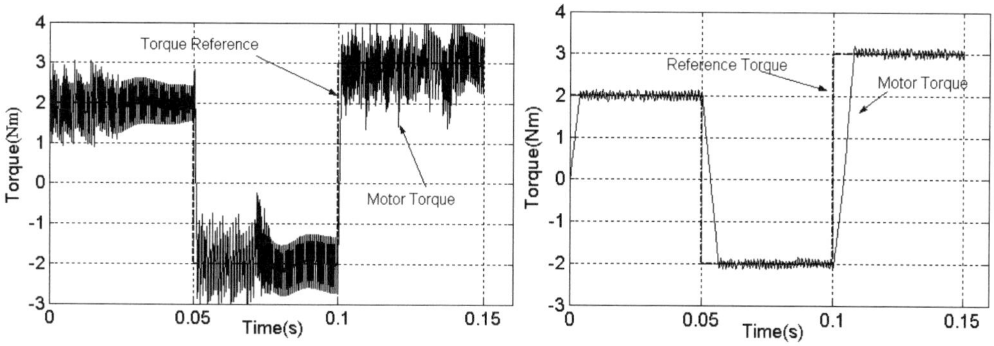

(a) (b)

Fig. 60. Motor torque dynamic under basic HDTC: (a) before (b) after connecting the AF

The motor steady state performance before and after applying the AF are shown in Fig. 61 to Fig. 64. Fig. 61-a and Fig. 61-b, show the phase voltage provided to the motor terminals before and after applying the filter topology respectively, observe the change of the waveform after applying the AF, it is clear that the phase voltage approaches sinusoidal waveform with almost free of voltage pulses appear in Fig. 61-a due to inverter switching. Better waveform can be obtained by increasing the active filter inductance L_F however, the cost and size of the AF will increase, and therefore suitable inductance value can be selected to achieve acceptable performance. Similar results have been provided by Yilmaz, (Yilmaz et al. 2000), however as compared to above result, their sinusoidal voltage waveform provided to the motor terminals is full of harmonic components.

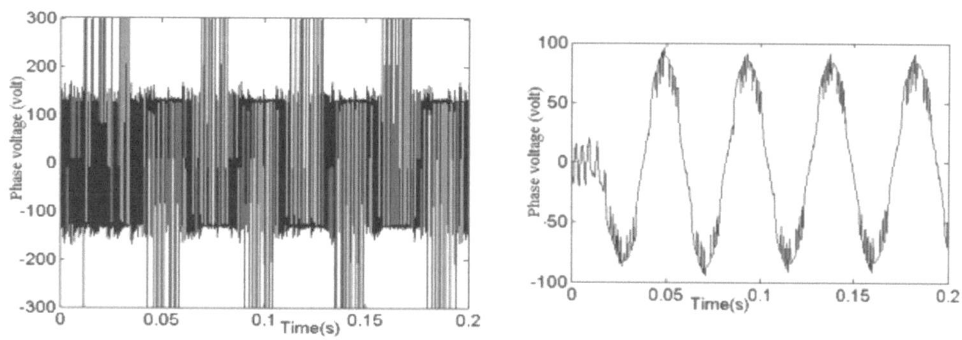

(a) (b)

Fig. 61. Starting motor phase voltage: (a) before (b) after connecting the AF topology

Fig. 62-a and Fig. 62-b show the response of the motor line currents under HDTC without and with the proposed filter topology respectively. In Fig. 62-a high distortion in line current can be observed, however the current waveform is smoother after applying the proposed filter topology. The reason of the high current distortion (ripples) is mainly due to the fact that switching of the inverter is only updated once at the sampling instances when the hysteresis controllers change state so, with existence of the proposed active filter a proper voltage is provided to the motor terminal which, in turn decreases current ripples.

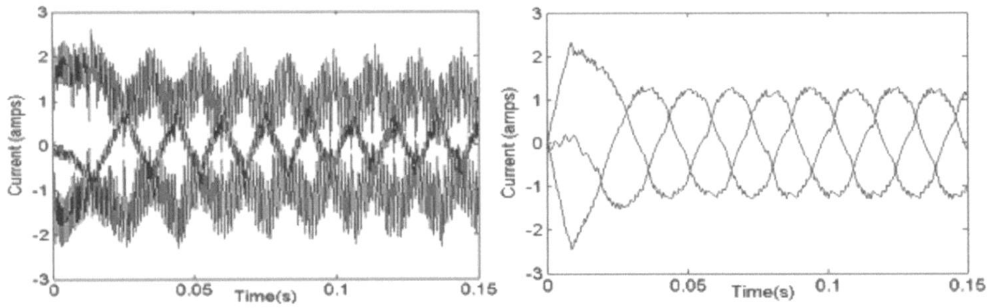

(a) (b)

Fig. 62. Motor line currents: (a) before (b) after applying the AF topology.

The torque response in Fig. 63 shows considerable reduction in torque ripples around the load torque when the proposed active filter is connected. The higher ripples of ±1.62Nm around the load torque in Fig. 63-a is mainly due to the existence of harmonic voltages provided to the motor terminals, so when the harmonics are reduced after insertion of the proposed filter topology the torque ripples is decreased down to ±0.1 Nm as shown in Fig. 63-b. The reduction in the torque ripples normally reflected in reduced motor mechanical vibration and hence reduced acoustic noise as well as smoother speed response as shown in Fig. 64.

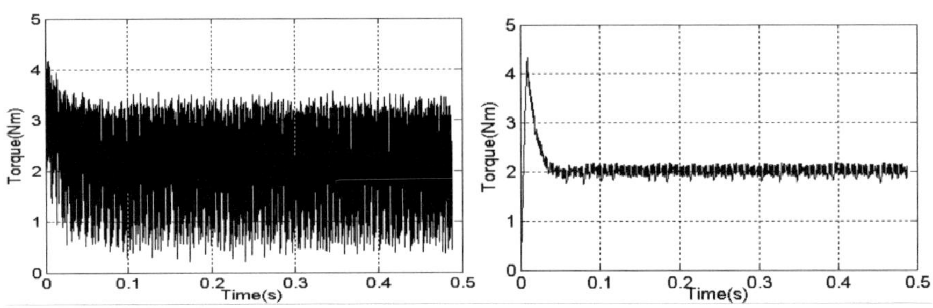

(a) (b)

Fig. 63. Steady state motor torque response under basic HDTC with 2.0 Nm as load torque (a) before (b) after connecting the AF topology

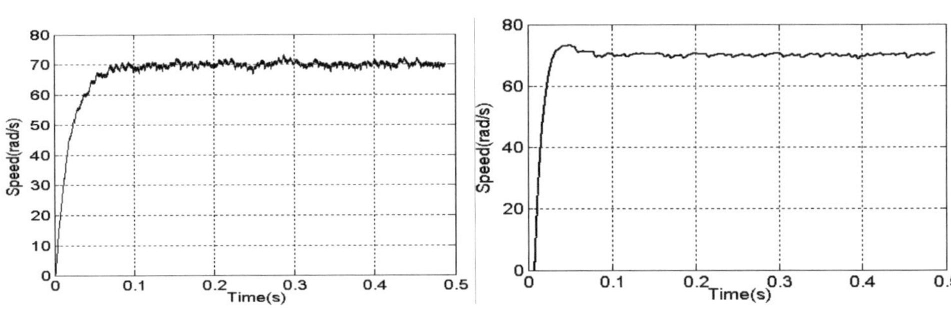

(a) (b)

Fig. 64. Rotor speed under basic HDTC (a)before (b)after applying the AF topology

The status of the phase voltage harmonics and EMI noise in the line currents before and after connecting the AF are shown in Fig. 65 to Fig. 66.
In Fig. 65-a the spectrum of the phase voltage before connecting the AF shows that disastrous harmonic voltages with THD of ~79% have widely scattered in the shown frequency range. These harmonic voltages if not cleared or reduced, it will result in parasitic ripples in motor developed torque and contribute to electromagnetic interference noise, so after connecting the AF, the THD is effectively reduced to less than 5% as in Fig. 65-b.

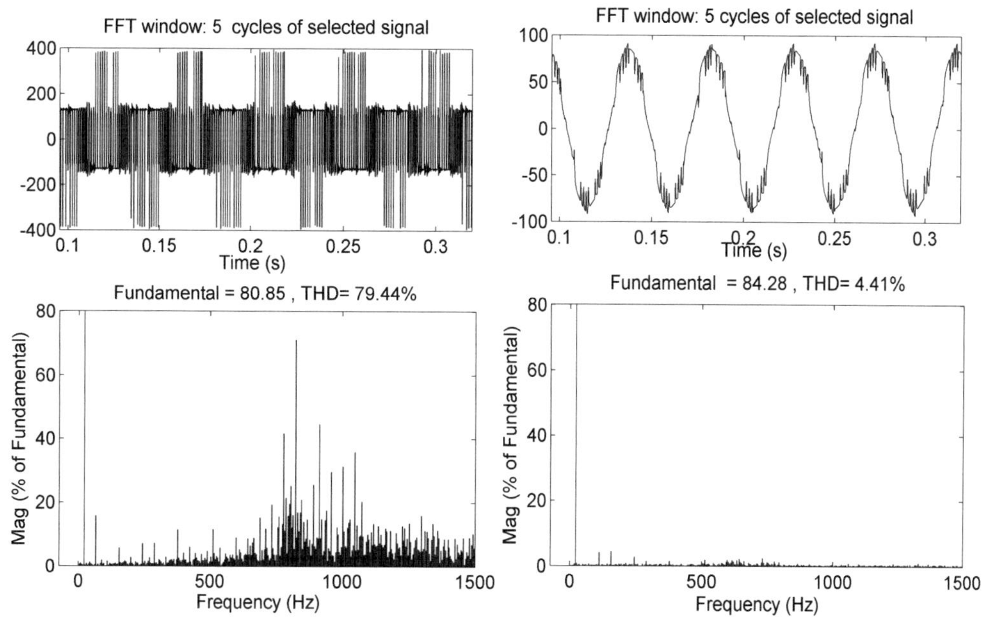

(a) (b)

Fig. 65. Phase-a voltage (upper) and it is spectrum (lower):
(a) Before connecting the AF topology (b) after connecting the AF topology

The EMI noise level before connecting the AF in Fig. 66-a shows a noise level of ~ 20dB at operating frequency, ~18dB at switching frequency (5KHz), and almost -40dBs for the most high frequencies (>0.2 MHz). These noise component frequencies have bad effect on the control system if not filtered. When the AF is connected the EMI noise level is tuned down to ~-18dB at operating frequency, ~-25dB at switching frequency and less than ~-60dBs for the most high frequencies as shown in Fig. 66-b.
From the results presented it can be seen that the steady state performance of the HDTC with the proposed filter topology is much better than the performance presented by Zhong(1997). This result can also be compared with experimental result presented by Tang et al (2004) though the effective average switching sampling time in that method is much less than the selected sampling period (150μs) and that due to the fact space vector modulation was used to drive the inverter.
The motor voltage waveform is better than that provided by Yilmaz, et al(2000), beside the filter topology presented by Yilmaz, et al (2000) is continuously required to be tuned when the switching frequency is changed. In addition in order to obtain acceptable sinusoidal

waveform, the resistor value used in the RLC loop is small, which involves larger current to flow through the loop composed of the RLC and the inverter which in turn causes over loading to the inverter elements.

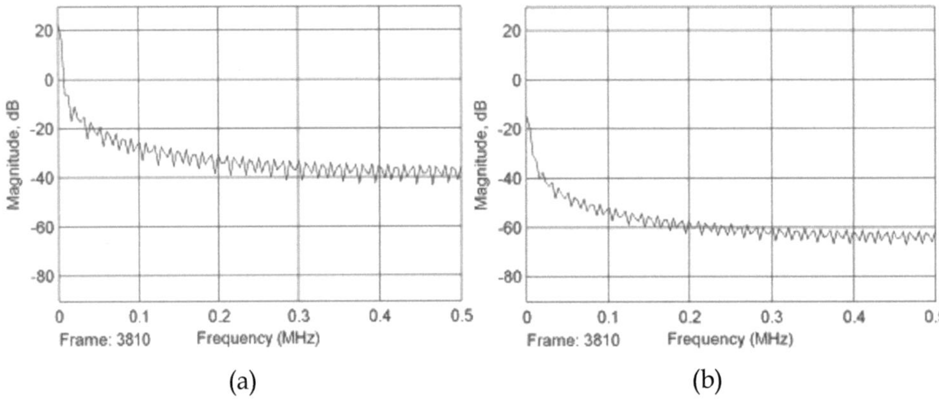

Fig. 66. EMI noise level: (a) before (b) after connecting the AF topology

7. References

Adam A. A. And Gulez K., (2009) "A New Sensorless Hysteresis Direct Torque Control Algorithm for PMSM with Minimum Torque Ripples", COMPEL, Vol.28, No.2, p.p. 437-453, April 2009.

Dariusz, S., Martin, P. K. And Frede, B., (2002),"DSP Based Direct Torque Control of Permanent Magnet Synchronous Motor Using Space Vector Modulation" Proceeding of the 2002 IEEE International Symposium on Industrial Electronics, ISIE 2002 , Vol. 3, 26-29 May , pp. 723-727.

Darwin, R., Morán, L., Dixon, W. J., and Espinoza, J. R,. (2003), "Improving Passive Filter Compensation Performance With Active Techniques," IEEE Transaction on Industrial Electronics, Vol. 50(1), pp. 161-170, Feb. 2003.

Depenbrock, M., (1984), "Direct Self-Control", U.S. Patent, No: 4678248, Oct. 1984.

Depenbrock, M., (1988), "Direct Self-Control of inverter-fed machine", IEEE Transactions on. Power Electronics Vol. 3, No.4, pp. 420-429, Oct. 1988.

Dirk, D., Jacobs, J., De Doncker, R. W. and Mall, H.G.,(2001), " A new Hybrid Filter to Dampen Resonances and Compensate Harmonic Currents in Industrial Power System With Power Factor Correction Equipment," IEEE Transaction on Industrial Electronics, Vol. 16(6), pp. 821-827, Nov. 2001.

Erik, P.,(1992), " Transient Effects in Application of PWM Inverters to Induction Motors", IEEE Transactions on Industry Application, Vol. 28, No. 5, pp. 1095-1101 Sept./ Oct. 1992.

French, C. and Acarnley, P., (1996), "Direct torque control of permanent magnet drives," IEEE Transactions on Industrial Applications., Vol. 32 Issue: 5, pp.1080–1088, Sept./Oct. 1996.

Gulez K., Adam A. A., Pastacı H. (2007), "Passive Filter Topology to Minimize Torque Ripples and Harmonic Noises in IPMSM Derived with HDTC", IJE-International Journal of Electronics, Vol. 94, No:1, p.p.23-33, Jan. (2007).

Gulez K., Adam A. A., Pastacı H. (2008) "Torque Ripples and EMI Noise Minimization in PMSM Using Active Filter Topology and Field Oriented Control", IEEE-Transactions on Industrial Electronics, Vol. 55, No. 1, Jan. (2008).

Hideaki, F., Takahiro, Y., and Hirofumi, A.,(2000), "A Hybrid active Filter For Damping of Harmonic Resonance in Industrial Power Systems," IEEE Transaction on Power Electronics, Vol. 15 (2) , pp. 215-222, Mar. 2000.

Holtz, J. and Springob, L.,(1996), "Identification and Compensation of Torque Ripple in High- Precision Permanent Magnet Motor Drives", IEEE Transactions on Industrial Electronics, Vol. 43, No. 2, April 1996, pp.309-320

Hugh, R., Juan, D. and Morán, L., (2003), "Active power filters as a solution to power quality problems in distribution networks", IEEE power & energy magazine Sept./Oct. 2003 pp. 32-40

Jahns, T. M. and Soong, W. L., (1996), "Pulsating torque minimization techniques for permanent magnet AC motor drives - a review," IEEE Transactions on Industrial Electronics, vol. 43, no. 2, pp. 321-330, Feb. 1996.

Jeong-seong, K., Shinji, D. and Muneaki, I.,(2002), "Improvement of IPMSM Sensor less control performance Using Fourier Transform and Repetitive control", IECON 02 Industrial Electronic Society Conference,5-7 Nov. 2002, IEEE, vol. 1 pp. 597-602

Luukko, J.,(2000), Direct Torque Control of Permanent Magnet Synchronous Machines - Analysis and Implementation, Diss. Lappeenranta University of Technology, Lappeenranta, Stockholm, 2000.

Satomi, H., Muneaki, I. and Takamasa, H., (2001), "Vibration Suppression Control Method for PMSM Utilizing Repetitive Control With Auto-tuning Function and Fourier Transform" IECON'01: The 27th Annual Conference of IEEE Industrial Electronics Society, 2001, pp 1673-1679.

Se-Kyo, C., Hyun-Soo, K. and Myun-Joong, Y., (1998),"A new Instantaneous Torque Control of PM Synchronous Motor for High-Performance Direct-Drive Applications", IEEE Transactions on Power Electronics Vol. 13, No. 3.

Springob, L. and Holtz, J., (1998),"High-Bandwidth Current Control for Torque-Ripple Compensation in PM Synchronous Machines", IEEE Transactions on Industrial Electronics, Vol. 45, NO. 5, October 1998, pp.713-721

Takahashi, I. and Naguchi, T.(1998), "A new quick-response and high efficiency control strategy of an induction motor," IEEE Transactions on Industrial Applications, vol. 34, No. 6 pp. 1246-1253, Nov./Dec. 1998.

Tan, Z. Y. and Li, M., (2001), "A Direct Torque Control of Induction Motor Based on Three Level Inverter" , IEEE, PESC'200, Vol. 2 pp. 1435-1439

Tang, L., Zhong, L., Rahman, M. F. and Hu, Y., (2004), "A Novel Direct Torque Controlled Interior Permanent Magnet Synchronous Machines Drive with Low Ripple in Flux and Torque and Fixed Switching Frequency", IEEE Transactions on Power Electronics Vol. 19, No. 2, Mar. 2004

Tang, L., Zhong, L, Rahman, M. F., and Hu, Y., (2001), " A novel Direct Torque Control for Interior Permanent Magnet Synchronous Machine Drive System with Low Ripple In torque and Flux-A Speed Sensor less Approach" IEEE, IAS, 13-18 Oct. 2002, vol. 1, pp.104-111.

Vas, P.,(1996), Electrical Machines and Drives- A Space-Vector Theory Approach, Oxford, USA, 1996.

Yilmaz, S., David, A. T., and Suhan, R., (2000), "New Inverter Output Filter Topology for PWM Motor Drives", IEEE Transaction on Power Electronics Vol. 15 No 6, pp. 1007-1017, Nov. 2000

Zhong, L., Rahman, M. F., Hu, W.Y. and Lim, K.W., (1997), "Analysis of direct torque control in permanent magnet synchronous motor drives," IEEE Transactions. on Power Electronics, vol. 12 Issue: 3, pp. 528 –536, May 1997.

Part 3

Special Controller Design and Torque Control of Switched Reluctance Machine

8

Switched Reluctance Motor

Jin-Woo Ahn, Ph.D
Kyungsung University
Korea

1. Introduction

Switched Reluctance Motors (SRM) have inherent advantages such as simple structure with non winding construction in rotor side, fail safe because of its characteristic which has a high tolerances, robustness, low cost with no permanent magnet in the structure, and possible operation in high temperatures or in intense temperature variations. The torque production in switched reluctance motor comes from the tendency of the rotor poles to align with the excited stator poles. The operation principle is based on the difference in magnetic reluctance for magnetic field lines between aligned and unaligned rotor position when a stator coil is excited, the rotor experiences a force which will pull the rotor to the aligned position. However, because SRM construction with doubly salient poles and its non-linear magnetic characteristics, the problems of acoustic noise and torque ripple are more severe than these of other traditional motors. The torque ripple is an inherent drawback of switched reluctance motor drives. The causes of the torque ripple include the geometric structure including doubly salient motor, excitation windings concentrated around the stator poles and the working modes which are necessity of magnetic saturation in order to maximize the torque per mass ratio and pulsed magnetic field obtained by feeding successively the different stator windings. The phase current commutation is the main cause of the torque ripple.

The torque ripple can be minimized through magnetic circuit design in a motor design stage or by using torque control techniques. In contrast to rotating field machines, torque control of switched reluctance machines is not based on model reference control theory, such as field-oriented control, but is achieved by setting control variables according to calculated or measured functions. By controlling the torque of the SRM, low torque ripple, noise reduction or even increasing of the efficiency can be achieved. There are many different types of control strategy from simple methods to complicated methods. In this book, motor design factors are not considered and detailed characteristics of each control method are introduced in order to give the advanced knowledge about torque control method in SRM drive.

1.1 Characteristic of Switched Reluctance Motor

The SRM is an electric machine that converts the reluctance torque into mechanical power. In the SRM, both the stator and rotor have a structure of salient-pole, which contributes to produce a high output torque. The torque is produced by the alignment tendency of poles. The rotor will shift to a position where reluctance is to be minimized and thus the inductance of the excited winding is maximized. The SRM has a doubly salient structure, but there are no windings or permanent magnets on the rotor [Lawrenson, 1980]. The rotor is basically a piece of steel (and laminations) shaped to form salient poles. So it is the only motor type with salient poles in both the rotor and stator. As a result of its inherent simplicity, the SRM promises a reliable and a low-cost variable-speed drive and will undoubtedly take the place of many drives now using the cage induction, PM and DC machines in the short future. The number of poles on the SRM's stator is usually unequal to the number of the rotor to avoid the possibility of the rotor being in a state where it cannot produce initial torque, which occurs when all the rotor poles are aligned with the stator poles. Fig.1 shows a 8/6 SRM with one phase asymmetric inverter. This 4-phase SRM has 8 stator and 6 rotor poles, each phase comprises two coils wound on opposite poles and connected in series or parallel consisting of a number of electrically separated circuit or phases. These phase windings can be excited separately or together depending on the control scheme or converter. Due to the simple motor construction, an SRM requires a simple converter and it is simple to control.

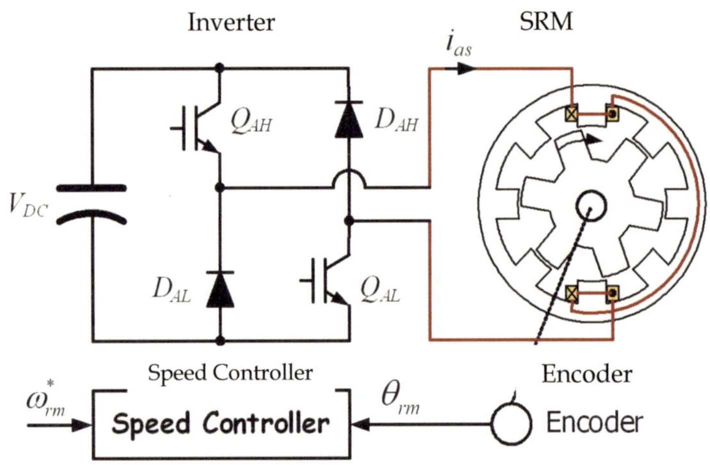

Fig. 1. SRM with one phase asymmetric inverter

The aligned position of a phase is defined to be the situation when the stator and rotor poles of the phase are perfectly aligned with each other ($\theta_1 - \theta_2$), attaining the minimum reluctance position and at this position phase inductance is maximum (L_a). The phase inductance decreases gradually as the rotor poles move away from the aligned position in either direction. When the rotor poles are symmetrically misaligned with the stator poles of a phase ($\theta_3 - \theta_s$), the position is said to be the unaligned position and at this position the phase has minimum inductance (L_u). Although the concept of inductance is not valid for a highly saturated machine like SR motor, the unsaturated aligned and unaligned incremental inductances are the two key reference positions for the controller. The relationship between inductance and torque production according to rotor position is shown in Fig. 2.

There are some advantages of an SRM compared with the other motor type. The SRM has a low rotor inertia and high torque/inertia ratio; the winding losses only appear in the stator because there is no winding in the rotor side; SRM has rigid structure and absence of permanent magnets and rotor windings; SRM can be used in extremely high speed application and the maximum permissible rotor temperature is high, since there are no permanent magnets and rotor windings [Miller, 1988].

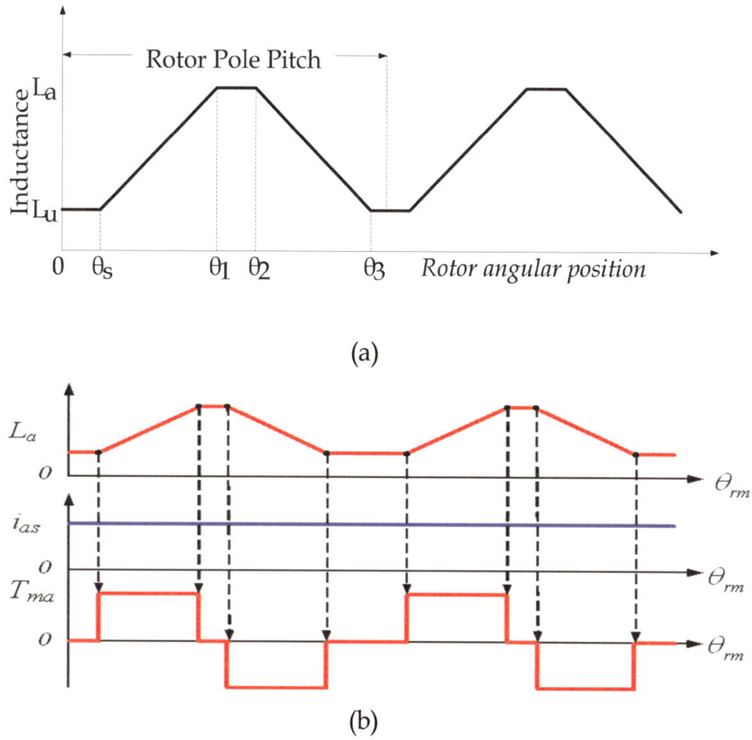

Fig. 2. (a) Inductance and (b) torque in SRM

Constructions of SRM with no magnets or windings on the rotor also bring some disadvantage in SRM. Since there is only a single excitation source and because of magnetic saturation, the power density of reluctance motor is lower than PM motor. The construction of SRM is shown in Fig. 3. The dependence on magnetic saturation for torque production, coupled with the effects of fringing fields, and the classical fundamental square wave excitation result in nonlinear control characteristics for the reluctance motor. The double saliency construction and the discrete nature of torque production by the independent phases lead to higher torque ripple compared with other machines. The higher torque ripple, and the need to recover some energy from the magnetic flux, also cause the ripple current in the DC supply to be quite large, necessitating a large filter capacitor. The doubly salient structure of the SRM also causes higher acoustic noise compared with other machines. The main source of acoustic noise is the radial magnetic force induced. So higher torque ripple and acoustic noise are the most critical disadvantages of the SRM.

The absence of permanent magnets imposes the burden of excitation on the stator windings and converter, which increases the converter kVA requirement. Compared with PM brushless machines, the per unit stator copper losses will be higher, reducing the efficiency and torque per ampere. However, the maximum speed at constant power is not limited by the fixed magnet flux as in the PM machine, and, hence, an extended constant power region of operation is possible in SRM.

The torque-speed characteristics of an SRM are shown in Fig. 4. Based on different speed ranges, the motor torque generation has been divided into three different regions: constant torque, constant power and falling power region.

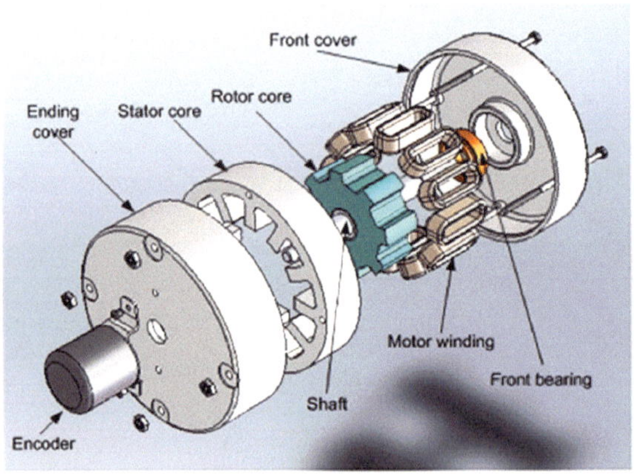

Fig. 3. The construction of SRM

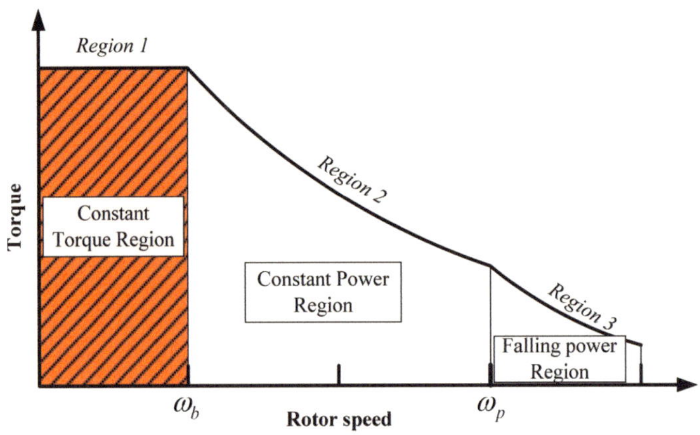

Fig. 4. The torque-speed of SRM

The base speed ω_b is the maximum speed at which maximum current and rated torque can be achieved at rated voltage. Below ω_b, the torque can be maintained constant or control the fat-top phase current. At lower speed, the phase current rises almost instantly after the phase switches turn-on since the back EMF is small at this time. So it can be set at any desired level by means of regulators (hysteresis or PWM controller). Therefore, the adjustment of firing angle and phase current can reduce noise and improve torque ripple or efficiency.

With speed increase, the back-EMF is increased. An advance turn-on angle is necessary to reach the desired current level before rotor and stator poles start to overlap. The desired current level depends on the speed and the load condition. At the same time, since no current chopping appears during the dwell angle, only the angle control can be used at this stage. So the torque cannot be kept constant and is falling linearly with the speed increase, resulting in a constant power production.

In the falling power region, as the speed increases, the turn-on angle cannot be advanced further. Because torque falls off more rapidly, the constant power cannot be maintained. As the speed grows, the tail current of the phase winding extends to the negative torque region.

The tail current may not even drop to zero. In the high speed operation, the continued conduction of current in the phase winding can increase magnitude of phase current and the power density can be increased.

1.2 Equivalent circuit of Switched Reluctance Motor

The equivalent circuit for SRM can be consisting of resistance and inductance with some condition. The effects of magnetic saturation, fringing flux around the pole corners, leakage flux, and the mutual coupling of phases are not considered. The linear analytical model of the SRM can be described by three differential equations, which can be classified as the voltage equation, the motional equation and the electromagnetic torque equation. The voltage equation is:

$$V = R.i + \frac{d\lambda(\theta,i)}{dt} \tag{1}$$

An equivalent circuit of the SRM is shown in Fig. 5. Where V is the applied phase voltage to phase, R is the phase resistance, and e is back-EMF. Ordinarily, e is the function of phase current and rotor position, and λ can be expressed as the product of inductance and winding current:

$$\lambda(\theta, i) = L(\theta, i) . i \tag{2}$$

And from (1) and (2), the function can be rewritten as:

$$V = R . i + \frac{d\lambda(\theta,i)}{di} . \frac{di}{dt} + \frac{d\lambda(\theta,i)}{d\theta} . \frac{d\theta}{dt} \tag{3}$$

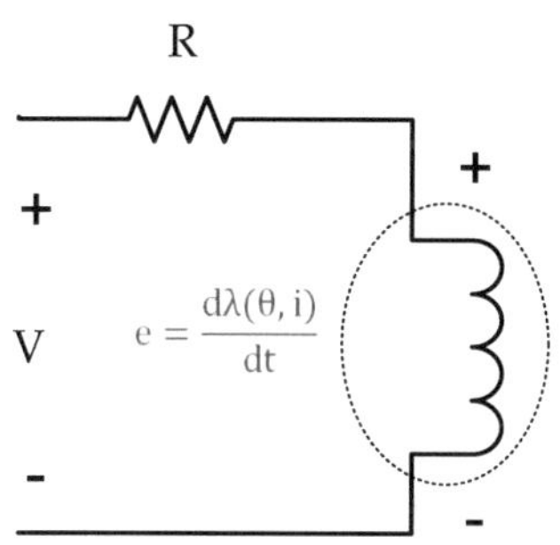

Fig. 5. Equivalent circuit of SR motor

For the electromechanical energy conversion, a nonlinear analysis takes account of the saturation of the magnetic circuit. Generally, the stored magnetic energy is defined as W_f and the co-energy is defined as W_c:

$$W_f = \int i d\psi \tag{4}$$

$$W_c = \int \psi \, di \tag{5}$$

The relationship between energy (W_f) and co-energy (W_c) as a function of flux and current shows in Fig. 6.

When rotor position matches the turn-on position, the phase switches are turned on; the phase voltage starts to build up phase current. At this time, one part of the input energy will

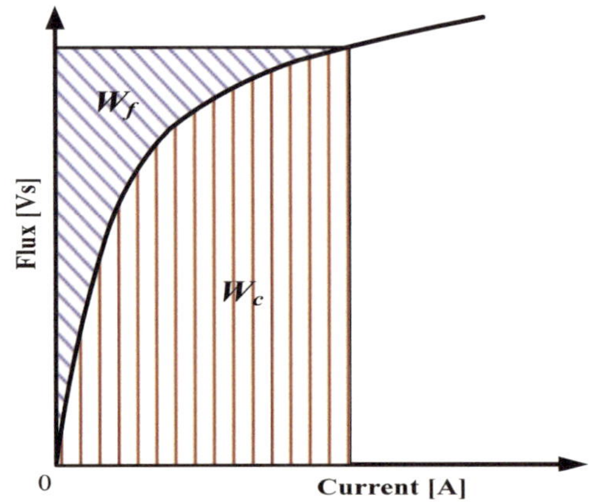

Fig. 6. Relationship between energy (W_f) and co-energy (W_c)

be stored in magnetic field. With the increasing inductance, the magnetic field energy will increase until turn-off angle. The other parts of input energy will be converted to mechanical work and loss. In Fig. 7, the flux of the SR motor operation is not a constant; nevertheless, uniform variation of the flux is the key point to obtain smoothing torque. W_1 is the mechanical work produced during the magnetization process, in other words, W_1 is co-energy in energy conversion. F+W_2 is magnetic field energy between turn-on and turn-off. During the derivation of the energy curve and the energy balance, constant supply voltage Vs and rotor speed ω are assumed.

When rotor position matches the turn-off position, phase switches are turned off. So the power source will stop to input energy. But magnetic field energy is $F+W_2$ at that moment. The magnetic field energy needs to be released, and then the phase current starts to feedback energy to power source. At this time, some of magnetic field energy, which is W_2, is converted into mechanical work and loss. The surplus of field energy F is feedback to the power source.

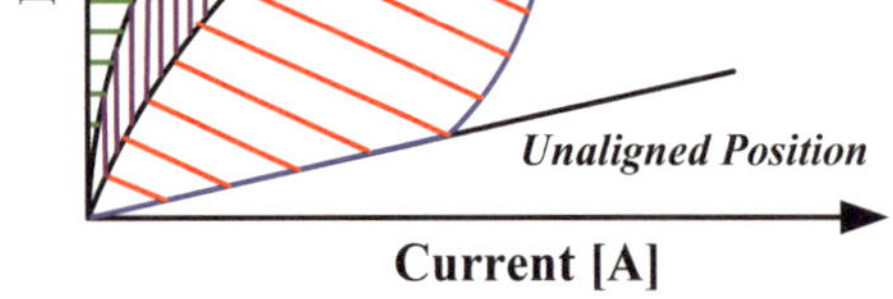

Fig. 7. Graphical interpretation of energy and co-energy for SR motor

The analytical answer of the current can be obtained from (3). The electromagnetic torque equation is:

$$T_e = \frac{\partial W\prime}{\partial \theta} = \frac{\partial W'(\theta+\Delta\theta) - \partial W'(\theta)}{\Delta\theta} \tag{6}$$

From (6), an analytical solution for the torque can be obtained. W' is the co-energy, which can be expressed as:

$$W' = \int_0^i \lambda \, di \tag{7}$$

And the motion equation is:

$$T_e = J\frac{\partial W}{\partial t} + D\omega + T_L \tag{8}$$

$$\omega = \frac{d\theta}{dt} \tag{9}$$

Where T_L, T_e, J, ω and D are load the electromagnetic torque, the rotor speed, the rotor inertia and the friction coefficient respectively.

The equations which have been mentioned above, can be combined together to build the simulation model for a SRM system. However, the function of inductance needs to be obtained by using a finite element method or by doing experiments with a prototype motor.

1.3 Torque control in Switch Reluctance Motor

The torque in SRM is generated toward the direction that the reluctance being to minimized. The magnitude of torque generated in each phase is proportional to the square of the phase current which controlled by the converter or drive circuit, and the torque control scheme. The drive circuit and torque control scheme directly affected to the performance and characteristic of the SRM. Many different topologies have emerged with a reduced number of power switch, faster excitation, faster demagnetization, high efficiency, high power factor and high power through continued research. Conventionally, there has always been a trade-off between gaining some of the advantages and losing some with each new topology.

The torque is proportional to the square of current and the slope of inductance. Since the torque is proportional to the square of current, it can be generated regardless of the direction of the current. And also because the polarity of torque is changed due to the slope of inductance, a negative torque zone is formed according to the rotor position. To have a motoring torque, switching excitation must be synchronized with the rotor position angle. As shown in Fig. 8, an inductance profile is classified into three regions, increasing ($\theta_{min1} \sim \theta_{max1}$), constant ($\theta_{max1} \sim \theta_{max2}$) and decreasing ($\theta_{max2} \sim \theta_{min2}$) period. If a constant exciting current flows through the phase winding, a positive torque is generated. When that is operated in inductance increasing period ($\theta_{min1} \sim \theta_{max2}$) and vice-versa in inductance decreasing ($\theta_{max2} \sim \theta_{min2}$).

In the case of a constant excitation, it cannot be generated any torque, because a positive torque and negative one are canceled out, and the shaft torque becomes zero. As a result, to achieve an effective rotating power, switching excitation must be synchronized with the inductance profile. In order to derive the phase current from (3), exact information about the inductance profile of the SRM is essential. In (10), the first term of the right side is voltage drops of winding resistance, the second term is the voltage drop of reactance and the last term is both the emf (electromotive magnetic force) and the mechanical output.

$$V = Ri + i(t)\frac{d\mathcal{L}(\theta,i)}{d\theta}\omega + \mathcal{L}(\theta,i)\frac{di(t)}{dt} \tag{10}$$

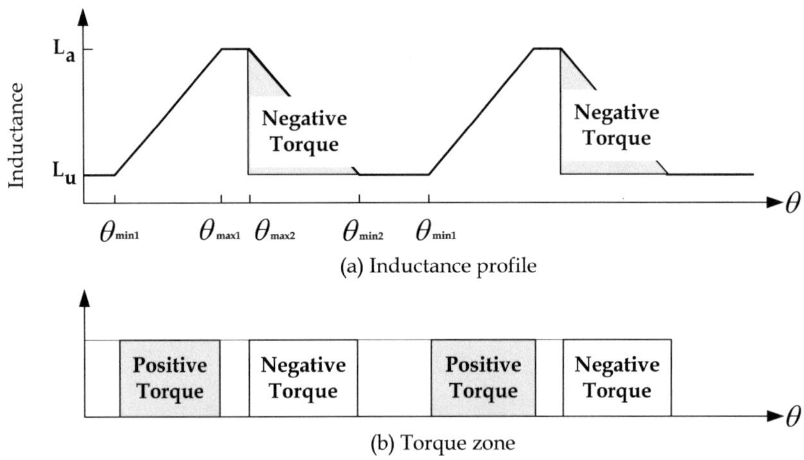

Fig. 8. (a)Inductance profile and (b) Torque zone

where, ω is the angular speed of the rotor.
In (10), the second in the right side can be considered as the back-emf; therefore, this term is expressed as:

$$e = \frac{d\mathcal{L}(\theta,i)}{d\theta}\omega i(t) = K\omega i(t) \tag{11}$$

$$\text{where, } K = \frac{d\mathcal{L}(\theta,i)}{d\theta} \tag{12}$$

As shown in (11), the back-emf equals to that of the DC motor. And also torque equation in (12) is equivalent with that of the DC series motor; therefore, the speed-torque of the magnetic energy in SRM is different from that of a mutual torque machine. And it operates more saturated level. The field energy in the magnetization curve is shown in Fig. 9.

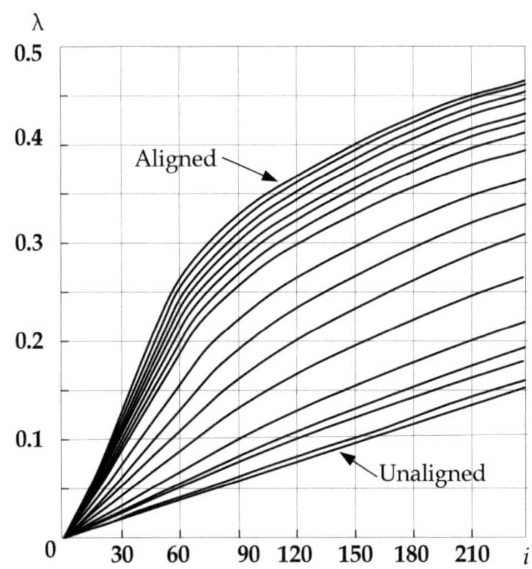

Fig. 9. Magnetizing curve and flux-linkage curve of SRM

It shows the magnetization curves from an aligned to an unaligned position. In SRM design, when poles of a rotor and a stator are aligned, the other phases are unaligned. In an aligned position, it has a maximum inductance with magnetically saturated easily. On the other hand, in an unaligned position it has a minimum inductance. As magnetic saturation is proportional to a rotor position, the magnetization curve according to the rotor position is an important factor to investigate the motor characteristics and to calculate the output power. The torque produced by a motor can be obtained by considering the energy variation. The generated torque is as:

$$T = \left[\frac{dW'}{d\theta}\right]_{i=const.} \tag{13}$$

where, w' means the co-energy, and it is given as:

$$W' = \int_0^i \lambda \, di \tag{14}$$

Under a constant phase current as shown in Fig. 10, when the rotor and total flux linkage are shifted from A to B, the SRM exchanges energy with the power source; thus, the stored field energy is also changed. The limitation to a constant current is that mechanical work done during the shifting region is exactly equal to the variation of co-energy. At a constant current, if the displacement between A and B is AB, the variation of energy received from the source can be expressed as:

$$\Delta W_e = ABCD \tag{15}$$

$$\Delta W_c = OBC - OAD \tag{16}$$

Then the mechanical work can be written as:

$$\Delta W_m = T\Delta\theta = \Delta W_e - \Delta W_c = OAB \tag{17}$$

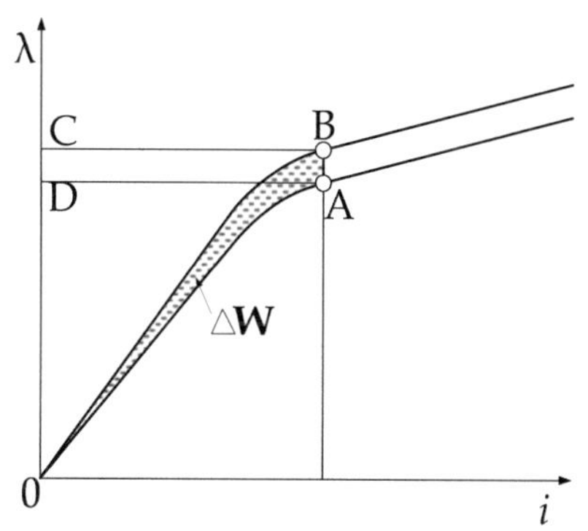

Fig. 10. Calculation of instant torque by the variation of co-energy at constant current

The above equation just shows the instantaneous mechanical output; therefore, in order to understand the characteristics of the motor, the average torque generated during an energy conversion cycle may be considered. The mechanical output is expressed as an area in an energy conversion curve (i-λ graph), the processes are separated with two stages as shown in Fig. 11.

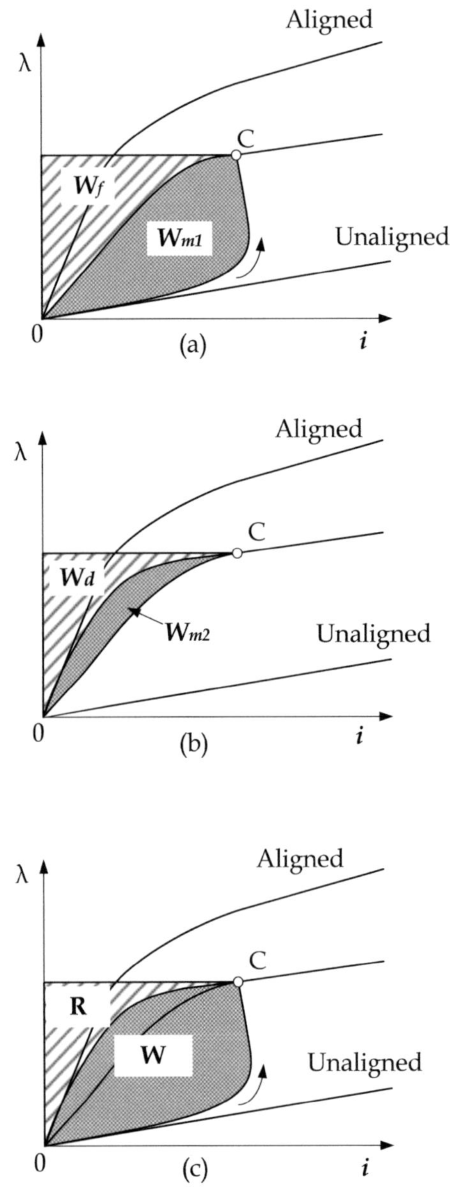

Fig. 11. Average torque (Energy conversion loop)

The total flux linkage is increased with phase current and inductance. Its operating area (i, λ) follows the curve between 0 and C as shown in Fig. 11(a). When the total flux linkage exists at point C, the mechanical work and stored energy between 0 and C becomes W_{m1} and W_f, respectively. Therefore, the total energy received from the source is summed up the mechanical work and the stored energy. On the other hand, when the demagnetizing voltage is applied at the point C, terminal voltage becomes negative; then current flows to the source through the diode. Its area follows the curve between C and 0 in Fig. 11(b). During process, some of the stored energy in SRM are appeared as a mechanical power;.

During the energy conversion, the ratio of supply and recovered energy considerably affects to the efficiency of energy conversion. To augment the conversion efficiency, the motor must be controlled toward to increase the ratio. Lawrenson [Lawrenson,1980]] proposed the energy ratio E that explains the usage ability of the intrinsic energy.

$$W = W_{m1} + W_{m2} \tag{18}$$

$$R = W_d = W_f - W_{m2} \tag{19}$$

The energy ratio is similar to the power factor in AC machines. However, because this is more general concept, it is not sufficient to investigate the energy flowing in AC machines. The larger energy conversion ratio resulted in decreasing a reactive power, which improves efficiency of the motor. In a general SRM control method, the energy conversion ratio is approximately 0.6 - 0.7.

$$E = \frac{W}{W+R} \tag{20}$$

In conventional switching angle control for an SRM, the switching frequency is determined by the number of stator and rotor poles.

$$f_e = \frac{1}{2} p_s p_r [Hz] \tag{21}$$

The general switching angle control has three modes, i.e., flat-topped current build-up, excitation or magnetizing, and demagnetizing. Each equivalent circuit is illustrated in Fig. 12.

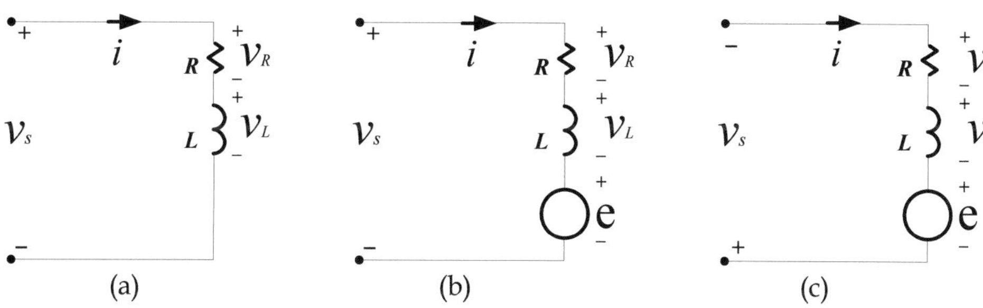

Fig. 12. Equivalent circuits when general switching angle control
(a) build-up mode (b) excitation mode (c) demagnetizing mode

Fig. 12(a) is a build-up mode for flat-topped current before inductance increasing. This mode starts at minimum inductance region. During this mode, there is no inductance variation; therefore, it can be considered as a simple RL circuit that has no back-emf. Fig. 12(b) shows an equivalent circuit at a magnetizing mode. In this mode, torque is generated from the built-up current. Most of mechanical torque is generated during this mode. A demagnetizing mode is shown in Fig. 12(c). During this mode, a negative voltage is applied to demagnetize the magnetic circuit not to generate a negative torque.

An additional freewheeling mode shown in Fig.13 is added to achieve a near unity energy conversion ratio. This is very effective under a light-load. By employing this mode, the energy stored is not returned to the source but converted to a mechanical power that is multiplication of phase current and back-emf. This means that the phase current is decreased by the back-emf.

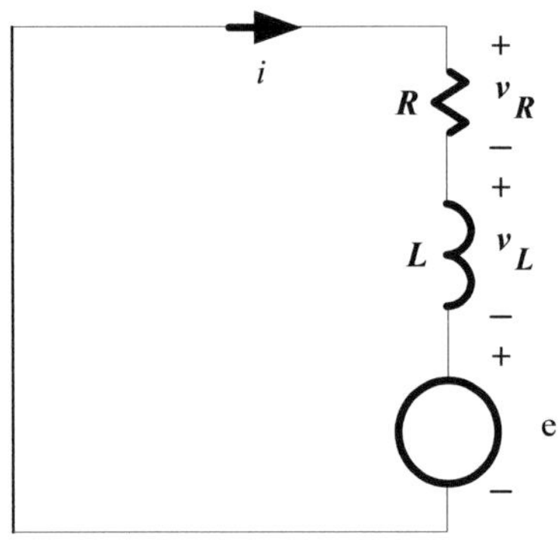

Fig. 13. Equivalent circuit of additional wheeling mode supplemented to conventional

If the increasing period of inductance is sufficiently large compared with the additional mode, the stored field energy in inductance can be entirely converted into a mechanical energy; then the energy conversion ratio becomes near unity.

1.4 Power converter for Switched Reluctance Motor

The selection of converter topology for a certain application is an important issue. Basically, the SRM converter has some requirements, such as:

- Each phase of the SR motor should be able to conduct independently of the other phases. It means that one phase has at least one switch for motor operation.
- The converter should be able to demagnetize the phase before it steps into the regenerating region. If the machine is operating as a motor, it should be able to excite the phase before it enters the generating region.

In order to improve the performance, such as higher efficiency, faster excitation time, fast demagnetization, high power, fault tolerance etc., the converter must satisfy some additional requirements. Some of these requirements are listed below.

Additional Requirements:

- The converter should be able to allow phase overlap control.
- The converter should be able to utilize the demagnetization energy from the outgoing phase in a useful way by either feeding it back to the source (DC-link capacitor) or using it in the incoming phase.
- In order to make the commutation period small the converter should generate a sufficiently high negative voltage for the outgoing phase to reduce demagnetization time.
- The converter should be able freewheel during the chopping period to reduce the switching frequency. So the switching loss and hysteresis loss may be reduced.
- The converter should be able to support high positive excitation voltage for building up a higher phase current, which may improve the output power of motor.
- The converter should have resonant circuit to apply zero-voltage or zero-current switching for reducing switching loss.

1.4.1 Basic Components of SR Converter

The block diagram of a conventional SRM converter is shown in Fig. 14. It can be divided into: utility, AC/DC converter, capacitor network, DC/DC power converter and SR motor.

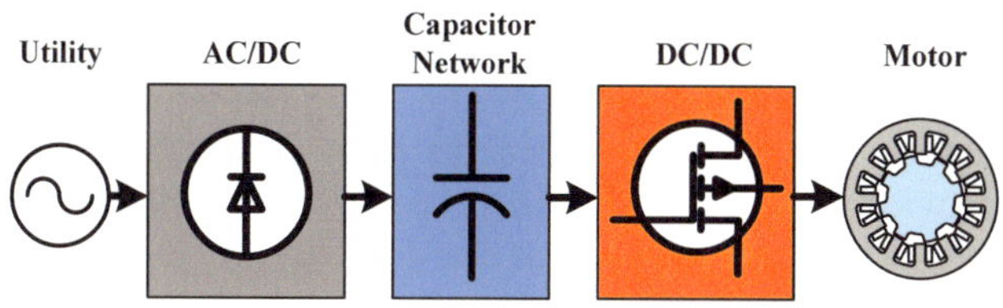

Fig. 14. Component block diagram of conventional SR drive

The converter for SRM drive is regarded as three parts: the utility interface, the front-end circuit and the power converter as shown in Fig. 15. The front-end and the power converter are called as SR converter.

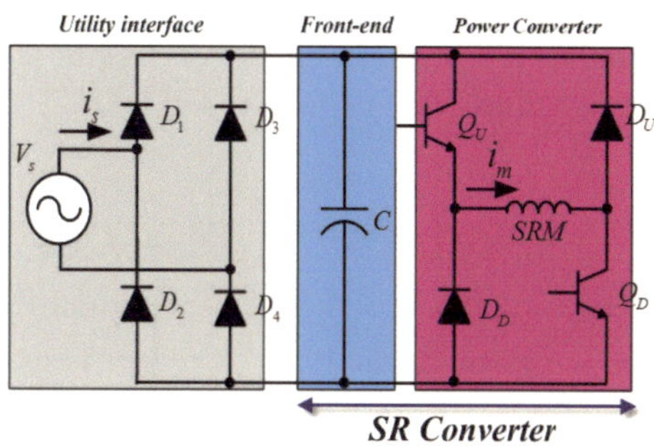

Fig. 15. Modules of SR Drive

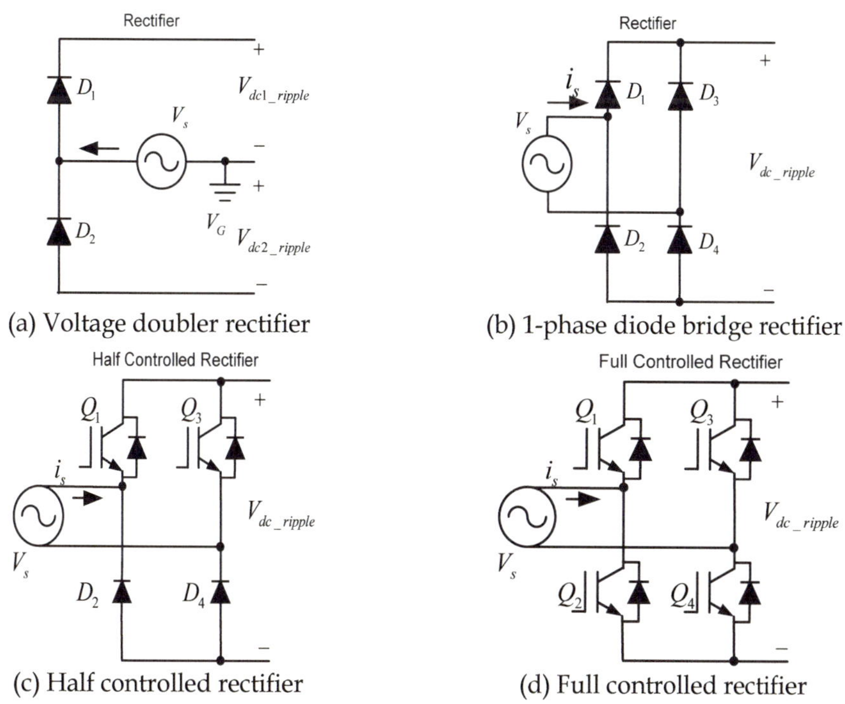

Fig. 16. Utility interface

A. Utility Interface

The main function of utility interface is to rectify AC to DC voltage. The line current input from the source needs to be sinusoidal and in phase with the AC source voltage. The AC/DC rectifier provides the DC bus for DC/DC converter. The basic, the voltage doubler and the diode bridge rectifier are popular for use in SR drives.

B. Front-end circuit

Due to the high voltage ripple of rectifier output, a large capacitor is connected as a filter on the DC-link side in the voltage source power converter. This capacitor gets charged to a value close to the peak of the AC input voltage. As a result, the voltage ripple is reduced to an acceptable valve, if the smoothing capacitor is big enough. However, during heavy load conditions, a higher voltage ripple appears with two times the line frequency. For the SR drive, another important function is that the capacitor should store the circulating energy when the phase winding returned to.

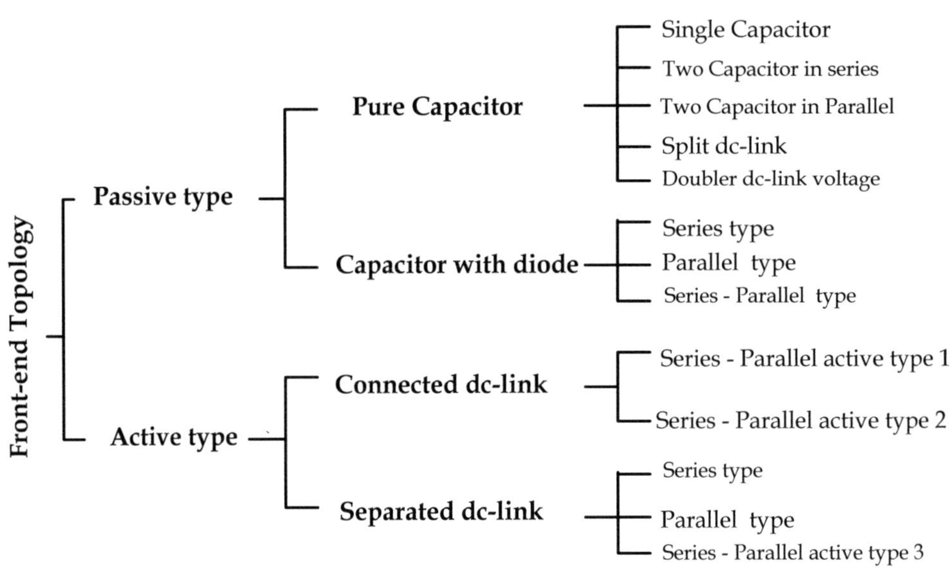

Fig. 17. Classification of capacitive type front-end topology

To improve performance of the SR drive, one or more power components are added. In this discussion, two capacitors networks are considered and no inductance in the front-end for reasonable implementation. Two types of capacitor network are introduced below: a two capacitors network with diodes and two capacitors with an active switch. The maximum boost voltage reaches two times the DC-link voltage.

The two capacitors network with diodes, which is a passive type circuit, is shown in Fig. 19. The output voltages of the series and parallel type front-ends are not controlled. Detailed characteristics are analyzed in Table 1.

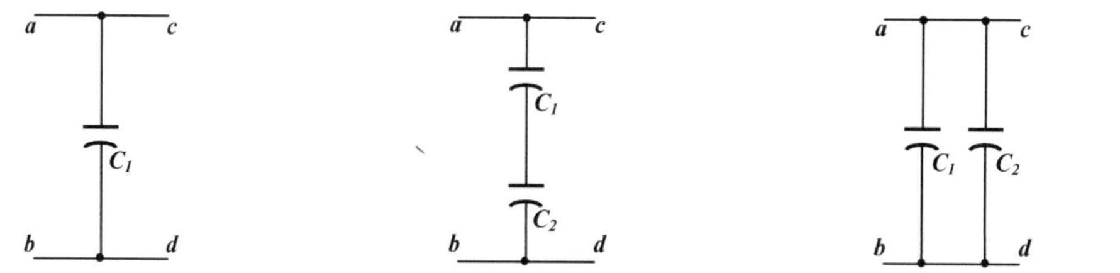

(a) Single cap. (b) Two cap. in series (c) Two cap. in parallel

(d) Split dc-link (e) Doublers dc-link voltage

Fig. 18. Pure capacitor network

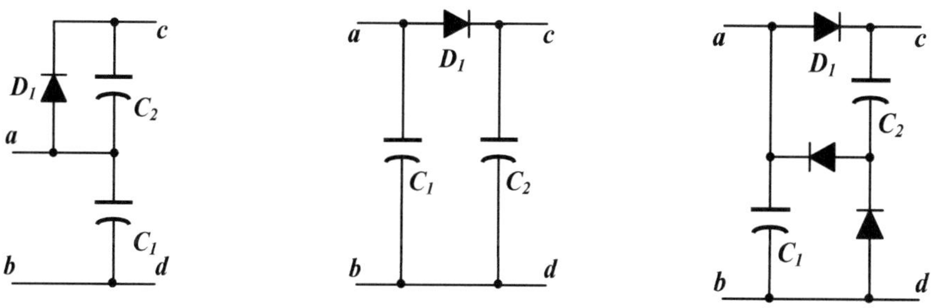

(a) Series type (b) Parallel type c) Series-parallel type

Fig. 19. Two capacitors network with diodes

Type	Series	Parallel	Series-parallel
No. of Capacitor	2	2	2
No. of Diode	1	1	3
V_{boost}	$V_{C1}+V_{C2}$	V_{C2}	$V_{C1}+V_{C2}$
V_{dc}	V_{DC}	V_{DC}	V_{DC}
Spec. Boost Capacitor	V_{DC}	V_{boost}	V_{DC}
Spec. Diode	V_{DC}	V_{DC}	V_{DC}

Table 1. Characteristics of two capacitor network with diodes

The active type of the two capacitors network connected to the DC-link, which is a two output terminal active boost circuit, is shown in Fig. 20 and Table 2.

(a) Series-parallel active type 1 (b) Series-parallel active type 2

Fig. 20. Active type of two capacitors network connected to DC-link

Type	Series-parallel 1	Series-parallel 2
No. of Capacitor	2	2
No. of Switch	1	1
No. of Diode	2	3
V_{boost}	$V_{C1}+V_{C2}$	V_{C2}
V_{demag}	$-(V_{C1}+V_{C2})$	$-(V_{C1}+V_{C2})$
Dc-link	V_{DC}	V_{DC}
Spec. Boost Capacitor	V_{DC}	V_{boost}
Spec. Diode	V_{DC}	V_{DC}

Table 2. Characteristics of active type of two capacitors connected to DC-link

The active type of two capacitors network separated to DC-link is shown in Fig. 21 and Table 3.

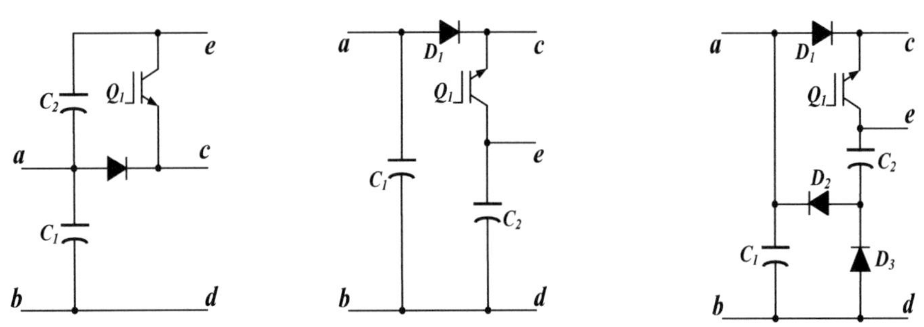

(a) Series type (b) Parallel type (c) Series-parallel active type3

Fig. 21. Active type of two capacitors network separated to DC-link

Type	Series	Parallel	Series-parallel type 3
No. of Capacitor	2	2	2
No. of Switch	1	1	1
No. of Diode	1	1	3
V_{boost}	$V_{C1}+V_{C2}$	V_{C2}	V_{C2}
V_{demag}	$-(V_{C1}+V_{C2})$	$-V_{C2}$	$-(V_{C1}+V_{C2})$
V_{dc}	V_{DC}	V_{DC}	V_{DC}
Spec. Capacitor	V_{DC}	V_{boost}	V_{C2}
Spec. Diode	V_{DC}	V_{DC}	V_{C2}

Table 3. Characteristics of active type of two capacitors separated to DC-link

C. Power converter

The power circuit topology is shown in Fig. 22 and Table 4. In this figure, five types of DC-DC converter are shown.

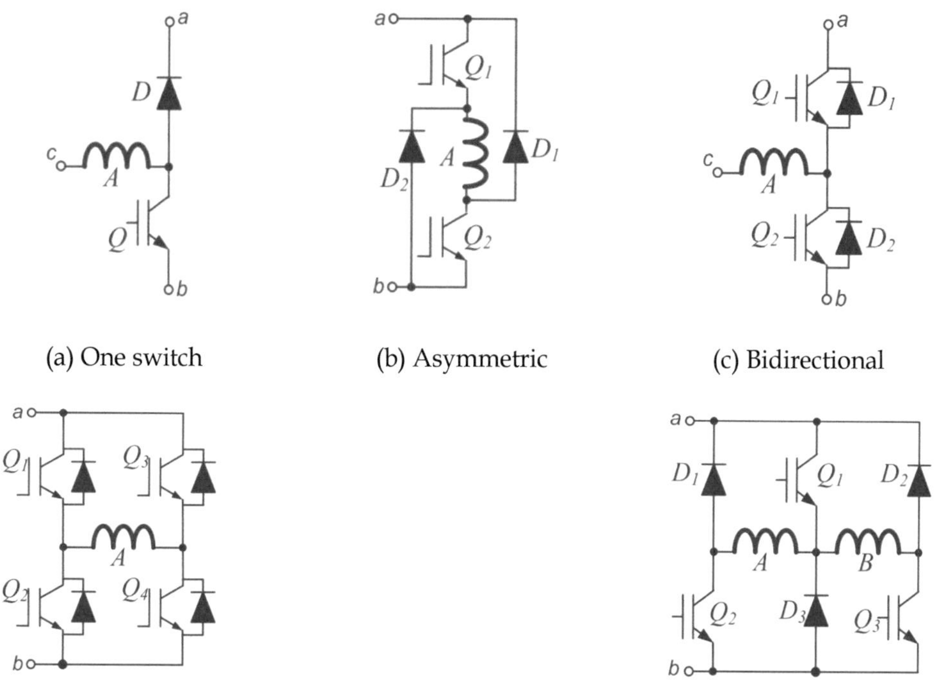

(a) One switch (b) Asymmetric (c) Bidirectional

(d) Full bridge (e) Shared switch

Fig. 22. Active type of two capacitors network separated to DC-link

Type	One switch	Asymetric	Bi-directional	Full bridge	Shared switch
No. Switch	1	2	2	4	3
No. Diode.	1	2	0	0	3
No. Phase	1	1	1	1	2
$V_{Excitation}$	V_{dc}	V_{dc}	$V_{dc}/2$	V_{dc}	V_{dc}
$V_{Demagnetitation}$	V_{dc}	V_{dc}	$V_{dc}/2$	V_{dc}	V_{dc}
Current Direction	Uni.	Uni.	Bi.	Uni.	Uni.

Table 4. Comparison of 5 types of DC/DC converter topology

1.4.2 Classification of SR converter

One of the well-known classifications of SRM converters only considering the number of power switches and diodes is introduced [miller,1990]. Different from the classification, a novel classification, which focuses on the characteristics of converters, is proposed in [Krishnan,2001].

A. SR converter by phase switch

The classification of power converter focuses on the number of power switches and diodes. These options have given way to power converter topologies with q, (q+1), 1.5q, and 2q switch topologies, where q is the number of motor phases. These configurations are classified and listed in Fig. 23 for easy reference. A two-stage power converter configuration which does not fit this categorization based on the number of machine phases is also included.

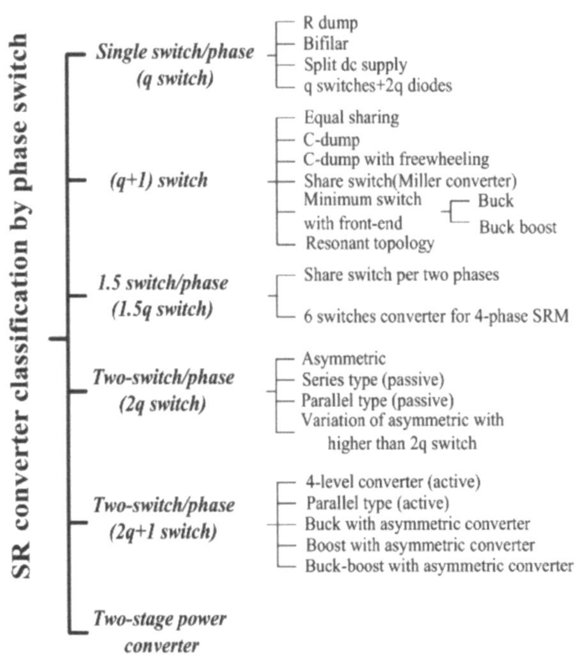

Fig. 23. SR converter classification by phase switch

All the converter topologies, except the two-stage power converter, assume that a DC voltage source is available for their inputs. This DC source may be from batteries or most usually a rectified AC supply with a filter to provide a stable DC input voltage to the SR converters.

Even though it is easy with the classification to find the number of semiconductors and the cost by counting the number of active components, it does not show important characteristics of a power converter, and the voltage ratings for the power switches and diodes are difficult to consider.

B. SR converter by commutation

Different converters which have the same number of switches may obtain different performance and characteristics. From this point of view, such a classification is not useful for finding the characteristic of an SRM converter.

The three types in the classification were presented as: extra commutation, half bridge and self commutation. In the extra commutation circuit, the capacitive, the magnetic and dissipative circuit is included. However, the distinction between three types in the classification is not clearly defined. Conventionally, the half bridge and the self commutation circuit also need a large capacitor in the front-end. They could also be classified as capacitive circuit. Moreover, the characteristic of circuits which contain one or more inductances is not shown in the classification.

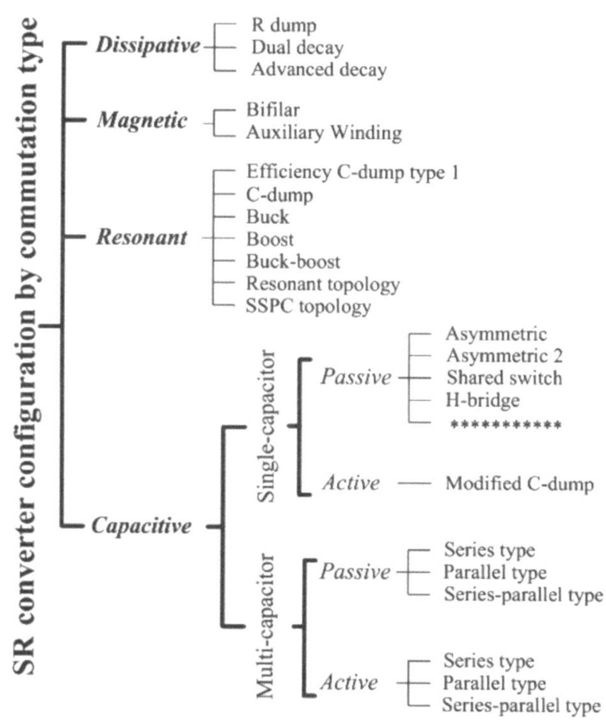

Fig. 24. SR converter configuration by commutation type

An SR converter configuration by commutation type is shows in Fig. 24. Based on the commutation type of the most of the returned or dissipated stored magnetic energy, the four major sorts are classified: dissipative, magnetic, resonant and capacitive type. Because the capacitive type is focused in this discussion, the capacitive converter category is split into several subclasses. The concepts for passive and active converters are introduced. The distinction between active and passive is determined by whether they include a controllable power switch or not.

1. Dissipative converter

The dissipative type dissipates some or all of the stored magnetic energy using a phase resistor, an external resistor or both of them. The remaining energy is transformed to mechanical energy. Therefore, none of the stored magnetic energy in the phase winding is returned to DC-link capacitor or source. The advantage of this type of converter is that it is simple; a low cost and has a low count of semiconductor components.

(a) R-dump (b) Zener-dump

Fig. 25. Two types of dissipative SR converter

2. Magnetic converter

The magnetic type is where the stored magnetic energy is transferred to a closely coupled second winding. Of course, that energy could be stored in DC-link capacitor or used to energize the incoming phase for multi-phase motors or use special auxiliary winding. The major advantage is a simple topology. The one switch per phase power circuit can be used. However, the potential rate of change of current is very high due to the stored magnetic energy is recovered by a magnetic manner. And the coupled magnetic phase winding which should be manufactured increases the weight of copper and cost of motor. Moreover, the power density of the motor is lower than that of the conventional ones.

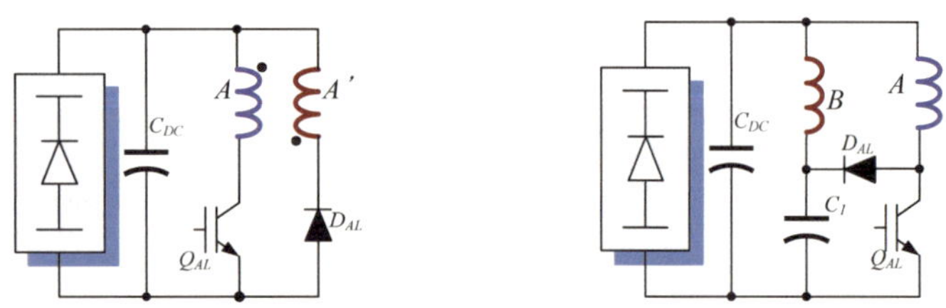

(a) Bifilar (b) Single controllable switch

Fig. 26. Two types of magnetic SR converter

3. Resonant converter

The resonant type has one or more external inductances for buck, boost or resonant purposes. Conventionally, the inductance, the diode and the power switch are designed as a snubber circuit. So, the dump voltage can be easily controlled, and the low voltage is easy to boost. In a special case, an inductance is used to construct a resonant converter. The major advantage is that the voltage of phase winding can be regulated by a snubber circuit. However, adding an inductance increases the size and cost of converter. The other additional components also increase the cost of converter. Three types of resonant type are shown in Fig. 27. All of them use a snubber circuit, which is composed by a power switch, a diode and an inductance.

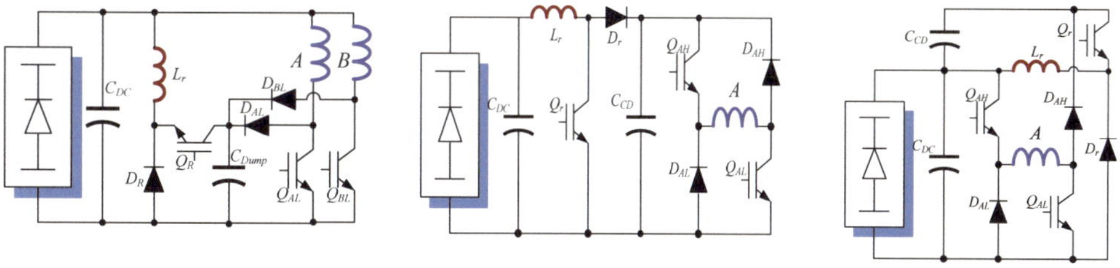

(a) C-Dump (b) Boost (c) High demagnetization

Fig. 27. Three types of resonant SR converter

4. Capacitive converter

The magnetic energy in the capacitive converters is fed directly back to the boost capacitor, the DC-link capacitor or both of the capacitors. Compared to the dissipative, magnetic, and resonant converters, one component is added in the main circuit. So, this component will increase the loss of the converter. Different from the other converters, the stored magnetic energy can easily be fed back using only the inductance of phase winding. Although the capacitor has an equivalent series resistance (ESR), the loss of ESR is lower than that of other converters. Therefore, the capacitive converter is more effective for use in SR drive.

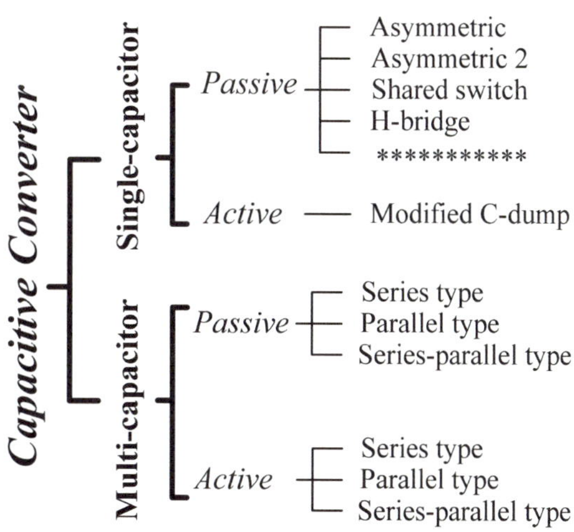

Fig. 28. Classification of capacitive SR converter

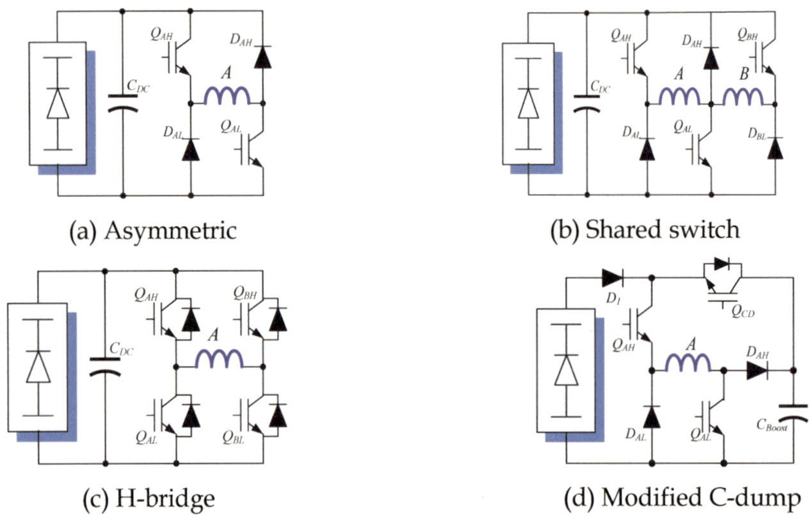

Fig. 29. Single capacitor type in capacitive SR converter

The capacitive converter can be divided two sorts: single capacitor and multi-capacitor type.

i. Single-capacitor converter

Single-capacitor converters have simple structure, which makes them very popular. Four single capacitor types are shown in Fig. 29. One capacitive converter has as a simple front-end as shown in Fig. 29(a)-(c). This capacitor should be large enough to remove the voltage ripple of the rectifier and store the magnetic energy. Since the DC-link capacitor voltage is uncontrollable during charging and discharging, this type of converter is defined as a passive converter. The modified C-dump converter is shown in Fig. 29(d). In this converter, the boost capacitor only stores the recovered energy to build up a boost voltage. Unfortunately, one power switch should be placed in front of the boost capacitor to control the voltage. Because the boost capacitor does not reduce the DC-link voltage from the rectifier, the fluctuating DC-link voltage is input directly to the phase winding. The boost capacitor has only to be big enough for the stored magnetic energy, so the size of this capacitor is smaller than that of conventional DC-link capacitor. The Single capacitor in capacitive converters simplifies the construction of the converter. However, the input voltage for the phase winding is kept fixed by the DC-link capacitor. If only a boost capacitor is used, the DC-link voltage is fluctuating, and one power switch is added to control the boost voltage. This extra switch may increase the cost of converter.

ii. Multi-capacitor converter

Multi-capacitor converters include two or more capacitors in the converter topology to obtain boost voltage. Extra capacitors may make the topology of converter more complex. In this discussion, different converter topologies, which include two capacitors, are considered. The different types of passive type front-ends are shown in Fig. 30. The passive converter with two capacitors in parallel type is in Fig. 30(a). Due to the direction of diode, the stored magnetic energy is only feed back to the boost capacitor. The maximum boost voltage can be obtained by a suitable size of the capacitor. Because the discharge of the boost capacitor is not controllable in the passive converter, the voltage of the boost capacitor is changed by the stored magnetic energy during different operating condition. When the phase switch is turned on, the voltage of the boost capacitor may fall very fast until the voltage reaches the DC-link voltage. Due to the non-linear characteristic of the SR motor, it is difficult to estimate advance angle or turn-on angle.

A passive converter with two capacitors in series is shown in Fig. 30(b). The stored magnetic energy charges the two capacitors in series. So, a part of the energy is stored in the boost capacitor to build up a boost voltage. It has the same advantage as for the parallel passive converter. However, the voltage rating of the boost capacitor is less than that of the parallel converter.

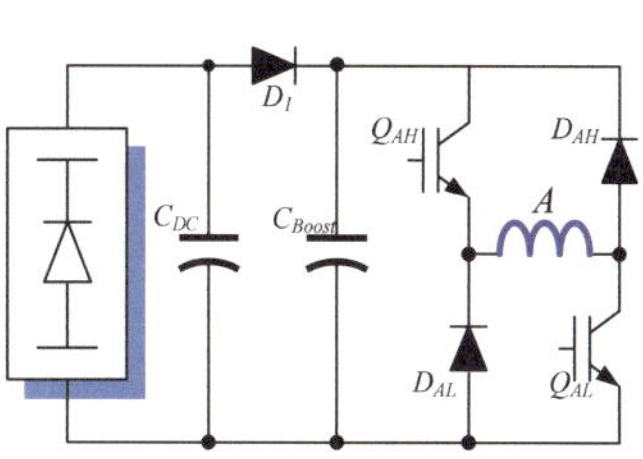

(a) Parallel type

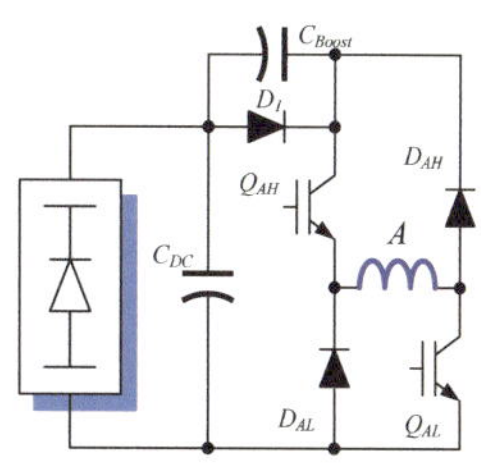

(b) Series type

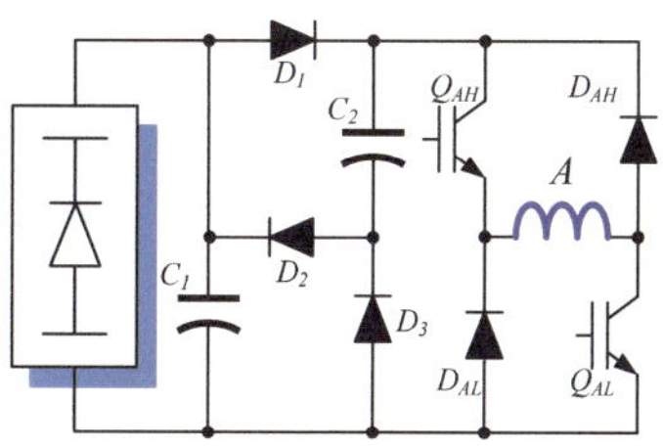

(c) Series-parallel type

Fig. 30. Passive boost converter with two capacitors

Another passive converter of two capacitors in series-parallel type is in Fig. 30(c). This converter is made of rectifier, the passive boost circuit and an asymmetric converter. The excitation voltage is the DC-link voltage, but the demagnetization voltage is twice of DC-link voltage. The high demagnetization voltage can reduce the tail current and negative torque; it could also extend the dwell angle to increase the output.

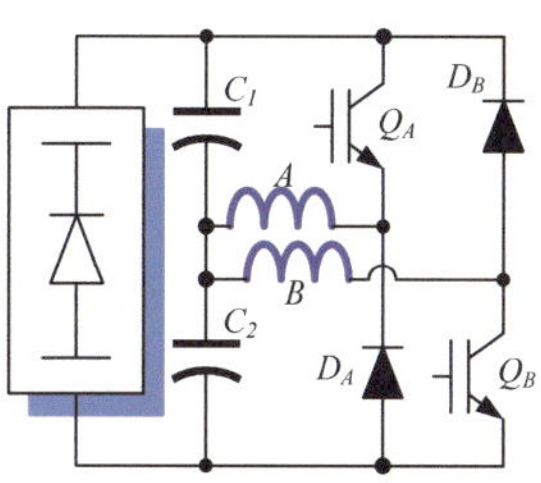

(a) Split dc-link type

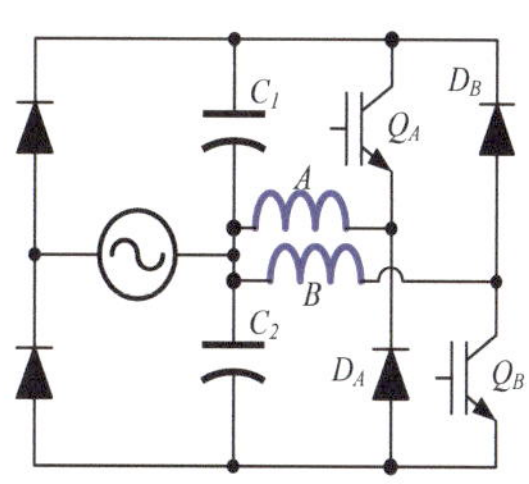

(b) Doublers dc-link voltage type

Fig. 31. other passive SR converter with series capacitor type

Other passive SR converter with two series capacitors is shown in Fig. 31. The front-end and DC-DC converter are same, but the bridge rectifier and the voltage doubling rectifier are connected. The split DC-link converter is shown in Fig. 31(a). The phase voltage of this converter is a half of DC-link voltage. The double dc-link voltage converter is shown in Fig. 31(b). The phase voltage is same to DC-link voltage. The main advantage of these two converters is that one switch and one diode per phase is used. However, the voltage rating of power switch and diode is the twice the input excitation voltage.

The active boost converter with two capacitors connected in parallel is shown in Fig. 32. The four active boost converters with two capacitors connected in parallel are introduced. To handle the charging of the capacitor in the beginning of the conduction period, one diode is needed to series or parallel with the power switch to protect the power switch. When parallel type 1 and 2 are used with the asymmetric converter, the maximum voltage rating of the power diode and the switch is the same as the desired boost voltage. While the diode is connected to the power switch, the boost capacitor is only charged by the stored magnetic energy. In the beginning, the voltage of the boost capacitor is increased from 0 to the desired value. For the parallel converter of type 2, a diode in parallel with the power switch is used, so the boost capacitor can be charged by the DC-link capacitor. Parallel converters of type 3 and 4 which belong to capacitor dump converters are shown in Fig. 32(c) and (d). If the demagnetization voltage is required to be the same to DC-link, the voltage rating of power diode and switch is at least twice of DC-link voltage.

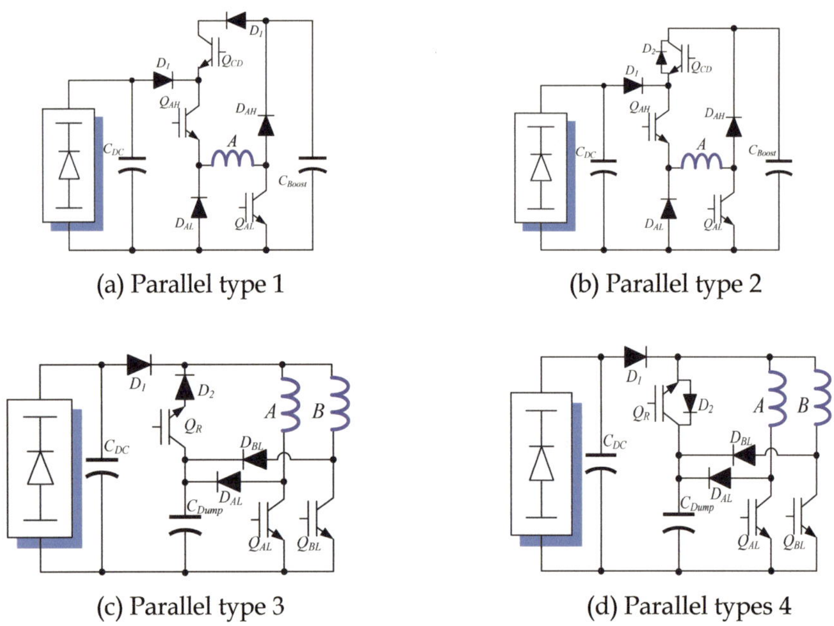

Fig. 32. Active boost converter with two capacitors connected in parallel

An active boost converter with two series connected capacitors is in Fig. 33(a). The stored magnetic energy charges the two series connected capacitors, so the boost voltage can be built up in the boost capacitor. The power switch Q_{cd} is used to control the boost voltage of the boost capacitor.

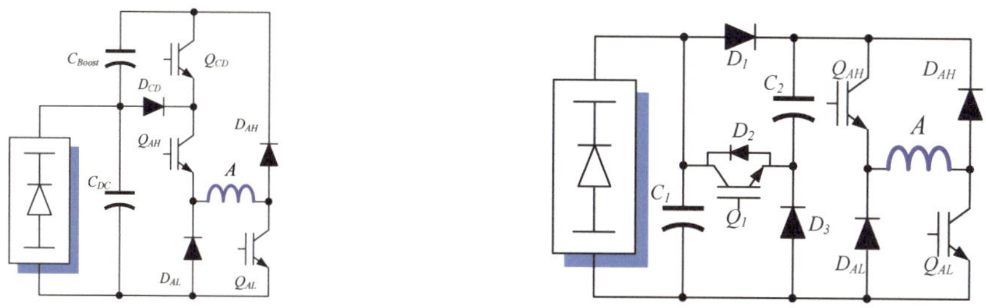

(a) Series capacitor type (b) Series-parallel capacitor type

Fig. 33. Active boost converter

An active boost converter with a series-parallel connection of the two capacitors is shown in Fig. 33(b). The active capacitor circuit added to the front-end consists of three diodes and one capacitor. This circuit combines a series-connected and a parallel-connected structure of two capacitors. Based on this active boost capacitor network, the two capacitors can be connected in series or parallel during different modes of operation. The operation mode of whole converter is presented in [Khrishnan,2001]. The fast excitation and demagnetization is easily obtained from the two series-connected capacitors. The stable voltage achieved with the two parallel-connected capacitors.

4 types of converter are compared in Table. 5. The converter with two capacitors connected in series or the converter with two capacitors connected in parallel may obtain a higher boost voltage than the series-parallel converter. However, an increased boost voltage may increase the cost of the converter. Since the series-parallel converter can limit the maximum voltage to twice the DC-link voltage, it is more stable and controllable.

	Asymmetric	2-capacitor in series type	2-capacitor in parallel type	2-capacitor in series-parallel
Vmax	Vdc	∞/2Vdc	∞/2Vdc	2Vdc
Vcontrol	No	Yes	Yes	optional
VC1_rate	Vdc	Vdc	Vdc	Vdc
VC2_rate	Vdc	∞/Vdc	∞/2Vdc	Vdc
No.Switch	2	3	3	3
No. Diode	2	3	3	4
Stability	Good	Normal	Normal	Good

Table 5. Comparison of 2-capacitor types

2. Torque control strategy

2.1 Angle control method

The switched reluctance drive is known to provide good adjustable speed characteristics with high efficiency. However, higher torque ripple and lack of the precise speed control are drawbacks of this machine. These problems lie in the fact that SR drive is not operated with an mmf current specified for dwell angle and input voltage. To have precise speed control with a high efficiency drive, SR drive has to control the dwell angle and input voltage instantaneously. The advance angle in the dwell angle control is adjusted to have high efficiency drive through efficiency test.

2.1.1 Switching angle control method

In SRM drive, it is important to synchronize the stator phase excitation with the rotor position; therefore, the information about rotor position is an essential for the proper switching operation. By synchronizing the appropriate rotor position with the exiting current in one phase; the optimal efficiency of SRM can be achieved. In this part various types of switching angle control method to achieve the optimal efficiency will be discussed.

A. Fixed angle switching method

Current source is a proper type to excite an SRM for its good feature of electromagnetic characteristics because it produces rectangular or flat-topped current and it is easy to control the torque production period. Therefore, it is considered as an ideal excitation method for switched reluctance machine but difficult and expensive to realize it.
To produce similar current shapes in voltage source, it is needed to regulate the supply voltage in the variable reluctance conditions. Usually PWM or chopper technique is used for this propose. But it is complex in its control circuit and increases loss. The other technique which is more simply in control is excitation voltage to form a flat-topped current by using fixed switching angle at various operation conditions. Fig. 34 shows excitation scheme with fixed switching angle control method.

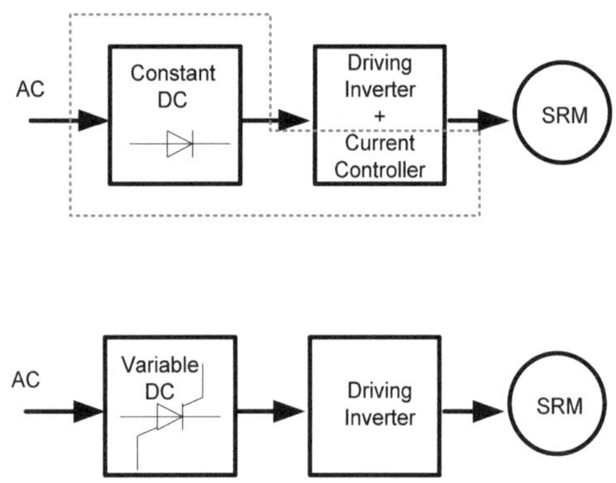

Fig. 34. Excitation scheme with fixed switching angle control method

In the fixed angle switching method, the turn-on angle and the turn-off angle of the main switches in the power converter are fixed; the triggering signals of the main switches are modulated by the PWM signal. The average voltage of phase winding could be adjusted by regulating the duty ratio of the PWM signal. So the output torque and the rotor speed of the motor are adjustable by regulating the phase winding average voltage.
Constant voltage source with current controller is substituted with variable voltage source to make the current flat-topped. Voltage equation of SRM for a phase is shown in (3). If winding resistance and magnetic saturation are ignored, an applied voltage to form a flat-topped current in the torque developed region is

$$V_c = KI_c\omega \tag{22}$$

Where V_c is amplitude of voltage, K is $dL/d\theta$, I_c is required current to balance load torque a, and ω is angular velocity. If magnetic saturation is considered, this equation is to be modified as

$$V' = \sigma K I_c \omega \tag{23}$$

Where σ is saturation factor. To calculate proper excitation voltage and switching angle for flat-topped current, let consider phase voltage and current as shown in Fig. 35. θ_{on} and θ_{off} are switching-on and switching-off angle, respectively. Phase current reaches to the desired value of current, I_c at θ_s, and become flat-topped current by this scheme, and the current decrease rapidly by reversing the applied voltage. θ_{off} is to be set in order to prevent the generation of negative torque. It can be divided into 3 regions to calculate the angles and voltages. In Region I and III, switching-on and switching-off angles are determined respectively. And in Region II, proper excitation voltage is calculated.

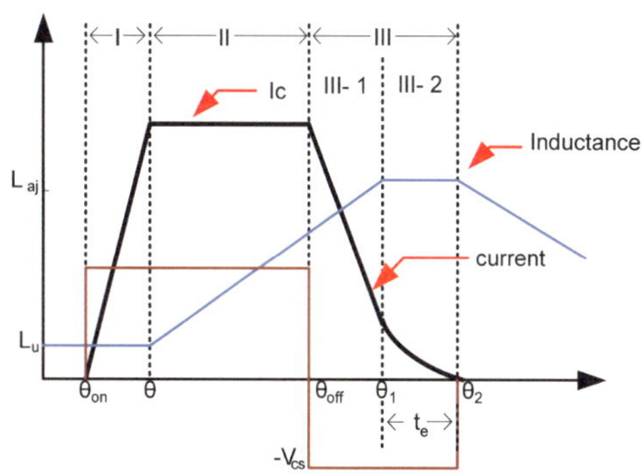

Fig. 35. Flat-topped phase current

- Region I : $\theta_{on} \le \theta \le \theta_s$ (switching-on angle determination)

θ_{on} is determined in this region. It is to ensure that current is to be settled to the desired value at θ_s. In this region, voltage equation becomes (24).

$$V_{cs} = Ri + L_u \frac{di}{dt} \tag{24}$$

Where L_u, is the minimum value of the inductance.
Required time, t_s to build up a phase current from 0 to I_c, which is the current to balance load torque, is derived from (23) and (24).

$$t_s = \frac{\theta_s - \theta_{on}}{\omega} = -\frac{L_u}{R} \ln(1 - \frac{R}{\sigma K \omega}) \tag{25}$$

Therefore, θ_{on} is

$$\theta_{on} = \theta_s - \frac{\omega L_u}{R} \ln(1 - \frac{R}{\sigma K \omega}) \tag{26}$$

θ_{on} is affected merely by saturation factor and not by speed variation except the range where speed is very low. Therefore, it can be fixed at the center of variation range of switching-on and compensate current build-up via applied voltage regulation for simple control.

- Region III : $\theta_{off} \le 0 \le \theta_t$ (Switching-off angle determination)

In this region, applied voltage must be reversed to accelerate current decay. It is divided into two sub-regions:

- Sub-region III-1
  Voltage and current equation are as follows.

$$-V_{cs} = Ri + L\frac{di}{dt} + \sigma K \omega i \tag{27}$$

$$i = I_c \left(2e^{-\frac{\sigma K\omega}{L}t} - 1\right) \tag{28}$$

  These equations are effective only during $\theta_{off} \leq \theta \leq \theta_1$

- Sub-region III-2
  In this region, the inductance has its maximum value L_{aj} and is constant. So, current is

$$i = -\frac{V_{cs}}{L_{aj}} t + I_0 \tag{29}$$

  Where I_0 is the current value at θ_1. This equation is effective during $\theta_1 \leq \theta \leq \theta_2$.

B. Advance angle control method

The SRM is controlled by input voltage, switch-on and switch-off angle. Switch-on and switch-off angle regulate the magnitude and shape of the current waveform. Also it results in affecting the magnitude and shape of the torque developed. To build up the current effectively with a voltage source, an advance switching before the poles meet is needed. The switch-on angle is one of the main factors to control the build-up currents. Therefore, this angle is controlled precisely to get optimal driving characteristics.

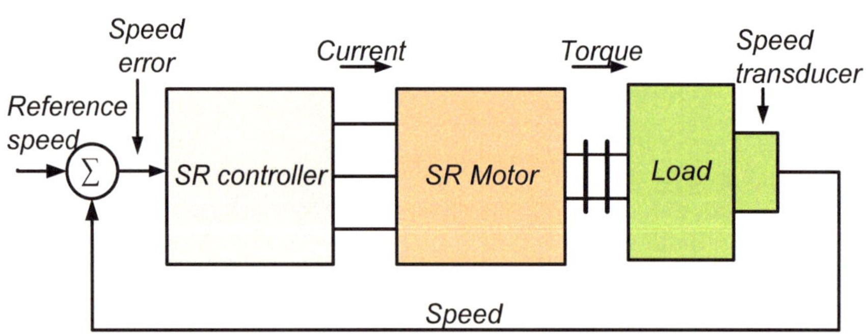

Fig. 36. Block diagram of advance angle control with feedback signal

In the real control system, control of advance angle which is controlled by variable load condition can be realized by simple feedback circuit using detecting load current. The block diagram of the advance angle control with a feedback signal shows in Fig.36.
The regulation of speed-torque characteristics of SRM drive is achieved by controlling advance angle and applied voltage. The advance angle is regulated to come up with the load variation in cooperation with the applied voltage.
The signal from the control loops is translated into individual current reference signal for each phase. The torque is controlled by regulating these currents. The feedback signal which is proportional to the phase detector is used to regulate the instantaneous applied voltage.

Variation relationship of torque with current or torque with rotor position must be compensated in the feed forward torque control algorithm. The relation between torques and current of a phase is shown in Fig. 37.

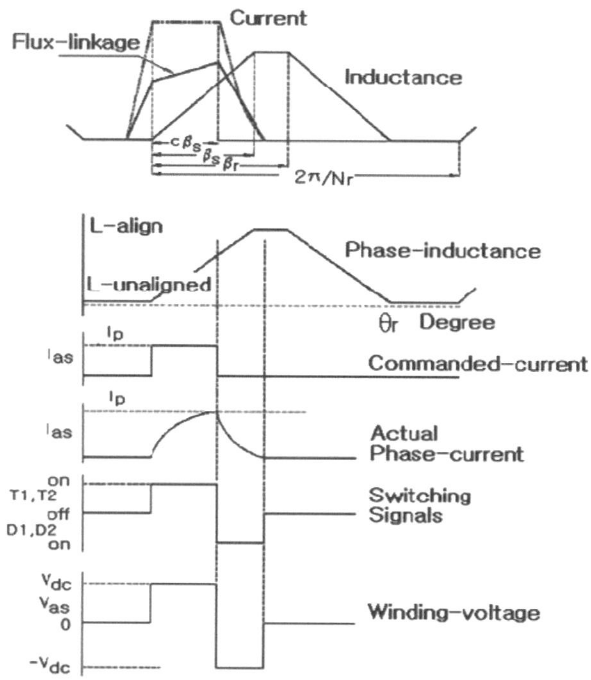

Fig. 37. Advance angle control

C. Switching-off angle control method

Control method of switch-off angle is introduced for variable load. The switching angle control method is based on two command signals for switching-on and switching-off angle independently. According to the motor speed and load condition, a proper switching-on angle θ_{on} is set at the cross point of negative slope of the sensor signal and the switching-on command signal V_{on} is

$$\theta_{on} = \left(1 - \frac{V_{on}}{V_{max}}\right)(\theta_0 - \theta_a) + \theta_a \tag{30}$$

The maximum switching-on angle is in the minimum inductance region. So, a fast build up of current is possible at the rated load. The minimum switching-on angle is in the increasing region of inductance. Therefore a smooth build up of current is possible at a light load with a smooth torque production. Similarly, the delay angle θ_{off} is set at the cross point of positive slope of the signal and the switching off command signal V_{off} as

$$\theta_{off} = \frac{V_{off}}{V_{max}}(\theta_d - \theta_0) + \theta_0 \tag{31}$$

In addition, the dwell angle is the interval of switching-on and switching-off angles, which takes the form

$$\theta_{dwell} = \theta_{off} - \theta_{on} \tag{32}$$

There are two types of control switch-off angle, one is constant torque angle (θ_{TQ}) control and the other is constant dwell angle (θ_{Dw}) control.

1. Constant torque angle control

Torque angle is the angle between the increasing of inductance to the switching-off angle. This control method is fixed the turn-off angle and turn-on angle is tuned for a fluctuation of speed and load by constant torque angle control method. The fluctuation of efficiency is small until rated power, but if the turn-on angle moves toward for an increase torque, even in the region of decreasing inductance, the current will flow and negative torque will be produced. Thus, the efficiency becomes reduced. Therefore, it is needed to find a proper position of turn-on angle and the phase current which determined by constant torque angle control method.

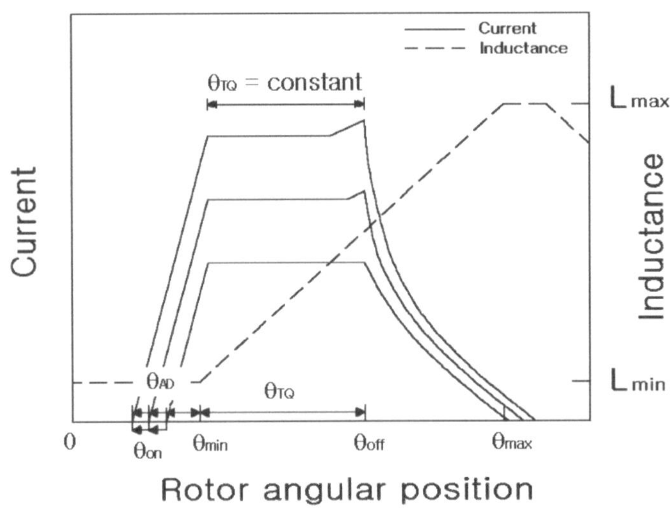

Fig. 38. Constant torque angle control

2. Constant dwell angle control

The constant dwell angle method controls the turn-on or turn-off angle by keep constant dwell angle (θ_{Dw}) for speed or output control. When turn-on angle is moved to keep the constant speed, effect of negative torque is regardless of speed and load. But because of the limits of rated power, it can be unstable to drive on overload. This method makes a control system simple and easy to avoid negative torque in the switching-off region. Fig. 39 shows the relation between current and rotor position in constant dwell angle (θ_{Dw}) control.

2.1.2 Single pulse control method

Torque production in SRM is not constant and it must be established from zero at every stroke. Each phase must be energized at the turn-on angle and switched off at the turn-off angle. In the low speed range, the torque is limited only by the current, which is regulated either by voltage-PWM or by instantaneous current. As the speed increases the back-EMF increases too, and there is insufficient voltage available to regulate the current; the torque can be controlled only by the timing of the current pulse. This control mode is called single-pulse mode.

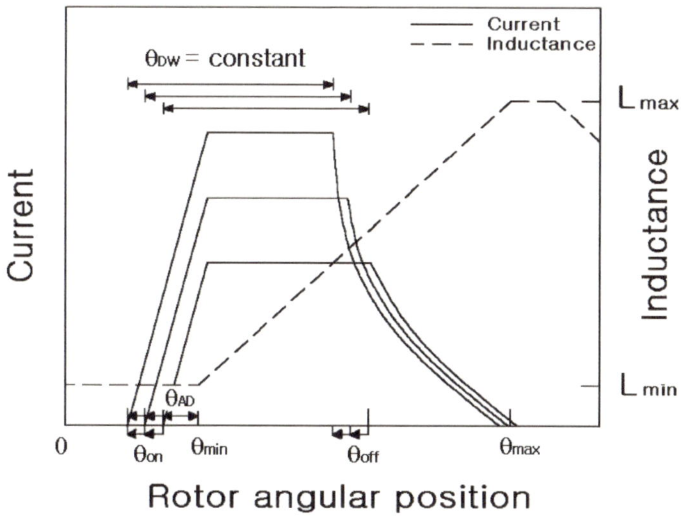

Fig. 39. Constant dwell angle (θ_{Dw}) control

In single pulse operation the power supply is kept switched on during the dwell angle and is switched off at the phase commutation angle. As there is no control of the current and a sharp increase of current, the amount of time available to get the desired current is short. Typically, single pulse operation is used at high mechanical speed with respect to the turn-on angle determined as a function of speed. Fig.40 shows the phase current in high speed region using an asymmetric converter. As shown in Fig. 40, SR drive is excited at θ_{on}

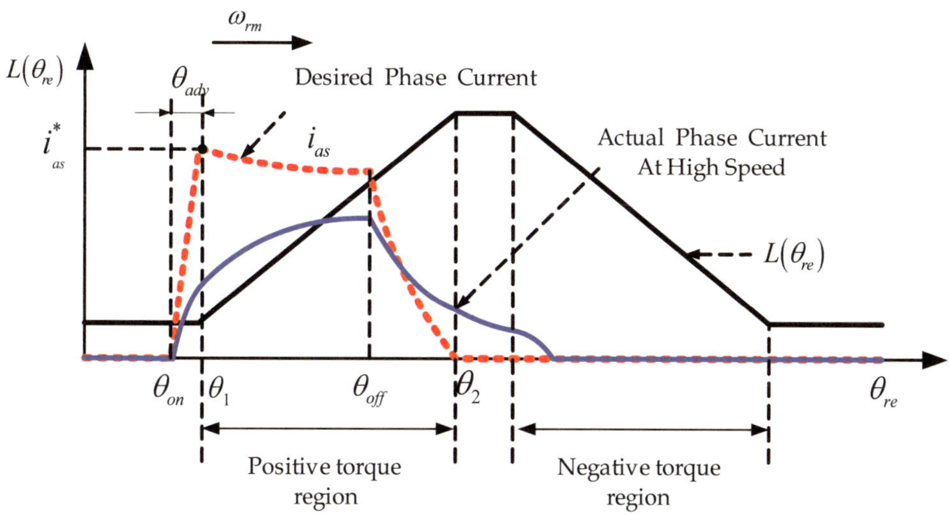

Fig. 40. Build-up of phase current in high speed region

position advanced as θ_{adv}, than the start point of positive torque region θ_1 in order to establish the sufficient torque current. The desired phase current shown as dash line in Fig. 40 is demagnetized at θ_{off} , and decreased as zero before the starting point of negative torque region θ_2 to avoid negative torque.
In order to secure enough time to build-up the desire phase current i_{as}^*, the advance angle θ_{adv} can be adjusted according to motor speed ω_m. From the voltage equations of SRM, the proper advance angle can be calculated by the current rising time as follows regardless of phase resistance at the turn-on position.

$$\Delta t = L(\theta_1).\frac{i_{abcs}^*}{V_{abcs}} \tag{33}$$

Where, i_{abcs}^* denotes the desired phase current of current controller and V_{abcs} is the terminal voltage of each phase windings. And the advance angle is determined by motor speed and (33) as follow

$$\theta_{adv} = \omega_m \,.\Delta t \tag{34}$$

As speed increase, the advance angle is to be larger and turn-on position may be advanced not to develop a negative torque. At the fixed turn-on position, the actual phase current denoted as solid line could not reach the desire value in high speed region as shown in Fig. 40. Consequently, the SRM cannot produce sufficient output torque. At the high speed region, turn-on and turn-off position are fixed and driving speed is changed. To overcome this problem, high excitation terminal voltage is required during turn-on region from θ_{on} to θ_1.

2.1.3 Dynamic angle control method

The dynamic angle control scheme is similar to power angle control in synchronous machine. When an SRM is driven in a steady-state condition, traces such as shown in Fig. 41(a) are produced. The switch-off instant is fixed at a preset rotor position. This may readily be done by a shaft mounted encoder. If the load is decreased, the motor is accelerated almost instantaneously. The pulse signal from a rotor encoder is advanced by this acceleration. This effect will reduce switch-off interval until the load torque and the developed torque balances [Ahn,1995]. Fig. 41(b) shows this action. On the contrary, if load is increased, the rotor will be decelerated and the switch-off instant will be delayed. The effect results in increasing the developed torque. Fig. 41(c) shows the regulating process of the dwell angle at this moment.
The principle of dynamic dwell angle is similar to PLL control. The function of the PLL in this control is to adjust the dwell angle for precise speed control. The phase detector in the PLL loop detects load variation and regulates the dwell angle by compares a reference signal (input) with a feedback signal (output) and locks its phase difference to be constant. Fig. 42 shows the block diagram of PLL in SR drive. It has a phase comparator, loop filter, and SRM drive.
The reference signal is a speed command and used for the switch-on signal. The output of the phase detector is used to control voltage through the loop filter. The switching inverter regulates switching angles. The output of phase detector is made by phase difference between reference signal and the signal of rotor encoder. It is affected by load variations. The dwell angle is similar to phase difference in a phase detector. To apply dynamic angle

control in an SR drive system, a reference frequency signals are used to switch-on, and the rotor encoder signal is used to switch-off similar to the function of a phase detector. The switch-off angle is fixed by the position of the rotor encoder. Therefore, the rotor encoder signal is delayed as load torque increased. This result is an increase of advance angle and initial phase current.

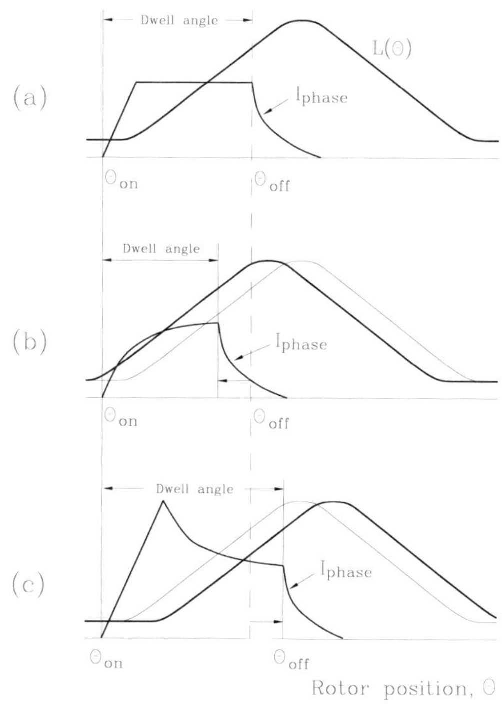

Fig. 41. Regulation of dwell angle according to load variation.
(a) steady-state. (b) load decreased. (c) load increased.

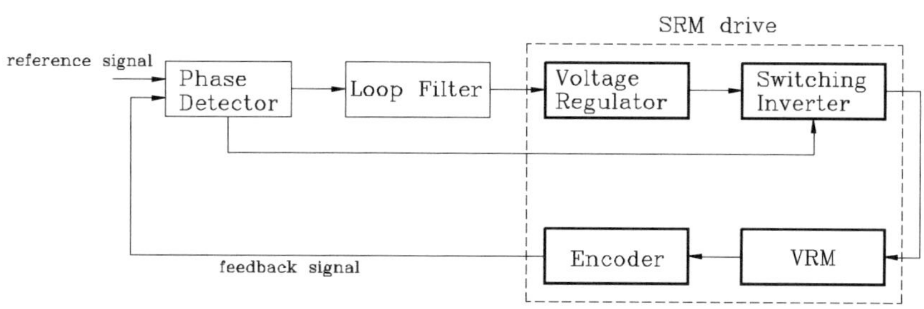

Fig. 42. Block diagram of PLL in SR drive.

2.2 Current control method

Control of the switched reluctance motor can be done in different ways. One of them is by using current control method. The current control method is normally used to control the torque efficiently. Voltage control has no limitation of the current as the current sensor is avoided, which makes it applicable in low-cost systems. Due to the development of microcontrollers, the different control loops have changed from analog to digital implementation, which allows more advanced control features. However, problems are still raised when designing high-performance current loop [miller,1990].

The main idea of current control method is timing and width of the voltage pulses. Two methods are too used in the current control, one is voltage chopping control method, and the other is hysteresis control method.

2.2.1 Voltage chopping control method

The voltage chopping control method compares a control signal $V_{control}$ (constant or slowly varying in time) with a repetitive switching-frequency triangular waveform or Pulse Width Modulation (PWM) in order to generate the switching signals. Controlling the switch duty ratios in this way allowed the average dc voltage output to be controlled. In order to have a fast built-up of the excitation current, high switching voltage is required. Fig. 43 shows an asymmetric bridge converter for SR drive. The asymmetric bridge converter is very popular for SR drives, consists of two power switches and two diodes per phase. This type of the SR drive can support independent control of each phase and handle phase overlap. The asymmetric converter has three modes, which are defined as magnetization mode, freewheeling mode, and demagnetization mode as shown in Fig. 44.

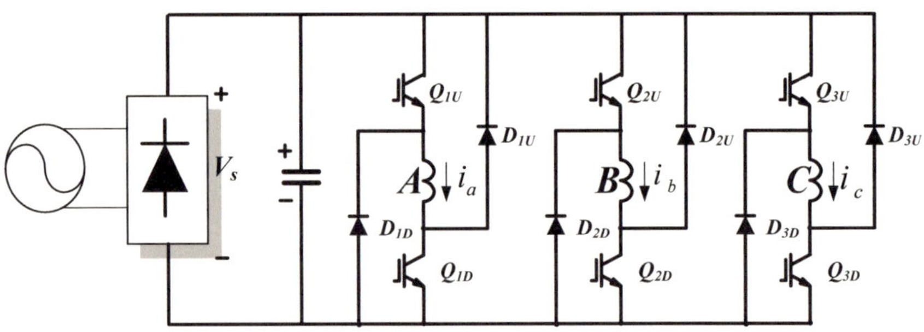

Fig. 43. Asymmetric bridge converter for SR drive

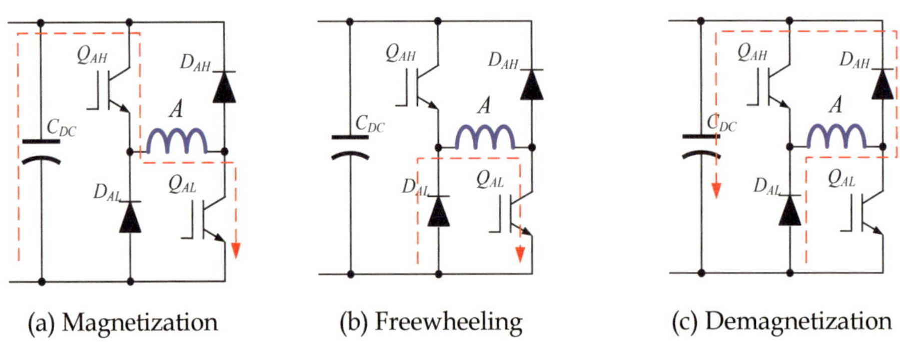

Fig. 44. Operation modes of asymmetric converter

From Fig. 44 (a) and (c), it is clear that amplitudes of the excitation and demagnetization voltage are close to terminal voltage of the filter capacitor. The fixed DC-link voltage limits the performance of the SR drive in the high speed application. On the other hand, the voltage chopping method is useful for controlling the current at low speeds. This PWM strategy works with a fixed chopping frequency. The chopping voltage method can be separated into two modes: the hard chopping and the soft chopping method. In the hard chopping method both phase transistors are driven by the same pulsed signal: the two transistors are switched on and switched off at the same time. The power electronics board is then easier to design and is relatively cheap as it handles only three pulsed signals. A disadvantage of the hard chopping operation is that it increases the current ripple by a large factor. The soft chopping strategy allows not only control of the current but a minimization of the current ripple as well. In this soft chopping mode the low side transistor is left on during the dwell angle and the high side transistor switches according to the pulsed signal. In this case, the power electronics board has to handle six PWM signals [Liang,2006].

2.2.2 Hysteresis control method

Due to the hysteresis control, the current is flat, but if boost voltage is applied, the switching is higher than in the conventional case. The voltage of the boost capacitor is higher in the two capacitor parallel connected converter. The hysteresis control schemes for outgoing and incoming phases are shown on the right side of Fig. 45.

Solid and dash lines denote the rising and falling rules, respectively. The y axis denotes phase state and the x axis denotes torque error (ΔT_{err}), which is defined as,

$$\Delta \mathrm{T}_{\mathrm{err}} = \mathrm{T}_{\mathrm{ref}} - \mathrm{T}_{\mathrm{est}} \tag{35}$$

The threshold values of torque error are used to control state variation in hysteresis controller. Compared to previous research, this method only has 3 threshold values (ΔE, *0* and -ΔE), which simplifies the control scheme. In order to reduce switching frequency, only one switch opens or closes at a time. In region 1, the incoming phase must remain in state 1 to build up phase current, and outgoing phase state changes to maintain constant torque. For example, assume that the starting point is (-1, 1), and the torque error is greater than 0. The switching states for the two phases will change to (0, 1). At the next evaluation period, the switching state will change to (1, 1) if torque error is more than ΔE and (-1, 1) if torque error is less than -ΔE. So the combinatorial states of (-1, 1), (0, 0) and (1, 1) are selected by the control scheme. The control schemes for region 2 and region 3 are shown in Fig. 45(b) and (c), respectively.

3. Advanced torque control strategy

There are some various strategies of torque control: one method is direct torque control, which uses the simple control scheme and the torque hysteresis controller to reduce the torque ripple. Based on a simple algorithm, the short control period can be used to improve control precision. The direct instantaneous torque control (DITC) and advanced DITC (ADITC), torque sharing function (TSF) method are introduced in this section.

3.1 Direct Instantaneous Torque Control (DITC)

The asymmetric converter is very popular in SRM drive system. The operating modes of asymmetric converter are shown in Fig. 46. The asymmetric converter has three states, which are defined as state 1, state 0 and state -1 in DITC method, respectively.

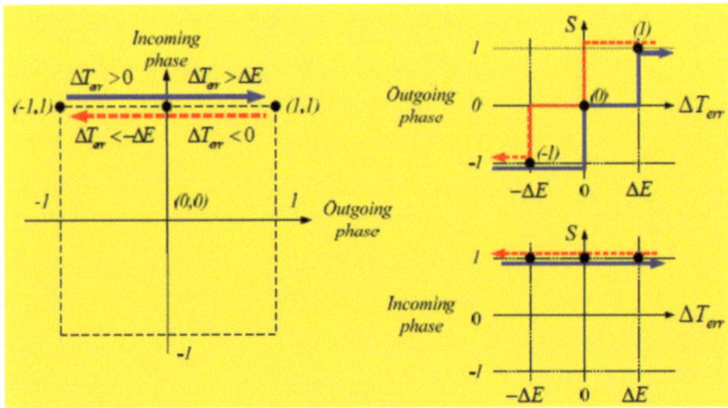

(a) Region 1

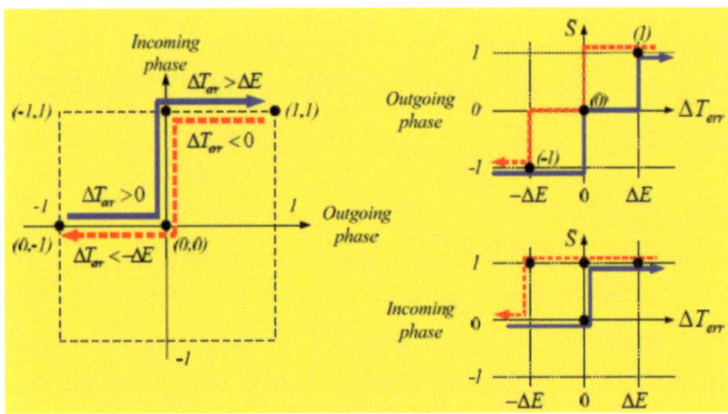

(b) Region 2

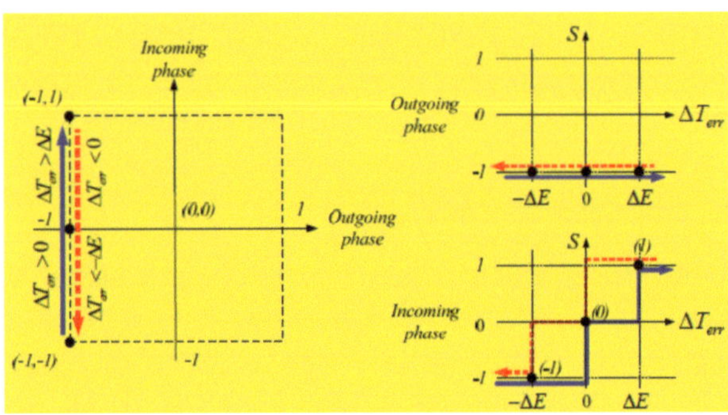

(c) Region 3

Fig. 45. The hysteresis control schemes for outgoing and incoming phases

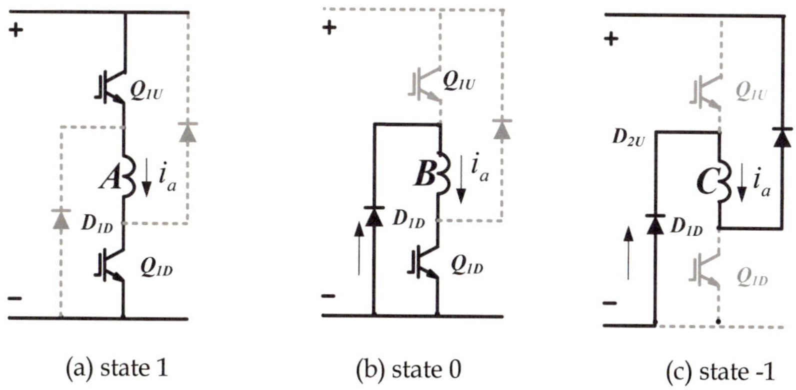

Fig. 46. 3 states in the asymmetric converter

In order to reduce a torque ripple, DITC method is introduced. By the given hysteresis control scheme, appropriate torque of each phase can be produced, and constant total torque can be obtained. The phase inductance has been divided into 3 regions shown as Fig. 47. The regions depend on the structure geometry and load. The boundaries of 3 regions are θ_{on1}, θ_1, θ_2 and θ_{on2} in Fig. 47. θ_{on1} and θ_{on2} are turn-on angle in the incoming phase and the next incoming phase, respectively, which depend on load and speed. The θ_1 is a rotor position which is initial overlap of stator and rotor. And θ_2 is aligned position of inductance in outgoing phase. Total length of these regions is 120 electrical degrees in 3 phases SRM. Here, let outgoing phase is phase A and incoming phase is phase B in Fig. 47. When the first region 3 is over, outgoing phase will be replaced by phase B in next 3 regions.

The DITC schemes of asymmetric converter are shown in Fig. 48. The combinatorial states of outgoing and incoming phase are shown as a square mesh. x and y axis denote state of outgoing and incoming phase, respectively. Each phase has 3 states, so the square mesh has 9 combinatorial states. However, only the black points are used in DITC scheme.

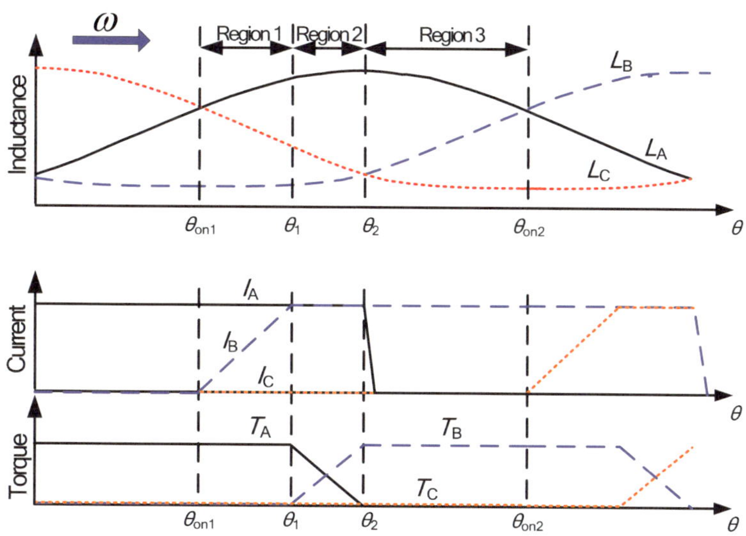

Fig. 47. Three regions of phase inductance in DITC method

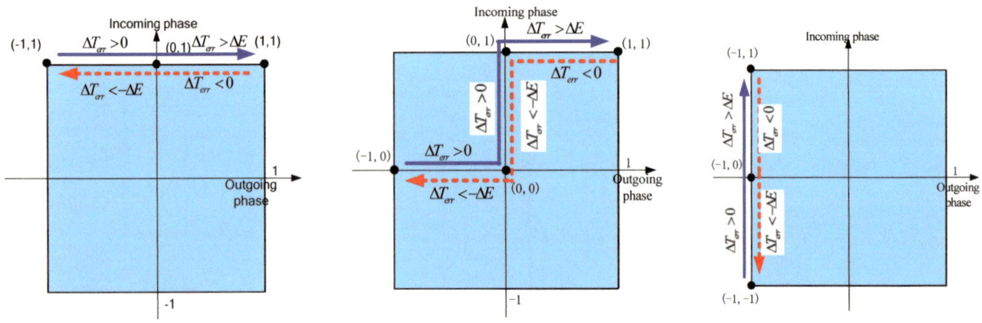

(a) region 1 (b) region 2 (c) region 3

Fig. 48. DITC scheme of asymmetric converter

Control diagram of DITC SR motor drive is shown in Fig. 49. The torque estimation block is generally implemented by 3-D lookup table according to the phase currents and rotor position. And the digital torque hysteresis controller which carries out DITC scheme generates the state signals for all activated machine phases according to torque error between the reference torque and estimated torque. The state signal is converted as switching signals by switching table block to control converter.

Through estimation of instantaneous torque and a simple hysteresis control, the average of total torque can be kept in a bandwidth. And the major benefits of this control method are its high robustness and fast toque response. The switching of power switches can be reduced.

However, based on its typical hysteresis control strategy, switching frequency is not constant. At the same time, the instantaneous torque cannot be controlled within a given bandwidth of hysteresis controller. The torque ripple is limited by the controller sampling time, so torque ripple will increase with speed increased.

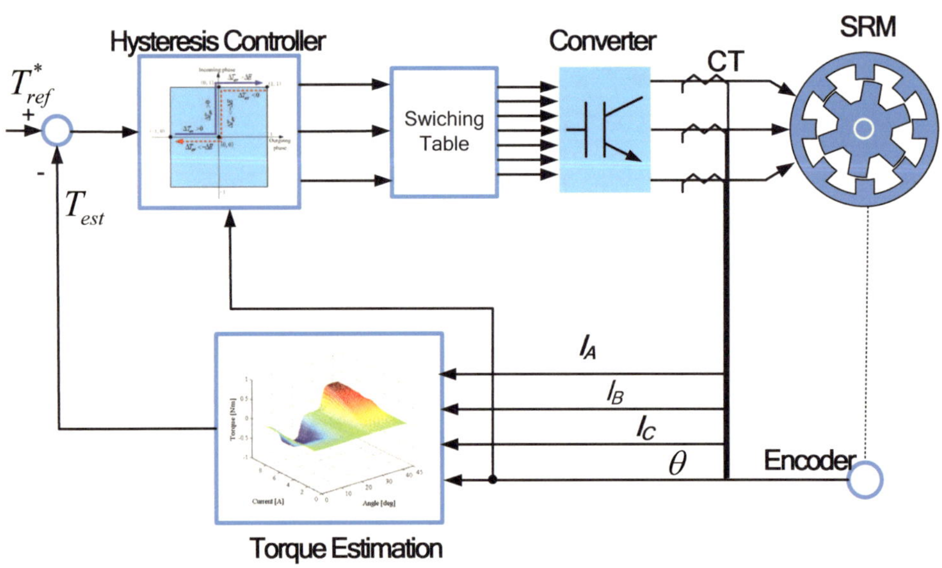

Fig. 49. Control diagram of DITC

3.2 Advanced Direct Instantaneous Torque Control (ADITC)

The conventional DITC method uses a simple hysteresis switch rules, so only one phase state is applied according to torque error at every sampling period. The toque variation with sampling time and speed under full dc-link voltage is shown in Fig. 50. In order to guarantee the torque ripple within a range, it has two methods: one is that reduces sampling time, which will increase the cost of hardware. Another is that control average voltage of phase winding in sampling time. PWM method can be used.

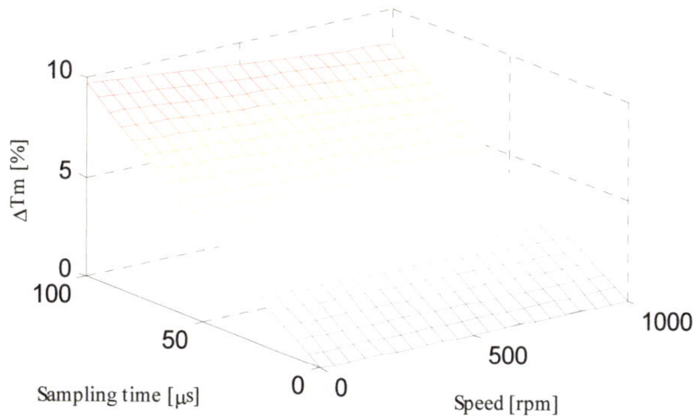

Fig. 50. Torque variation with sampling time and speed

ADITC combines the conventional DITC and PWM method. The duty ratio of the phase switch is regulated according to the torque error and simple control rules of DITC. Therefore, the sampling time of control can be extended, which allows implementation on low cost microcontrollers.

ADITC is improved from the conventional DITC, so the divided region of phase inductance is similar to DITC method. The control scheme of ADITC is shown in Fig. 51, $D_{t(k)}$ means incoming phase, $D_{t(k-1)}$ means outgoing phase. X-axis denotes torque error, and y-axis denotes switching state of $D_{t(k)}$ and $D_{t(k-1)}$.

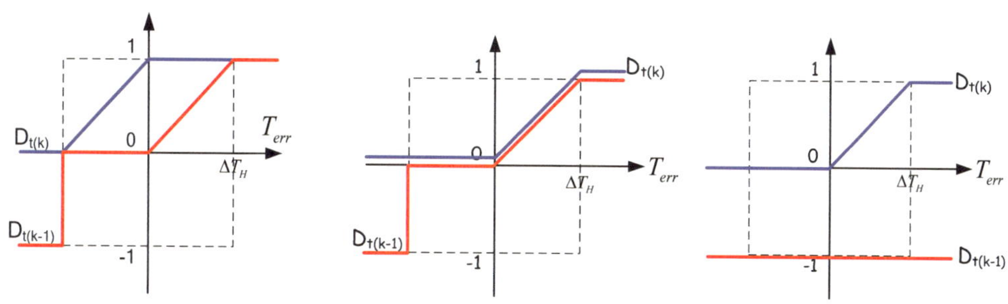

(a) Region 1 (b) Region 2 (c) Region

Fig. 51. ADITC scheme of asymmetric converter

Profit from the effect of PWM, the average voltage of phase winding can be adjusted from 0 to V_{dc} in one sampling time. And the hysteresis rule is removed from the control scheme. Now, the current state can select the phase state between state 0 and 1 by duty ratio of PWM.

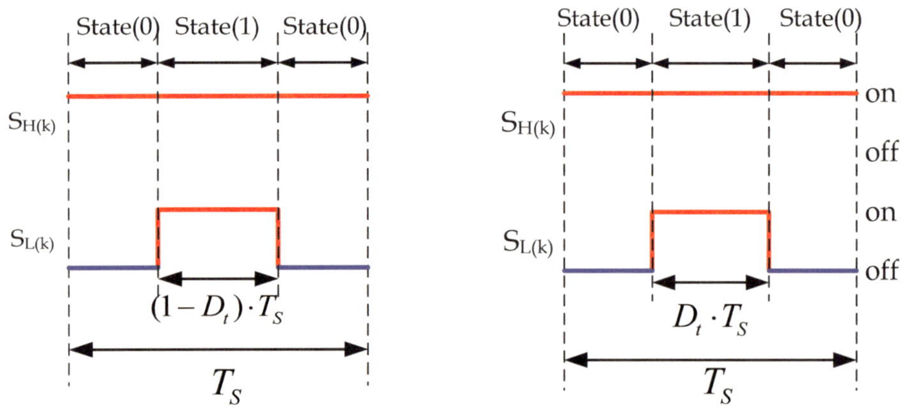

(a) Incoming phase (b) outgoing phase

Fig. 52. Switching modes of incoming and outgoing phase

The duty ratio of switching modes is decided by the torque error as shown in Fig. 52, and D_t is expressed as follows:

$$D_t = \text{Abs}\,(T_{err})/\Delta T_H \tag{36}$$

Where, T_{err} is torque error, ΔT_H is torque error bandwidth. The control block diagram of ADITC is similar to Fig. 53. The hysteresis controller is replaced by Advanced DITC controller, and the PWM generator is added.

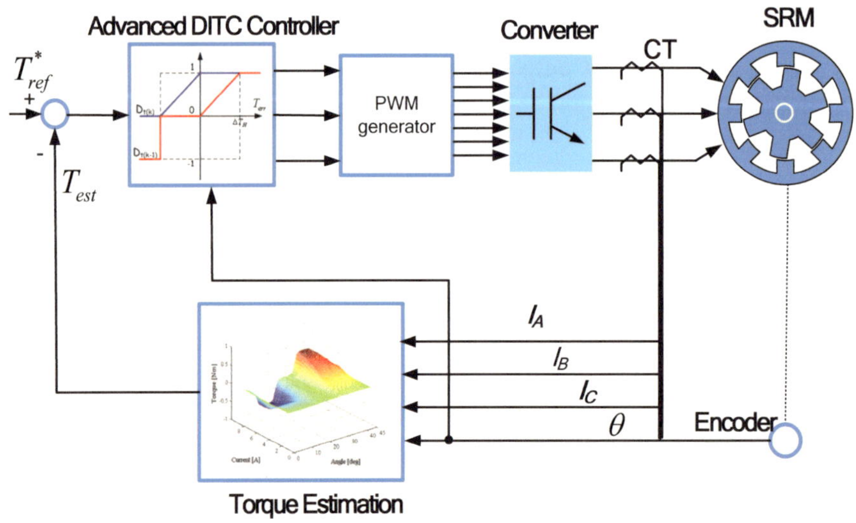

Fig. 53. Control diagram of ADITC

ADITC method can adjust average phase voltage to control variety of phase current in one sampling time, which can extend the sampling time and obtain smaller torque ripple than conventional DITC. However, PWM generator is added, and the switching frequency of

ADITC is double of DITC's with uniform sampling time in the worst case. So the switching loss and EMC noise are increased in ADITC method.

3.3 Torque sharing control

Another control method to produce continuous and constant torque is indirect torque control, which uses the complicated algorithms or distribution function to distribute each phase torque and obtain current command. And then, the current controller is used to control phase torque by given current command. The linear, cosine and non linear logical torque sharing function (TSF) are introduced.

Among them, the simple but powerful method is torque sharing function (TSF). The TSF method uses the pre-measured non-linear torque characteristic, and simply divided torque sharing curve is used for constant torque generation. Besides the direct torque control method, another method is indirect torque control. TSF is simple but powerful and popular method among the indirect torque control method. It simply divided by torque sharing curve that is used for constant torque generation. And the phase torque can be assigned to each phase current to control smoothing torque. But phase torque has relationship of square current. So the current ripple should keep small enough to generate smooth torque. So the frequency of current controller should be increased.

Fig. 54 shows the torque control block diagram with TSF method. The input torque reference is divided into three-phase torque command according to rotor position. Torque references of each phase are changed to current command signal in the "Torque-to-Current" block according to rotor position. Since the output torque is determined by the inductance slope and phase current, and the inductance slope is changed by rotor position, so the reference currents of each phase is determined by the target torque and rotor position. The switching rule generates an active switching signal of asymmetric converter according to current error and hysteresis switching tables.

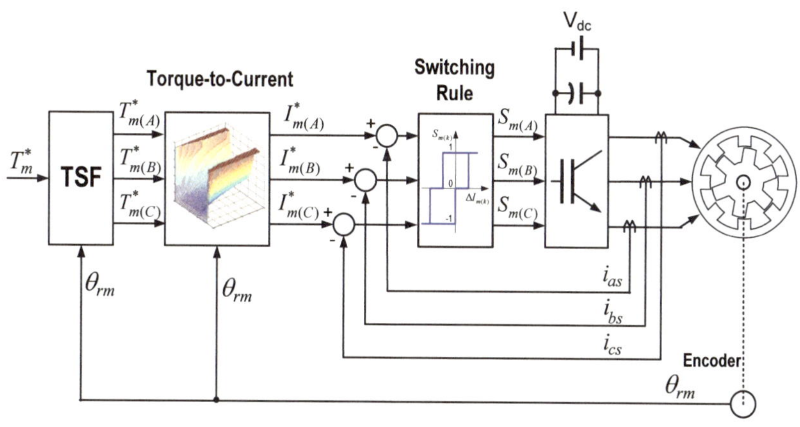

Fig. 54. The torque control block diagram with TSF method

In the over-lap region of inductances, the two-phase currents generate the output torque together. A simple torque sharing curves are studied for constant torque generation in the commutation region such as linear and cosine function.

Fig. 55 shows the inductance profiles of three-phase SRM, cosine and linear TSF curves. As shown in Fig. 55, region 2 denotes the one phase activation area. Region 1 and region 3 are two phases activation area explained as the commutation region. In one phase activation region, TSF is constant in every torque sharing functions. But TSF is different in the commutation regions. The linear TSF has constant slope of torque in commutation region. This method is simple, but it is very difficult to generate the linear torque slope in the commutation region due to the non-linear inductance characteristics.

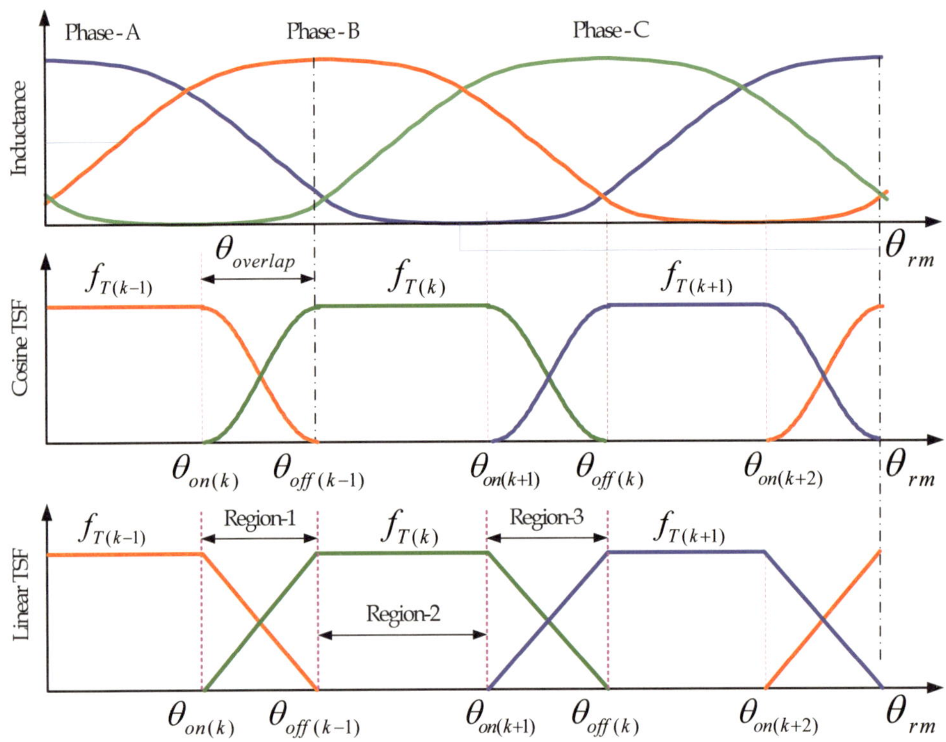

Fig. 55. Phase inductances and cosine, and linear TSF curves

The cosine TSF uses the cosine function in commutation region as shown in Fig. 55. The cosine function is relatively simple and it is similar to the non-linear inductance characteristics. But the non-linear characteristic of SRM is very complex, so cosine torque function can not be satisfied in the aspect of torque ripple and efficiency.
In the cosine TSF, the TSF of each phase in the commutation region are defined as follow

$$f_{T(k)} = \frac{1}{2}\left[1 - \cos\left(\frac{\theta_{rm} - \theta_{on(k)}}{\theta_{overlap}}\pi\right)\right] \tag{37}$$

$$f_{T(k-1)} = 1 - f_{T(k)} \tag{38}$$

$$f_{T(k+1)} = 0 \tag{39}$$

And the linear TSF method, the TSF of each phase can be obtained as follow

$$f_{T(k)} = \frac{\theta_{rm} - \theta_{on(k)}}{\theta_{overlap}} \tag{40}$$

$$f_{T(k-1)} = 1 - f_{T(k)} \tag{41}$$

$$f_{T(k+1)} = 0 \tag{42}$$

These two TSFs are very simple, but they can not consider nonlinear phenomena of the SRM and torque dip is much serious according to rotor speed. For the high performance torque control, a novel non-linear torque sharing function is suitable to use. In order to reduce torque ripple and to improve efficiency in commutation region, the TSF uses a non-linear current distribution technique at every rotor position. And the torque sharing function can be easily obtained by the current coordinates of each rotor position. In the commutation region, the total torque reference is divided by two-phase torque reference.

$$T_m^* = T_{m(k)}^* + T_{m\,(k+1)}^* \tag{43}$$

In the equation, the subscripts *k+1* denotes the incoming phase and k denotes outgoing phase. The actual torque can be obtained by inductance slope and phase current. So the torque equation can be derived as follows.

$$T_m^* = \frac{{I_{m(k)}^*}^2}{a^2} + \frac{{I_{m(k+1)}^*}^2}{b^2} \tag{44}$$

where,
$$a = \sqrt{\frac{2}{\partial L_{(k)}/\partial L_{(m)}}},\ b = \sqrt{\frac{2}{\partial L_{(k+1)}/\partial L_{(m)}}} \tag{45}$$

This equation is same as ellipse equation. In order to generate a constant torque reference, current references of the outgoing and incoming phases is placed on the ellipse trajectory in the commutation region. And the aspect of the ellipse and its trajectory is changed according to rotor position, inductance shape and the reference torque. Since the TSFs uses a fixed torque curve such as linear and cosine, the outgoing phase current should keep up the reference. And the actual current should remain higher level around rotor aligned position. Fig. 56 shows each phase current reference and actual phase torque for constant torque production according to rotor position. As shown in Fig. 56, the actual torque profile has non-linear characteristics around match position of rotor and stator position. So the current reference of each phase for constant torque generation is changed according to the rotor position and the amplitude of the torque reference. However, the actual phase current is limited by the performance of a motor and a drive. And the actual torque can not be satisfied the torque reference around the aligned position due to the non-linear torque characteristics shown as Fig. 56. If the current of outgoing phase is increased as a limit value of the motor, the actual torque is decreased after D_k position. And the actual torque of incoming phase can not be satisfied at the start position of the commutation due to the same reason. In order to generate the constant torque from A_k to G_{k+1}, the outgoing and incoming current reference should be properly selected so that the total torque of each phase is remained as constant value of T_m^*.

In order to reduce the commutation region, the outgoing phase current should be decreased fast, and the incoming phase current should be increased fast with a constant torque generation. At the starting point of commutation, the incoming phase current should be increased from zero to A_{k+1} point, and the end of the commutation, the outgoing phase current should be decreased from G_k point to zero as soon as possible shown in Fig. 56.

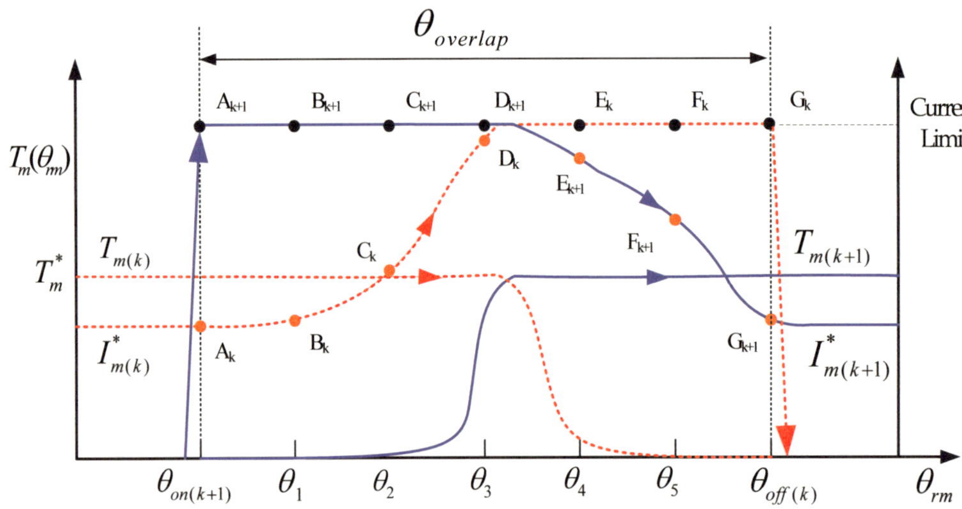

Fig. 56. Phase current and actual torque trajectory for constant torque production during phase commutation

In order to reduce the torque ripple and increase the operating efficiency, a non-linear TSF is based on minimum changing method. One phase current reference is fixed, and the other phase current reference is changed to generate constant torque during commutation. Fig. 57 shows the basic principle of the non-linear TSF commutation method.

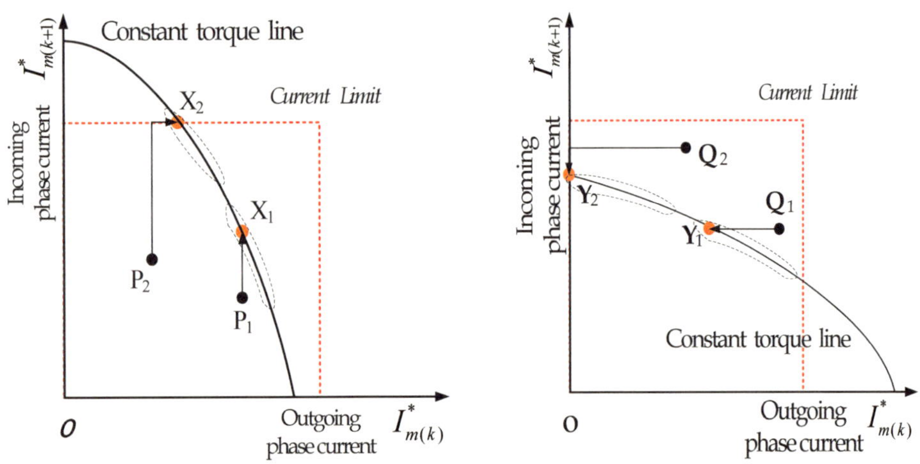

(a) In case of $T_m < T_m^*$ (b) In case of $T_m > T_m^*$

Fig. 57. Basic principle of the commutation method based on minimum changing

In this method, the incoming phase current is changed to a remaining or an increasing direction to produce the primary torque. And the outgoing phase current is changed to a remaining or a decreasing direction to produce the auxiliary torque. In case of $T_m < T_m^*$, the outgoing phase current is fixed, and the incoming phase current is increased to reach the constant torque line from P_1 to X_1 shown as Fig. 57(a). If the incoming phase current is limited by the current limit and the actual torque is under the reference value, the auxiliary torque is generated by the outgoing phase current from P_2 to X_2 shown as Fig. 57(a). In case of $T_m > T_m^*$, the incoming phase current is fixed, and the outgoing phase current is decreased to reach the constant torque line from Q_1 to Y_1 shown as Fig. 57(b), because the incoming phase current is sufficient to generate the reference torque. If the outgoing phase current is reached to zero, and the actual torque is over to reference value, the incoming phase current is decreased from Q_2 to Y_2 shown as Fig. 57(b). This method is very simple, but the switching number for torque control can be reduced due to the minimum number changing of phase. As the other phase is fixed as the previous state, the torque ripple is dominated by the one phase switching. Especially, the outgoing phase current is naturally decreased when the incoming phase current is sufficient to produce the torque reference. The demagnetization can be decreased fast, and the tail current which generates negative torque can be suppressed.

Table 6 shows the logical TSF, and the Fig. 58 is the ideal current trajectory during commutation region. In Fig. 58, the ellipse curves are current trajectory for constant torque at each rotor position under commutation.

In case of $T_m < T_m^*$		when	In case of $T_m > T_m^*$		when
$T_{m(k+1)}^*$	$T_m^* - T_{m(k)}$	$I_{m(k+1)}^* < I_{max}$	$T_{m(k)}^*$	$T_m^* - T_{m(k+1)}$	$I_{m(k)}^* > 0$
	$T_{m(k+1)}$* *At current limit*	$I_{m(k+1)}^* > I_{max}$		0	$I_{m(k)}^* < 0$
$T_{m(k)}^*$	$T_m^* - T_{m(k+1)}^*$		$T_{m(k+1)}^*$	$T_{m(k+1)}$	$I_{m(k)}^* > 0$
				$T_m^* - T_{m(k)}$	$I_{m(k)}^* < 0$

Table 6. The logical TSF in commutation region.

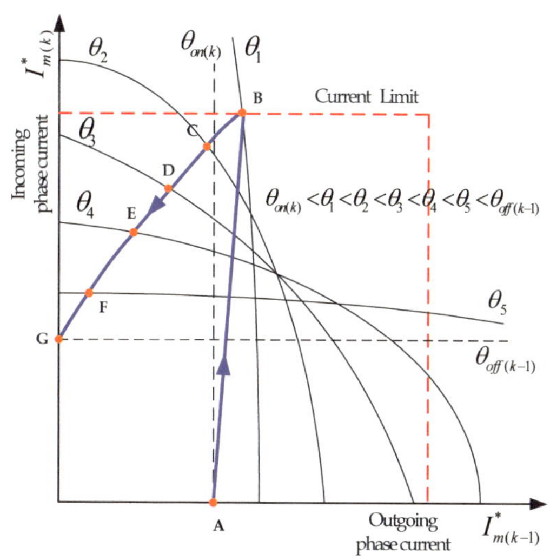

Fig. 58. The ideal current trajectory at commutation region

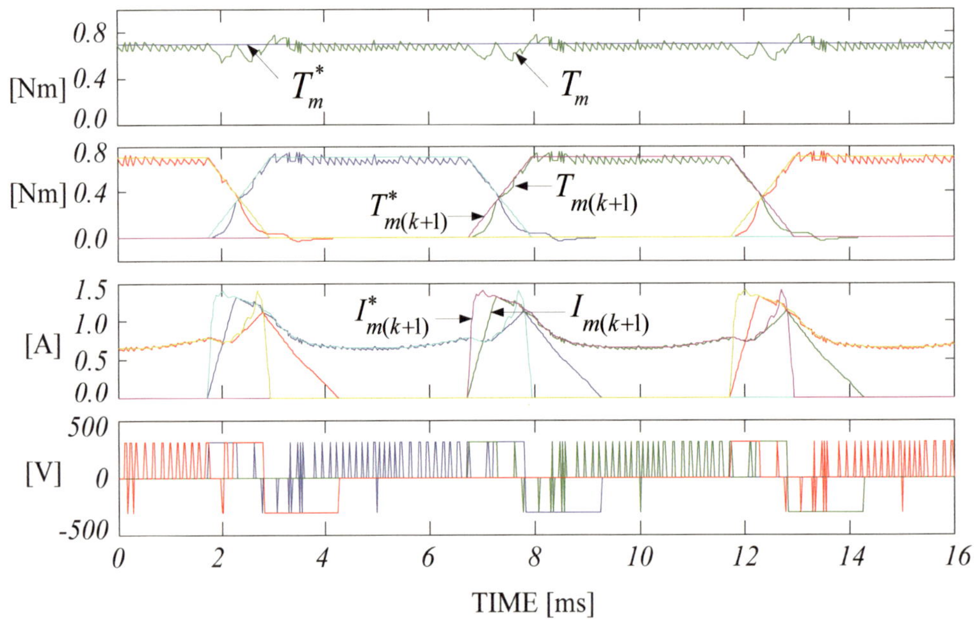

(a) Linear TSF

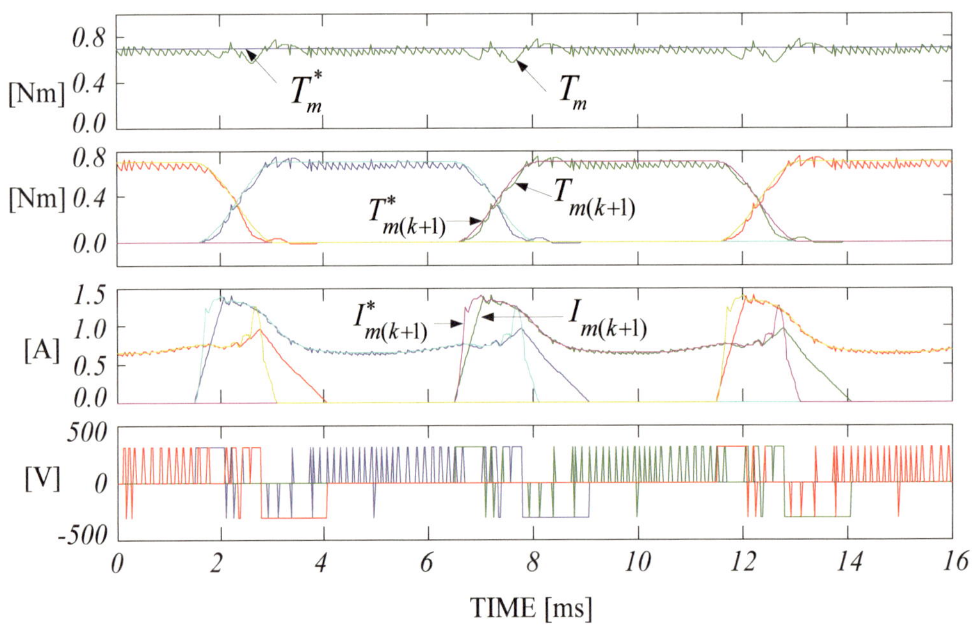

(b) Cosine TSF

Fig. 59. Simulation result at 500 rpm with rated torque

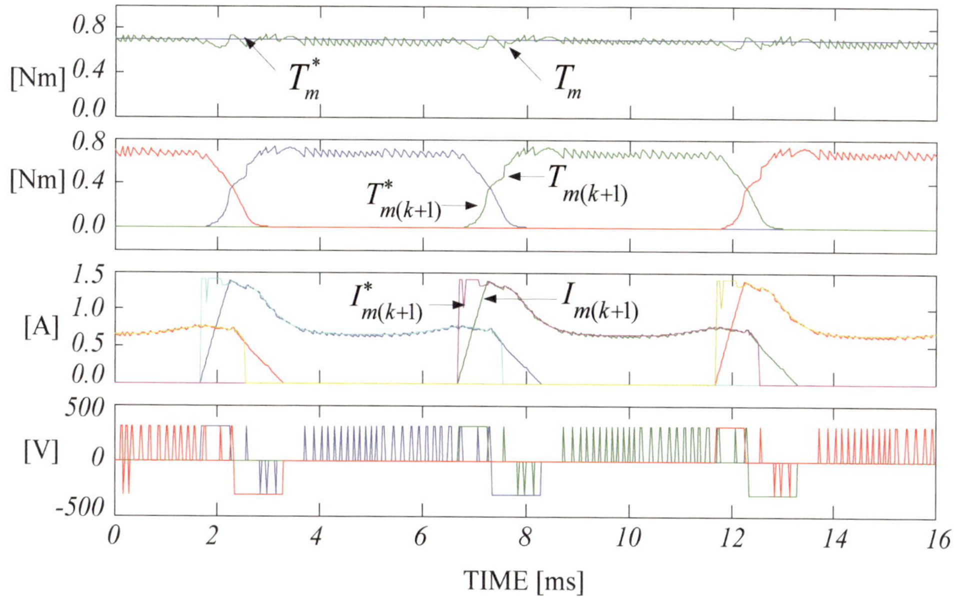

(c) non-linear Logical TSF

Fig. 59. Simulation results at 500rpm with rated torque (continued)

In order to verify the non-linear TSF control scheme, computer simulations are executed and compared with conventional methods. Matlab and simulink are used for simulation. Fig. 59 shows the simulation comparison results at 500[rpm] with rated torque reference. The simulation results show the total reference torque, actual total torque, reference phase torque, actual phase torque, reference phase current, actual phase current and phase voltage, respectively. As shown in Fig. 59, torque ripple is linear TSF > cosine TSF > the logical TSF.

Fig. 60 shows the actual current trajectory in the commutation region. In the conventional case, the cross over of the outgoing and incoming phase is serious and two-phase current are changed at each rotor position. But the cross over is very small and one-phase current is changed at each rotor position in the logical TSF method.

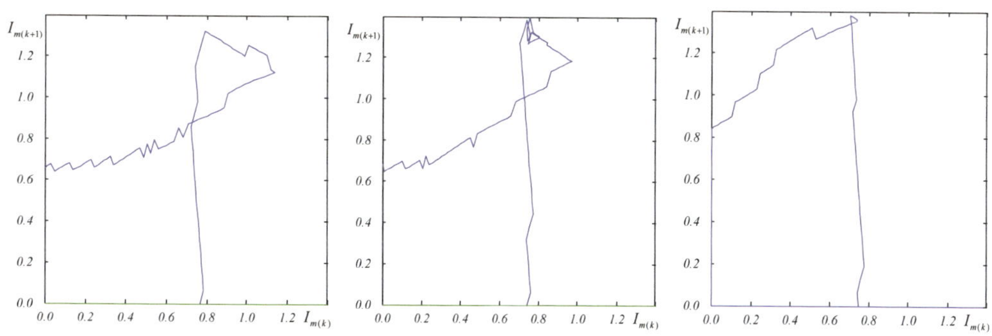

(a) Linear TSF (b) Cosine TSF (c) logical TSF

Fig. 60. The current trajectory for constant torque production in commutation region

Fig. 61 shows the experimental setup. The main controller is designed by TMS320F2812 from TI(Texas Instruments) and phase current and voltage signals are feedback to 12bit ADC embedded by DSP. The rotor position and speed is obtained by 512ppr optical encoder. At every 1.6[ms], the rotor speed is calculated from captured encoder pulse by QEP function of DSP.

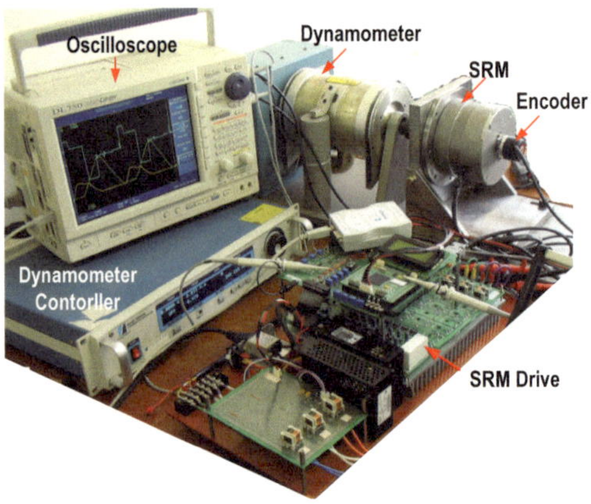

Fig. 61. The experimental configuration

Fig. 62, 63 and 64 show the experimental results in case of linear TSF, cosine TSF and the non-linear logical TSF at 500rpm, respectively. Torque ripple can be reduced in case of the TSF method due to the minimum phase changing.

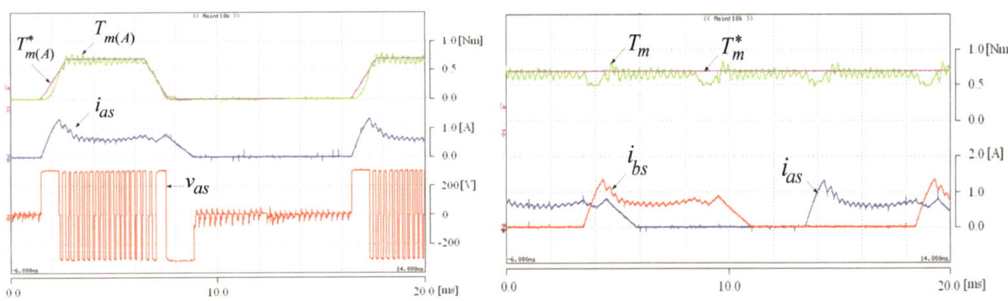

(a) (b)

Fig. 62. Experimental results in linear TSF(at 500[rpm])
(a) Reference torque, actual torque, phase current and terminal voltage
(b) Total reference torque, actual torque and phase currents

Fig. 65 shows experimental results at 1200rpm. As speed increase, torque ripple is increased due to the reduction of the commutation time. However, the control performance is much improved in this case.

Fig. 66 shows efficiency of the logical control schemes. In the low speed range, the TSF control scheme has about 5% higher efficiency than that of the conventional ones with low torque ripple. In high speed range, the actual efficiency is similar to all other control method due to the short commutation time. But the practical torque ripple can be reduced than other two control schemes shown in simulation and experimental results.

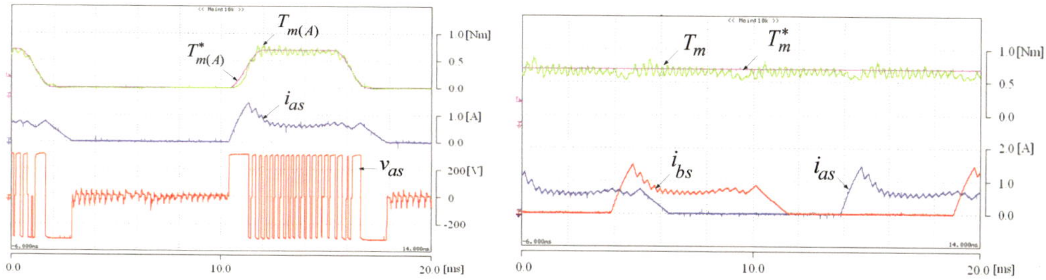

(a) (b)

Fig. 63. Experimental results in cosine TSF(at 500[rpm])
(a) Reference, actual torque, phase current and terminal voltage
(b) Total reference torque, actual torque and phase currents

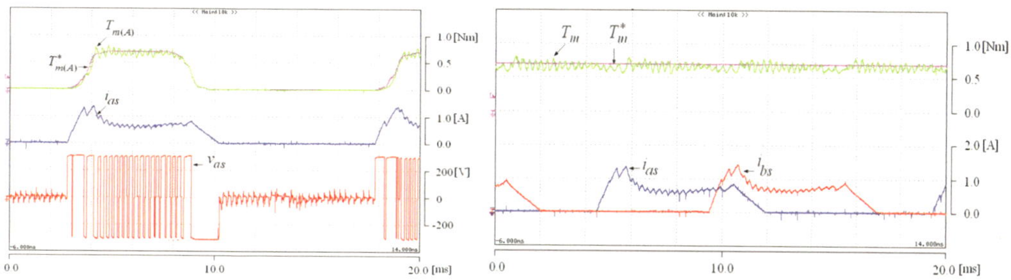

(a) (b)

Fig. 64. Experimental results in case of the non-linear logical TSF(at 500[rpm])
(a) Reference, actual torque, phase current and terminal voltage
(b) Total reference torque, actual torque and phase currents

4. Conclusion

The torque production in switched reluctance motor structures comes from the tendency of the rotor poles to align with the excited stator poles. However, because SRM has doubly salient poles and non-linear magnetic characteristics, the torque ripple is more severe than these of other traditional motors. The torque ripple can be minimized through magnetic circuit design or drive control. By controlling the torque of the SRM, low torque ripple, noise reduction or even increasing of the efficiency can be achieved. There are many different types of control methods. In this chapter, detailed characteristics of each control method are introduced in order to give the advanced knowledge about torque control method in SRM drive.

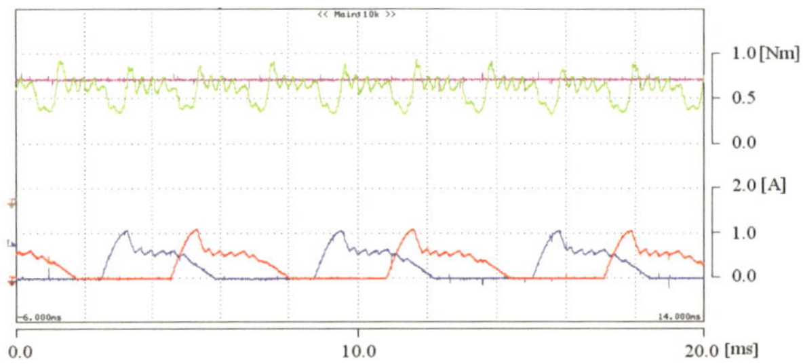

(a) Reference torque, total torque and phase currents in linear TSF

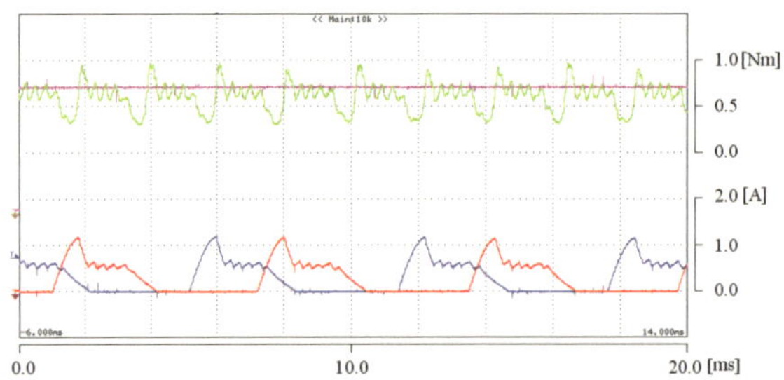

(b) Reference torque, total torque and phase currents in cosine TSF

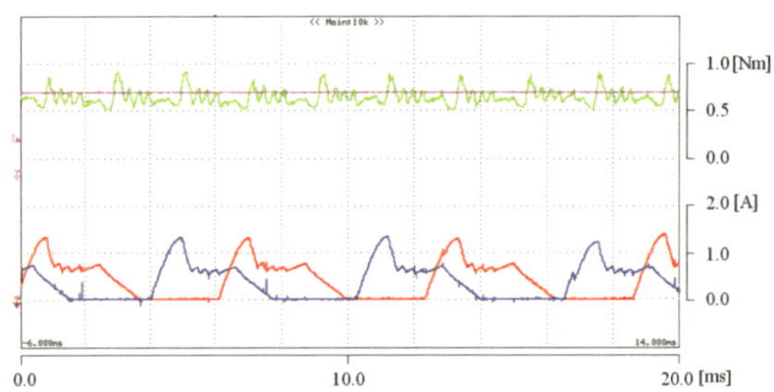

(c) Reference torque, total torque and phase currents in non-linear logical TSF

Fig. 65. Experimental results at 1200rpm with rated torque

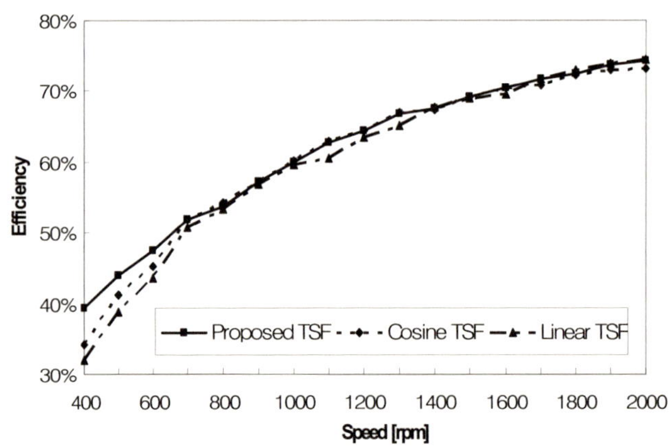

Fig. 66. Efficiency comparison

5. References

A. Chiba, K. Chida and T. Fukao, "Principles and Characteristics of a Reluctance Motor with Windings of Magnetic Bearing," in *Proc. PEC Tokyo*, pp.919-926, 1990.

Bass, J. T., Ehsani, M. and Miller, T. J. E ; "Robust torque control of a switched reluctance motor without a shaft position sensor," *IEEE Transactions*, Vol.IE-33, No.33, 1986, 212-216.

Bausch, H. and Rieke, B.; "Speed and torque control of thyristorfed reluctance motors." *Proceedings ICEM*, Vienna Pt.I, 1978, 128.1-128.10. Also : "Performance of thyristorfed electric car reluctance machines." *Proceedings ICEM*, Brussels E4/2.1-2.10

Byrne, J. V. and Lacy, J.G.; "Characteristics of saturable stepper and reluctance motors." *IEE Conf.* Publ. No.136,Small Electrical Machines, 1976, 93-96.

Corda, J. and Stephenson, J. M., "Speed control of switched reluctance motors," *International Conference on Electrical Machines*, Budapest, 1982.

Cossar, C. and miller, T.J.E., "Electromagnetic testing of switched reluctance motors," *International Conference on Electrical Machines*, Manchester, September 15-17, 1992, 470-474.

Davis, R. M., "A Comparison of Switched Reluctance Rotor Structures," *IEEE Trans. Indu. Elec.*, Vol.35, No.4, pp.524-529, Nov. 1988.

D.H. Lee, J. Liang, Z.G. Lee, J.W. Ahn, "A Simple Nonlinear Logical Torque Sharing Function for Low-Torque Ripple SR Drive", Industrial Electronics, *IEEE Transactions* on, Vol. 56, Issue 8, pp.3021-3028, Aug. 2009.

D.H. Lee, J. Liang, T.H. Kim, J.W. Ahn, "Novel passive boost power converter for SR drive with high demagnetization voltage", International Conference on Electrical Machines and Systems, 2008, pp.3353-3357, 17-20 Oct. 2008.

D.H. Lee, T.H. Kim, J.W. Ahn, " Pressure control of SR Driven Hydraulic Oil-pump Using Data Based PID Controller", *Journal of Power Electronics* Vol.9, September 2009.

D.S. Schramm, B.W. Williams, and T.C. Green; "Torque ripple reduction of switched reluctance motors by phase current optimal profiling", in *Proc. IEEE PESC' 92*, Vol. 2, Toledo, Spain, pp.857-860, 1992 .

Harris, M. R. and Jahns, T. M., "A current-controlled switched reluctance drive for FHP applications," *Conference on Applied Motion Control*, Minneapolis, June 10-12 , 1986.

Ilic-Spong, M., Miller, T. J. E., MacMinn, S. R. and Thorp, J. S., "Instantaneous torque control of electric motor drives," *IEEE Transactions*, Vol.IA-22, 1987, 708-715.

J.W. Ahn, Se.G. Oh, J.W. Moon, Y.M. Hwang; "A three-phase switched reluctance motor with two-phase excitation", Industry Applications, *IEEE Transactions* on, Vol. 35, Issue 5, pp.1067-1075, Sept.-Oct. 1999.

J.W. Ahn, S. G. Oh, and Y. M. Hwang, "A Novel Control Scheme for Low Cost SRM Drive, " in *Proc. IEEE/ISIE '95*, July 1995, Athens, pp. 279-283.

J.W. Ahn, S.G. Oh, " DSP Based High Efficiency SR Drive with Precise Speed Control", *PESC '99, june 27*, Charleston, south Carolina.

J.W. Ahn, "Torque Control Strategy for High Performance SR Drive", *Journal of Electrical Engineering & Technology(JEET)*, Vol.3. No.4. 2008, pp.538-545.

J.W. Ahn , S. G. Oh, C. U. Kim, Y. M. Hwang, "Digital PLL Technique for Precise Speed Control for SR Drive," in *Proc. IEEE/PESC'99*, Jun./Jul. 1999, Charleston, pp.815-819

J.M. Stephenson; J. Corda, "Computation of Torque and Current in Doubly-Salient Reluctance Motors from Nonlinear Magnetization Data", *Proceedings IEE*, Vol. 126, pp.393-396, May 1979.

J. N.Liang, Z. G. Lee, D. H. Lee, J. W. Ahn, " DITC of SRM Drive System Using 4-Level Converter " , *Proceedings of ICEMS 2006*, Vol. 1, 21-23 Nov. 2006

J. N. Liang, S.H. Seok, D.H. Lee, J.W. Ahn, "Novel active boost power converter for SR drive" *International Conference on Electrical Machines and Systems*, 2008, pp.3347-3352, 17-20 Oct. 2008.

Lawrenson, P.J.et al; "Variable-speed switched reluctance motors." *Proceedings IEE*. Vol.127, Pt.B 253-265,1980.

M. Stiebler, G. Jie; "A low Voltage switched reluctance motor with experimentally optimized control", *Proceedings of ICEM '92*, Vol. 2, pp. 532-536, Sep. 1992.

Miller, T. J. E., Bower, P. G., Becerra, R. and Ehsani, M., "Four- quadrant brushless reluctance motor drive," *IEE Conference on Power Electronics and Variable Speed Drives*, London, 1988.

Pollock, C. and Willams, B. W.; "Power convertor circuit for switched reluctance motors with the minimum number of switches," *IEE Proceedings-B*, Vol.137, 1990, No.6.

R. Krishnan; "*Switched Reluctance Motor Drives: Modeling, Simulation, Analysis, Design, and Applications*", CRC Press, 2001

R. Orthmann, H.P. Schoner; "Turn-off angle control of switched reluctance motors for optimum torque output", *Proceedings of EPE '93*, Vol. 6, pp.20-55, 1993.

Stephenson, J.M. and El-Khazendar, M.A., "Saturation in doubly salient reluctance motors," *IEE Proceedings-B*, Vol.136, No.1, 1989, 50-58.

T. Skvarenina; "*The Power Electronics Handbook*", CRC Press, 2002

T.J.E. Miller, M. McGilp, "Nonlinear theory of the switched reluctance motor for rapid computer-aided design", *IEE Proceedings B (Electric Power Applications)*, Vol. 137, No. 6, pp.337-347, Nov. 1990.

Unnewehr, L. E. and Koch, W. H.; "An axial air-gap reluctance motor for variable-speed applications." *IEEE Transactions*, 1974, PAS-93, 367-376.

Vukosavic, S. and Stefanovic, V. R., "SRM inverter topologies : a comparative evaluation," *IEEE IAS Annual Meeting*, Conf. Record, Seattle, WA, 1990.

Wallace, R. S. and Taylor, D. G., "Low torque ripple switched reluctance motors for direct-drive robotics," *IEEE Transactions on Robotics and Automation*, Vol.7, No.6, 1991, 733-742.

Wallace, R. S. and Taylor, D. G., "A balanced commutator for switched reluctance motors to reduce torque ripple," *IEEE Transactions on Power Electronics*, October 1992.

9

Controller Design for Synchronous Reluctance Motor Drive Systems with Direct Torque Control

Tian-Hua Liu
Department of Electrical Engineering,
National Taiwan University of Science and Technolog
Taiwan

1. Introduction

A. Background

The synchronous reluctance motor (SynRM) has many advantages over other ac motors. For example, its structure is simple and rugged. In addition, its rotor does not have any winding or magnetic material. Prior to twenty years ago, the SynRM was regarded as inferior to other types of ac motors due to its lower average torque and larger torque pulsation. Recently, many researchers have proposed several methods to improve the performance of the motor and drive system [1]-[3]. In fact, the SynRM has been shown to be suitable for ac drive systems for several reasons. For example, it is not necessary to compute the slip of the SynRM as it is with the induction motor. As a result, there is no parameter sensitivity problem. In addition, it does not require any permanent magnetic material as the permanent synchronous motor does.

The sensorless drive is becoming more and more popular for synchronous reluctance motors. The major reason is that the sensorless drive can save space and reduce cost. Generally speaking, there are two major methods to achieve a sensorless drive system: vector control and direct torque control. Although most researchers focus on vector control for a sensorless synchronous reluctance drive [4]-[12], direct torque control is simpler. By using direct torque control, the plane of the voltage vectors is divided into six or twelve sectors. Then, an optimal switching strategy is defined for each sector. The purpose of the direct torque control is to restrict the torque error and the stator flux error within given hysteresis bands. After executing hysteresis control, a switching pattern is selected to generate the required torque and flux of the motor. A closed-loop drive system is thus obtained.

Although many papers discuss the direct torque control of induction motors [13]-[15], only a few papers study the direct torque control for synchronous reluctance motors. For example, Consoli et al. proposed a sensorless torque control for synchronous reluctance motor drives [16]. In this published paper, however, only a PI controller was used. As a result, the transient responses and load disturbance responses were not satisfactory. To solve the problem, in this chapter, an adaptive backstepping controller and a model-reference adaptive controller are proposed for a SynRM direct torque control system. By using the proposed controllers, the transient responses and load disturbance rejection capability are obviously improved. In addition, the proposed system has excellent tracking ability. As to the authors best knowledge, this is the first time that the adaptive backstepping controller and model reference adaptive controller have been used in the direct torque control of synchronous reluctance motor drives. Several experimental results validate the theoretical analysis.

B. Literature Review

Several researchers have studied synchronous reluctance motors. These researchers use different methods to improve the performance of the synchronous reluctance motor drive system. The major categories include the following five methods:

1. Design and manufacture of the synchronous reluctance motor

The most effective way to improve the performance of the synchronous reluctance motor is to design the structure of the motor, which includes the rotor configuration, the windings, and the material. Miller et al. proposed a new configuration to design the rotor configuration. By using the proposed method, a maximum L_d / L_q ratio to reach high power factor, high torque, and low torque pulsations was achieved [17]. In addition, Vagati et al. used the optimization technique to design a rotor of the synchronous reluctance motor. By applying the finite element method, a high performance, low torque pulsation synchronous reluctance motor has been designed [18]. Generally speaking, the design and manufacture of the synchronous reluctance motor require a lot of experience and knowledge.

2. Development of Mathematical Model for the synchronous reluctance motor

The mathematical model description is required for analyzing the characteristics of the motor and for designing controllers for the closed-loop drive system. Generally speaking, the core loss and saturation effect are not included in the mathematical model. However, recently, several researchers have considered the influence of the core loss and saturation. For example, Uezato et al. derived a mathematical model for a synchronous reluctance motor including stator iron loss [19]. Sturtzer et al. proposed a torque equation for synchronous reluctance motors considering saturation effect [2]. Stumberger discussed a parameter measuring method of linear synchronous reluctance motors by using current, rotor position, flux linkages, and friction force [20]. Ichikawa et al. proposed a rotor estimating technique using an on-line parameter identification method taking into account magnetic saturation [5].

3. Controller Design

As we know, the controller design can effectively improve the transient responses, load disturbance responses, and tracking responses for a closed-loop drive system. The PI controller is a very popular controller, which is easy to design and implement. Unfortunately, it is impossible to obtain fast transient responses and good load disturbance responses by using a PI controller. To solve the difficulty, several advanced controllers have been developed. For example, Chiang et al. proposed a sliding mode speed controller with a grey prediction compensator to eliminate chattering and reduce steady-state error [21]. Lin et al. used an adaptive recurrent fuzzy neural network controller for synchronous reluctance motor drives [22]. Morimoto proposed a low resolution encoder to achieve a high performance closed-loop drive system [7].

4. Rotor estimating technique

The sensorless synchronous reluctance drive system provides several advantages. For example, sensorless drive systems do not require an encoder, which increases cost, generates noise, and requires space. As a result, the sensorless drive systems can reduce costs and improve reliability. Several researchers have studied the rotor estimating technique to realize a sensorless drive. For example, Lin et al. used a current-slope to estimate the rotor position and rotor speed [4]. Platt et al. implemented a sensorless vector controller for a synchronous reluctance motor [9]. Kang et al. combined the flux-linkage estimating method and the high-frequency injecting current method to achieve a sensorless rotor position/speed drive system [23]. Ichikawa presented an extended EMF model and initial position estimation for synchronous motors [10].

5. Switching strategy of the inverter for synchronous reluctance motor

Some researchers proposed the switching strategies of the inverter for synchronous reluctance motors. For example, Shi and Toliyat proposed a vector control of a five-phase synchronous reluctance motor with space vector pulse width modulation for minimum switching losses [24].

Recently, many researchers have created new research topics for synchronous reluctance motor drives. For example, Gao and Chau present the occurrence of Hopf bifurcation and chaos in practical synchronous reluctance motor drive systems [25]. Bianchi, Bolognani, Bon, and Pre propose a torque harmonic compensation method for a synchronous reluctance motor [26]. Iqbal analyzes dynamic performance of a vector-controlled five-phase synchronous reluctance motor drive by using an experimental investigation [27]. Morales and Pacas design an encoderless predictive direct torque control for synchronous reluctance machines at very low and zero speed [28]. Park, Kalev, and Hofmann propose a control algorithm of high-speed solid-rotor synchronous reluctance motor/generator for flywheel-based uniterruptible power supplies [29]. Liu, Lin, and Yang propose a nonlinear controller for a synchronous reluctance drive with reduced switching frequency [30]. Ichikawa, Tomita, Doki, and Okuma present sensorless control of synchronous reluctance motors based on extended EMF models considering magnetic saturation with online parameter identification [31].

2. The synchronous reluctance motor

In the section, the synchronous reluctance motor is described. The details are discussed as follows.

2.1 Structure and characteristics

Synchronous reluctance motors have been used as a viable alternative to induction and switched reluctance motors in medium-performance drive applications, such as: pumps, high-efficiency fans, and light road vehicles. Recently, axially laminated rotor motors have been developed to reach high power factor and high torque density. The synchronous reluctance motor has many advantages. For example, the synchronous reluctance motor does not have any rotor copper loss like the induction motor has. In addition, the synchronous reluctance motor has a smaller torque pulsation as compared to the switched reluctance motor.

2.2 Dynamic mathematical model

In synchronous d-q reference frame, the voltage equations of the synchronous reluctance motor can be described as

$$v_{qs} = r_s i_{qs} + p\lambda_{qs} + \omega_r \lambda_{ds} \tag{1}$$

$$v_{ds} = r_s i_{ds} + p\lambda_{ds} - \omega_r \lambda_{qs} \tag{2}$$

where v_{qs} and v_{ds} are the q-axis and the d-axis voltages, r_s is the stator resistance, i_{qs} is the q-axis equivalent current, i_{ds} is the d-axis equivalent current, p is the differential operator, λ_{qs} and λ_{ds} are the q-axis and d-axis flux linkages, and ω_r is the motor speed. The flux linkage equations are

$$\lambda_{qs} = (L_{ls} + L_{mq})\ i_{qs} \tag{3}$$

$$\lambda_{ds} = (L_{ls} + L_{md})\ i_{ds} \tag{4}$$

where L_{ls} is the leakage inductance, and L_{mq} and L_{md} are the q- axis and d-axis mutual inductances. The electro-magnetic torque can be expressed as

$$T_e = \frac{3}{2}\ \frac{P_0}{2}\ (L_{md} - L_{mq})\ i_{ds}\ i_{qs} \tag{5}$$

where T_e is the electro-magnetic torque of the motor, and P_0 is the number of poles of the motor. The rotor speed and position of the motor can be expressed as

$$p\,\omega_{rm} = \frac{1}{J}(T_e - T_l - B\,\omega_{rm}) \tag{6}$$

and

$$p\,\theta_{rm} = \omega_{rm} \tag{7}$$

where J is the inertia constant of the motor and load, T_l is the external load torque, B is the viscous frictional coefficient of the motor and load, θ_{rm} is the mechanical rotor position, and ω_{rm} is the mechanical rotor speed. The electrical rotor speed and position are

$$\omega_r = \frac{P_0}{2}\omega_{rm} \tag{8}$$

$$\theta_r = \frac{P_0}{2}\theta_{rm} \tag{9}$$

where ω_r is the electrical rotor speed, and θ_r is the electrical rotor position of the motor.

2.3 Steady-state analysis

When the synchronous reluctance motor is operated in the steady-state condition, the d-q axis currents, i_d and i_q, become constant values. We can then assume $x_q = \omega_e L_{qs}$ and $x_d = \omega_e\ L_{ds}$, and derive the steady-state d-q axis voltages as follows:

$$v_d = r_s i_d - x_q i_q \tag{10}$$

$$v_q = r_s i_q + x_d i_d \tag{11}$$

The stator voltage can be expressed as a vector V_s and shown as follows

$$V_s = v_q - j v_d \tag{12}$$

Now, from equations (10) and (11), we can solve the d-axis current and q-axis current as

$$i_d = \frac{r_s v_d + x_q v_q}{r_s^2 + x_d x_q} \tag{13}$$

and

$$i_q = \frac{r_s v_q - x_d v_d}{r_s^2 + x_d x_q} \tag{14}$$

By substituting equations (13)-(14) into (5), we can obtain the steady-state torque equation as

$$T_e = \frac{3}{2}\frac{P}{2}\frac{1}{\omega_e}\frac{x_d - x_q}{(r_s^2 + x_d x_q)^2}[(r_s x_q v_q^2) - (r_s x_d v_d^2) + (r_s^2 - x_d x_q) v_d v_q] \tag{15}$$

According to (15), when the stator resistance r_s is very small and can be neglected, the torque equation (15) can be simplified as

$$T_e = \frac{3}{2}\frac{P}{2}\frac{1}{\omega_e}\frac{x_d - x_q}{2 x_d x_q} V_s^2 \sin(2\delta) \tag{16}$$

The output power is

$$\begin{aligned} P &= T_e \frac{\omega_e}{(\frac{P}{2})} \\ &= \frac{3}{2}\frac{x_d - x_q}{2 x_d x_q} V_s^2 \sin(2\delta) \end{aligned} \tag{17}$$

where P is the output power, and δ is the load angle.

3. Direct torque control

3.1 Basic principle

Fig. 1 shows the block diagram of the direct torque control system. The system includes two major loops: the torque-control loop and the flux-control loop. As you can observe, the flux and torque are directly controlled individually. In addition, the current-control loop is not required here. The basic principle of the direct torque control is to bound the torque error and the flux error in hysteresis bands by properly choosing the switching states of the inverter. To achieve this goal, the plan of the voltage vector is divided into six operating sectors and a suitable switching state is associated with each sector. As a result, when the voltage vector rotates, the switching state can be automatically changed. For practical implementation, the switching procedure is determined by a state selector based on pre-calculated look up tables. The actual stator flux position is obtained by sensing the stator voltages and currents of the motor. Then, the operating sector is selected. The resolution of the sector is 60 degrees for every sector. Although the direct torque is very simple, it shows good dynamic performance in torque regulation and flux regulation. In fact, the two loops on torque and flux can compensate the imperfect field orientation caused by the parameter variations. The disadvantage of the direct torque control is the high frequency ripples of the torque and flux, which may deteriorate the performance of the drive system. In addition, an advanced controller is not easy to apply due to the large torque pulsation of the motor.

In Fig.1, the estimating torque and flux can be obtained by measuring the a-phase and the b-phase voltages and currents. Next, the speed command is compared with the estimating speed to compute the speed error. Then, the speed error is processed by the speed controller to obtain the torque command. On the other hand, the flux command is compared to the estimated flux. Finally, the errors ΔT_e and $\Delta\lambda_s$ go through the hysteresis controllers and the switching table to generate the required switching states. The synchronous reluctance motor rotates and a closed-loop drive system is thus achieved. Due to the limitation of the scope of this paper, the details are not discussed here.

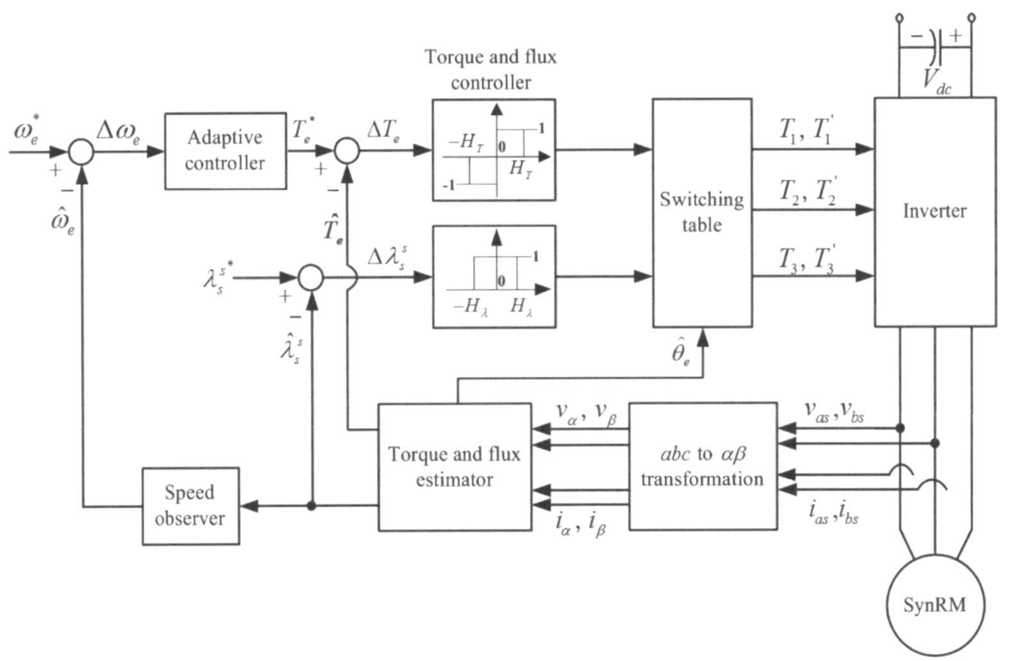

Fig. 1. The block diagram of the direct torque control system

3.2 Controller design

The SynRM is easily saturated due to its lack of permanent magnet material. As a result, it has nonlinear characteristics under a heavy load. To solve the problem, adaptive control algorithms are required. In this paper, two different adaptive controllers are proposed.

A. Adaptive Backstepping Controller

From equation (6), it is not difficult to derive

$$\begin{aligned}\frac{d}{dt}\omega_r &= \frac{1}{J_m}\left[T_e - T_L - B_m\omega_r\right] \\ &= A_1T_e + A_2T_L + A_3\omega_r\end{aligned} \tag{18}$$

and

$$A_1 = \frac{1}{J_m} \tag{19}$$

$$A_2 = -\frac{1}{J_m} \tag{20}$$

$$A_3 = -\frac{B_m}{J_m} \tag{21}$$

Where A_1, A_2, A_3 are constant parameters which are related to the motor parameters. In the real world, unfortunately, the parameters of the SynRM can not be precisely measured and are varied by saturated effect or temperature. As a result, a controller designer should consider the problem. In this paper, we proposed two control methods. The first one is an adaptive backstepping controller. In this method, we consider the parameter variations and external load together. Then

$$\frac{d}{dt}\omega_r = A_1T_e + A_3\omega_r + (A_2T_L + \Delta A_1T_e + \Delta A_2\ T_L + \Delta A_3\omega_r) = A_1T_e + A_3\omega_r + d \tag{22}$$

and

$$d = (A_2T_L + \Delta A_1T_e + \Delta A_2\ T_L + \Delta A_3\omega_{rm}) \tag{23}$$

where $\Delta A_1, \Delta A_2, \Delta A_3$ are the variations of the parameters, and d is the uncertainty including the effects of the parameter variations and the external load.
Define the speed error e_2 as

$$e_2 = \omega_{rm}^* - \omega_{rm} \tag{24}$$

Taking the derivation of both sides, it is easy to obtain

$$\dot{e}_2 = \dot{\omega}_{rm}^* - \dot{\omega}_{rm} \tag{25}$$

In this paper, we select a Lyapunov function as

$$\begin{aligned}V &= \frac{1}{2}e_2^2 + \frac{1}{2}\frac{1}{\gamma}\tilde{d}^2 \\ &= \frac{1}{2}e_2^2 + \frac{1}{2}\frac{1}{\gamma}\left(d - \hat{d}\right)^2\end{aligned} \tag{26}$$

Taking the derivation of equation (26), it is easy to obtain

$$\begin{aligned}\dot{\mathrm{V}} &= e_2\dot{e}_2 + \frac{1}{\gamma}\tilde{d}\dot{\tilde{d}} \\ &= e_2\dot{e}_2 + \frac{1}{\gamma}\tilde{d}\left(\dot{d} - \dot{\hat{d}}\right) \\ &= e_2\dot{e}_2 - \frac{1}{\gamma}\tilde{d}\dot{\hat{d}}\end{aligned} \tag{27}$$

By substituting (25) into (27) and doing some arrangement, we can obtain

$$\begin{aligned}\dot{\mathrm{V}} &= e_2\left(\dot{\omega}_{rm}^{*} - \mathrm{A}_1 T_e - \mathrm{A}_3\omega_{rm} - d\right) - \frac{1}{\gamma}\tilde{d}\dot{\hat{d}} \\ &= e_2\left(\dot{\omega}_{rm}^{*} - \mathrm{A}_1 T_e - \mathrm{A}_3\omega_{rm} - \tilde{d} - \hat{d}\right) - \frac{1}{\gamma}\tilde{d}\dot{\hat{d}}\end{aligned} \tag{28}$$

Assume the torque can satisfy the following equation

$$T_e = \frac{1}{\mathrm{A}_1}\left(\dot{\omega}_{rm}^{*} - \mathrm{A}_3\omega_{rm} - \hat{d} + \mathrm{M}e_2\right) \tag{29}$$

Substituting (29) into (28), we can obtain

$$\dot{\mathrm{V}} = -\mathrm{M}e_2^{\,2} - \tilde{d}e_2 - \frac{1}{\gamma}\tilde{d}\dot{\hat{d}} \tag{30}$$

From equation (30), it is possible to cancel the last two terms by selecting the following adaptive law

$$\dot{\hat{d}} = -\gamma e_2 \tag{31}$$

In equation (31), the convergence rate of the $\hat{d}$ is related to the parameter γ. By submitting (31) into (30), we can obtain

$$\dot{\mathrm{V}} = -\mathrm{M}e_2^{\,2} \le 0 \tag{32}$$

From equation (32), we can conclude that the system is stable; however, we are required to use Barbalet Lemma to show the system is asymptotical stable [32]-[34].
By integrating equation (32), we can obtain

$$\int_0^{\infty} \dot{\mathrm{V}} d\tau = \mathrm{V}(\infty) - \mathrm{V}(0) < \infty \tag{33}$$

From equation (33), the integrating of parameter $e_2^{\,2}$ of the equation (32) is less than infinite. Then, $e_2(t) \in \mathrm{L}_\infty \cap \mathrm{L}_2$, and $\dot{e}_2(t)$ is bounded. According to Barbalet Lemma, we can conclude [32]-[34]

$$\lim_{t\to\infty} e_2(t) = 0 \tag{34}$$

The block diagram of the proposed adaptive backstepping control system is shown in Fig. 2, which is obtained from equations (29) and (31).

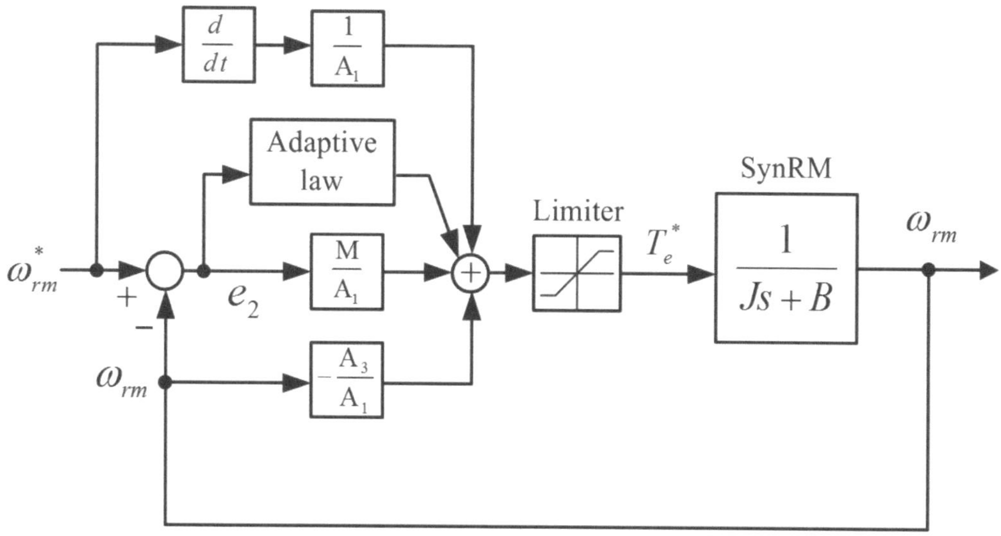

Fig. 2. The adaptive backstepping controller.

B. Model-Reference Adaptive Controller

Generally speaking, after the torque is applied, the speed of the motor incurs a delay of several micro seconds. As a result, the transfer function between the speed and the torque of a motor can be expressed as:

$$\frac{\omega_{rm}}{T_e} = \frac{1/J_m}{\left(s + B_m/J_m\right)} e^{-\tau s} \tag{35}$$

Where τ is the delay time of the speed response. In addition, the last term of equation (35) can be described as

$$e^{-\tau s} \cong \frac{1}{1+\tau s} \cong \frac{1/\tau}{s+1/\tau} \tag{36}$$

Substituting (36) into (35), one can obtain

$$\frac{\omega_{rm}}{T_e} = \frac{1/J_m}{(s + B_m/J_m)} \frac{1/\tau}{(s+1/\tau)} = \frac{b_0}{s^2 + a_1 s + a_0} \tag{37}$$

where

$$a_1 = \left(\frac{B_m}{J_m} + \frac{1}{\tau}\right) \tag{38a}$$

$$a_0 = \frac{B_m}{J_m \tau} \tag{38b}$$

$$b_0 = \frac{1}{J_m \tau} \tag{38c}$$

Equation (37) can be described as a state-space representation:

$$\begin{bmatrix} \dot{x}_1 \\ \dot{x}_2 \end{bmatrix} = \begin{bmatrix} 0 & 1 \\ -a_0 & -a_1 \end{bmatrix} \begin{bmatrix} x_1 \\ x_2 \end{bmatrix} + \begin{bmatrix} 0 \\ b_0 \end{bmatrix} u \tag{39a}$$

$$y = \begin{bmatrix} 1 & 0 \end{bmatrix} \begin{bmatrix} x_1 \\ x_2 \end{bmatrix} \tag{39b}$$

Where $x_1 = \omega_{rm} = y_p$, $x_2 = \dot{\omega}_{rm}$, and $u = T_e$. Next, the equations (39a) and (39b) can be rewritten as :

$$\dot{X}_p = A_p X_p + B_p u \tag{40a}$$

and

$$y_p = C_p^T X_p \tag{40b}$$

where

$$X_p = \begin{bmatrix} x_1 \\ x_2 \end{bmatrix} \tag{41a}$$

$$A_p = \begin{bmatrix} 0 & 1 \\ -a_0 & -a_1 \end{bmatrix} \tag{41b}$$

$$B_p = \begin{bmatrix} 0 \\ b_0 \end{bmatrix} \tag{41c}$$

$$C_p{}^T = \begin{bmatrix} 1 & 0 \end{bmatrix} \tag{41d}$$

After that, we define two state variables w_1 and w_2 as:

$$\dot{w}_1 = -hw_1 + u \tag{42}$$

and

$$\dot{w}_2 = -hw_2 + y_p \tag{43}$$

The control input u can be described as

$$u = Kr + Q_1 w_1 + Q_2 w_2 + Q_0 y_p \tag{44}$$

$$= \theta^T \phi$$

where

$$\theta^T = \begin{bmatrix} K & Q_1 & Q_2 & Q_0 \end{bmatrix}$$

and

$$\phi = \begin{bmatrix} r & w_1 & w_2 & y_p \end{bmatrix}^T$$

where γ is the reference command. Combining (40a),(42), and (43), we can obtain a new dynamic equation as

$$\begin{bmatrix} \dot{X}_p \\ \dot{w}_1 \\ \dot{w}_2 \end{bmatrix} = \begin{bmatrix} A_p & 0 & 0 \\ 0 & -h & 0 \\ C_p^T & 0 & -h \end{bmatrix} \begin{bmatrix} X_p \\ w_1 \\ w_2 \end{bmatrix} + \begin{bmatrix} B_p \\ 1 \\ 0 \end{bmatrix} u \tag{45}$$

Substituting (44) into (45), we can obtain

$$\begin{bmatrix} \dot{X}_p \\ \dot{w}_1 \\ \dot{w}_2 \end{bmatrix} = \begin{bmatrix} A_p + B_p Q_0 C_p^T & B_p Q_1 & B_p Q_2 \\ Q_0 C_p^T & -h + Q_1 & Q_2 \\ C_p^T & 0 & -h \end{bmatrix} \begin{bmatrix} X_p \\ w_1 \\ w_2 \end{bmatrix} + \begin{bmatrix} B_p K \\ K \\ 0 \end{bmatrix} r \tag{46}$$

Define $\tilde{K} = K - K^*$, $\tilde{Q}_1 = Q_1 - Q_1^*$, $\tilde{Q}_2 = Q_2 - Q_2^*$, $\tilde{Q}_0 = Q_0 - Q_0^*$
Then, equation (46) can be rearranged as

$$\begin{bmatrix} \dot{X}_p \\ \dot{w}_1 \\ \dot{w}_2 \end{bmatrix} = \begin{bmatrix} A_p + B_p Q_0^* C_p^T & B_p Q_1^* & B_p Q_2^* \\ Q_0^* C_p^T & -h + Q_1^* & Q_2^* \\ C_p^T & 0 & -h \end{bmatrix} + \begin{bmatrix} B_p K^* \\ K^* \\ 0 \end{bmatrix} r + \begin{bmatrix} B_p \\ 1 \\ 0 \end{bmatrix} \tilde{\theta}^T \phi \tag{47}$$

Where $\tilde{\theta}^T = \begin{bmatrix} \tilde{K} & \tilde{Q}_1 & \tilde{Q}_2 & \tilde{Q}_0 \end{bmatrix}$ is the parameter errors. It is possible to rearrange equation (47) as a simplified form

$$\dot{X}_c = A_m X_c + \begin{bmatrix} B_p K^* \\ K^* \\ 0 \end{bmatrix} r + B_m \tilde{\theta}^T \phi \tag{48}$$

and

$$Y_c = C_m^T X_c \tag{49}$$

where

$$X_c = \begin{bmatrix} X_p \\ w_1 \\ w_2 \end{bmatrix},\ A_m = \begin{bmatrix} A_p + B_p Q_0^* C_p^T & B_p Q_1^* & B_p Q_2^* \\ Q_0^* C_p^T & -h + Q_1^* & Q_2^* \\ C_p^T & 0 & -h \end{bmatrix},$$

$$B_m^T = \begin{bmatrix} B_p & 1 & 0 \end{bmatrix},\ C_m^T = \begin{bmatrix} C_p^T & 0 & 0 \end{bmatrix}.$$

After that, the referencing model of the closed-loop system can be described as :

$$\dot{X}_m = A_m X_m + B_m K^* r \tag{50}$$

and

$$Y_m = C_m^T X_m \tag{51}$$

where $X_m^T = \begin{bmatrix} X_p^* & w_1^* & w_2^* \end{bmatrix}$ is the vector of the state variables, and Y_m is the output of the referencing model. Now, we define the derivation of the state variable error and the output error as:

$$\dot{e} = \dot{X}_c - \dot{X}_m \tag{52}$$

and

$$e_1 = Y_c - Y_m \tag{53}$$

Substituting (50)-(51) into (52) and (53), one can obtain

$$\dot{e} = A_m e + B_m \tilde{\theta}^T \phi \tag{54a}$$

and

$$e_1 = C_m^T e \tag{54b}$$

By letting $\overline{B}_m = B_m K^*$, it is not difficult to rearrange (54a) as

$$\dot{e} = A_m e + \overline{B}_m \frac{1}{K^*} \tilde{\theta}^T \phi \tag{55a}$$

Combining (54b) and (55a), one can obtain

$$\dot{e}_1 = \frac{1}{K^*} C_m^T (sI - A_m)^{-1} \overline{B}_m \tilde{\theta}^T \phi \tag{56}$$

It is essential that the degree of the referencing model equal the uncontrolled plant. As a result, equation (55a) has to be revised as [12]:

$$\dot{e} = A_m e + \overline{B}_{m1} \frac{1}{K^*} \tilde{\theta}^T \overline{\phi} \tag{57a}$$

where $\overline{B}_{m1} = \overline{B}_m L_{(s)}$, $\overline{\phi} = L_{(s)}^{-1} \phi$, $L_{(s)} = s + F$; $F > 0$ '
After that, we can obtain

$$\dot{e}_1 = \frac{1}{K^*} C_m^T (sI - A_m)^{-1} \overline{B}_{m1} \tilde{\theta}^T \overline{\phi} \tag{58}$$

Now, selecting a Lyapunov function as

$$V = \frac{1}{2} e^T P_m e + \frac{1}{2} \tilde{\theta}^T \Gamma^{-1} \tilde{\theta} \left| \frac{1}{K^*} \right| \tag{59}$$

where P_m is a symmetry positive real matrix, and Γ is a positive real vector.
The matrix P_m satisfies the following two equations:

$$A_m^T P_m + P_m A_m = -Q \tag{60}$$

and

$$P_m \overline{B}_{m1} = C_m^T \tag{61}$$

where Q is a symmetry positive real matrix . Taking the derivation of equation (59) and substituting (60), (61) into the derivation equation, we can obtain

$$\begin{aligned} \dot{V} &= \frac{-1}{2} e^T Q e + e^T P_m \overline{B}_{m1} \frac{1}{K^*} \tilde{\theta}^T \overline{\phi} + \tilde{\theta}^T \Gamma^{-1} \dot{\tilde{\theta}} \left| \frac{1}{K^*} \right| \\ &= \frac{-1}{2} e^T Q e + e_1 \frac{1}{K^*} \tilde{\theta}^T \overline{\phi} + \tilde{\theta}^T \Gamma^{-1} \dot{\tilde{\theta}} \left| \frac{1}{K^*} \right| \end{aligned} \tag{62}$$

It is possible to select the adaptive law as

$$\dot{\theta} = -\mathrm{sgn}\left(\frac{1}{K^*}\right) \Gamma e_1 \overline{\phi} \tag{63}$$

where $\mathrm{sgn}\left(\frac{1}{K^*}\right) = \frac{|K^*|}{K^*}$, substituting (63) into (62), we can obtain :

$$\dot{V} = \frac{-1}{2} e^T Q e \le 0 \tag{64}$$

Next, by using Barbalet Lemma, we can obtain that the system is asymmetrical and

$$\lim_{t \to \infty} e_1(t) = 0 \tag{65}$$

Finally, we can obtain

$$u_p = L_{(s)}\theta^T L_{(s)}^{-1}\phi = L_{(s)}\theta^T\overline{\phi}$$
$$= \dot{\theta}^T\overline{\phi} + \theta^T\dot{\overline{\phi}} + F\theta^T\overline{\phi} = \dot{\theta}^T\overline{\phi} + \theta^T\phi \qquad (66)$$

The block diagram of the model-reference control system is shown in Fig. 3, which includes referencing model, adaptive controller, and adaptive law.

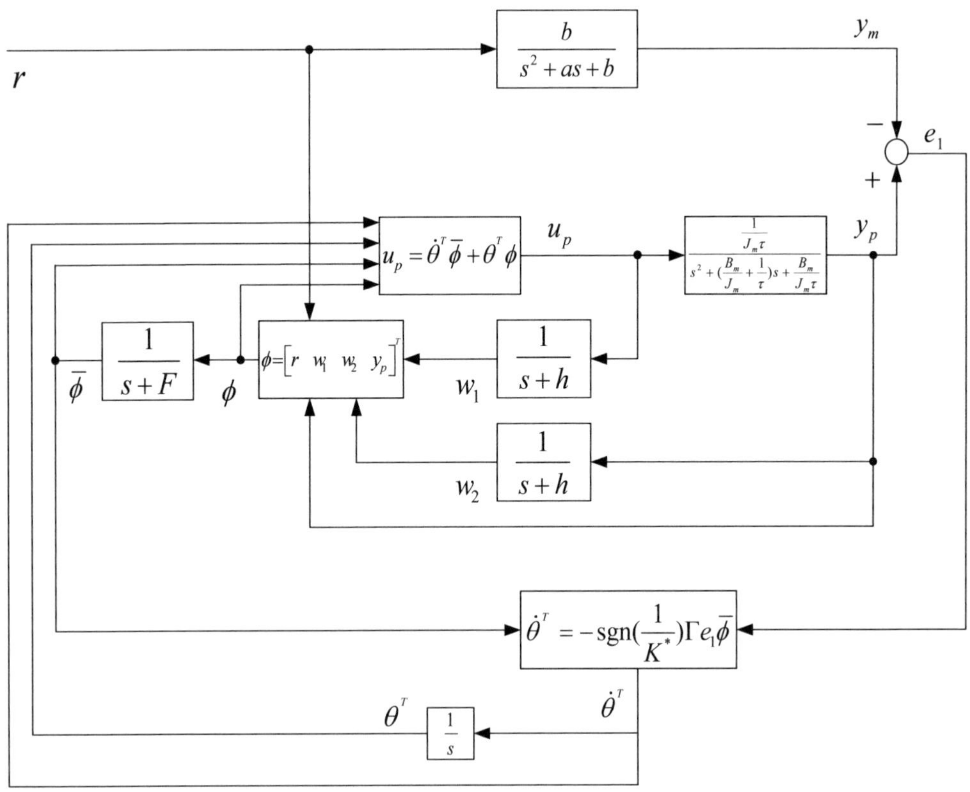

Fig. 3. The block diagram of the model reference adaptive controller.

4. Implementation

The implemented system is shown in Fig. 4. The system includes two major parts: the hardware circuits and the software programs. The hardware circuits include: the synchronous reluctance motor, the driver and inverter, the current and voltage sensors, and the A/D converters. The software programs consist of the torque estimator, the flux estimator, the speed estimator, the adaptive speed controller, and the direct torque control algorithm. As you can observe, the most important jobs are executed by the digital signal processor; as a result, the hardware is quite simple. The rotor position can be obtained by stator flux, which is computed from the stator voltages and the stator currents. The digital signal processor outputs triggering signals every 50 μs; as a result, the switching frequency of the inverter is 20 kHz. In addition, the sampling interval of the speed control loop is 1 ms although the adaptive controllers are quite complicated. The whole drive system, therefore, is a multi-rate fully digital control system.

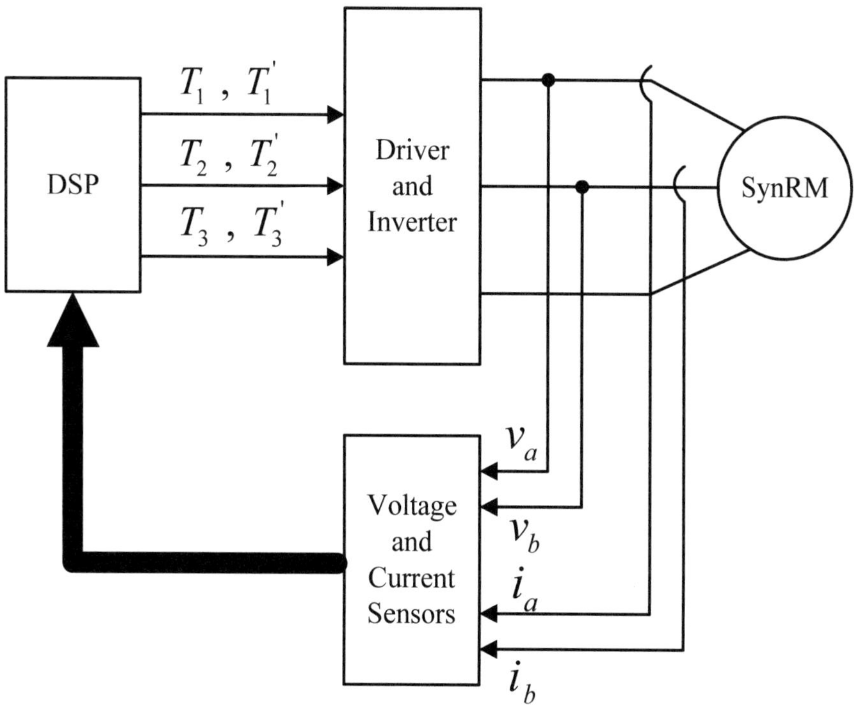

Fig. 4. The implemented system.

A. Hardware Circuits

The hardware circuits of the synchronous reluctance drive system includes the major parts. The details are discussed as follows.

a. The delay circuit of the IGBT triggering signals.

Fig. 5 shows the proposed delay circuit of the IGBT triggering signals. The delay circuit is designed to avoid the overlapping period of the turn-on interval of the upper IGBT and the lower IGBT for the inverter. Then, the inverter can avoid a short circuit. In this paper, the delay time of the delay circuit is set as 2 μs . To achieve the goal, two integrated circuit chips are used: 74LS174 and 74LS193. The basic idea is described as follows. First, the digital signal processor sends a clock signal to 74LS193. The time period of the clock is 62.5 μs . The 74LS193 executes the dividing frequency function and finally generates a clock signal with a 0.5 μs period. After that, the 74LS193 sends it into the $\underline{CLK}$ pin of 74LS174. The 74LS174 provides 6 series D-type flip-flop to generate a 3 μs delay. Finally, an AND gate is used to make a 3 μs for a rising-edge triggering signal but not a falling-edge triggering signal.

b. The driver of the IGBTs

The power switch modules used in the paper are IGBT modules, type 2MBI50-120. Each module includes two IGBTs and two power diodes. The driver of the IGBT is type EX-B840, made by Fuji company. The detailed circuit of the driver for an IGBT is shown in Fig. 6. In Fig. 6, the EX-B840, which is a driver, uses photo-couple to convert the control signal into a

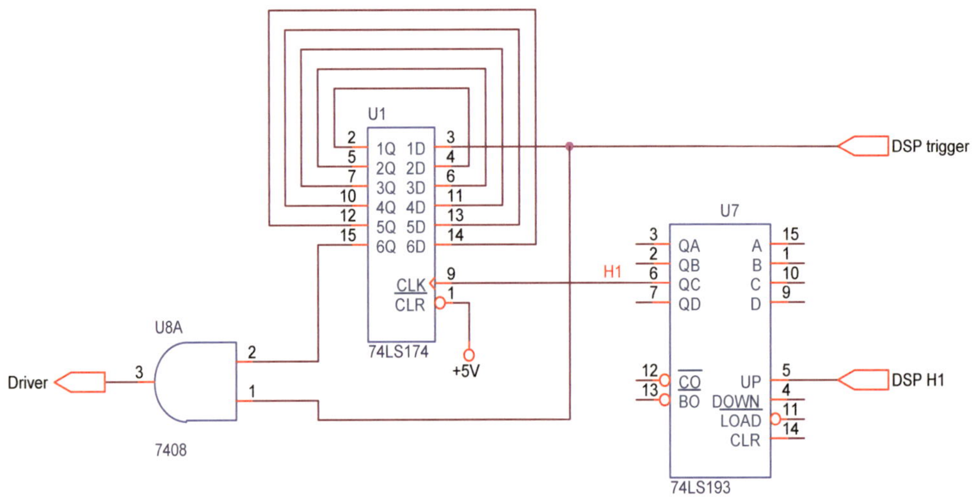

Fig. 5. The delay circuit of the IGBT triggering signals.

triggering signal for an IGBT. In addition, the EX-B840 provides the isolation and over-current protection as well.

When the control signal is "High", the photo transistor is turned on. Then, the photo diode is conducted. A 15V can across the gate and emitter of the IGBT to turn on the IGBT. On the other hand, when the control signal is "Low", the photo transistor is turned off. As a result, the photo diode is cut off. A -5V can across the gate and emitter of the IGBT to make IGBT turn off immediately.

The protection of the IGBT is included in Fig. 6. When the IGBT has over-current, the voltage across the collector and emitter of the IGBT is obviously dropped. After the 6-pin of EX-B840 detects the dropped voltage, the 5-pin of the EX-B840 sends a "Low" voltage to the photo diode. After that, the photo diode is opened, and a -5V across the gate and emitter of the IGBT is sent to turn off the IGBT.

c. The snubber circuit

The snubber circuit is used to absorb spike voltages when the IGBT is turned off. As we know, the synchronous reluctance motor is a kind of inductive load. In Fig. 7, when the upper leg IGBT T is turned off, the low leg IGBT T' cannot be turned on immediately due to the required dead-time, which can avoid short circuits. A new current path to keep the current continuous flow is required. The new current path includes the fast diode D and the snubber capacitor C_s. So, the current can flow through the fast diode D and the capacitor C_S, and then stores its energy into the capacitor C_S. On the other hand, when the IGBT is turned on in next time interval, the stored energy in the capacitor C_S can flow through the resistance R_s and the IGBT T. Finally, the energy dissipates in the resistance R_s. By suitably selecting the parameter C_s and R_s, a snubber circuit with satisfactory performance can be obtained.

d. The current detecting circuit

The current detecting circuit is used to measure the stator current of the synchronous reluctance motor, and can be shown in Fig. 8. The Hall current sensor, typed LP-100, is used to sense the stator current of the motor and to provide the isolation between the power stage and the control circuit. The primary side of the LT-100P can measure 0 to 100 A with a bandwidth of 100 kHz. The basic principle is discussed as follows. The primary of the LT-100P has 5 turns. As a result, when 1A flows into the primary side, the secondary side of the LT-100P can generate 5mA. The current flows from the M pin of the LP-100P to the 0.1 $K\Omega$ resistance, and then provides 0.5V voltage drop. As a result, in this chapter, for every 1A primary current, the circuit can output 0.5V. A low-pass filter is designed to eliminate the high-frequency noise.

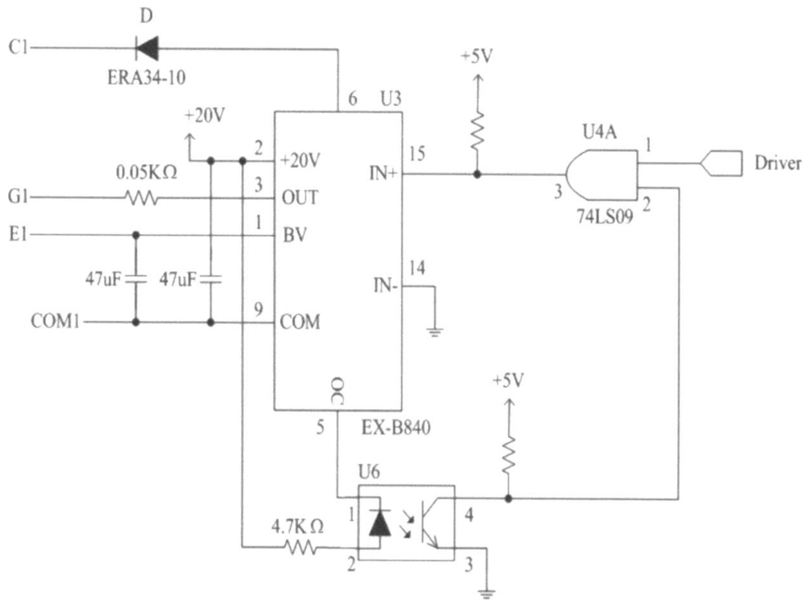

Fig. 6. The circuit of the driver for IGBTs.

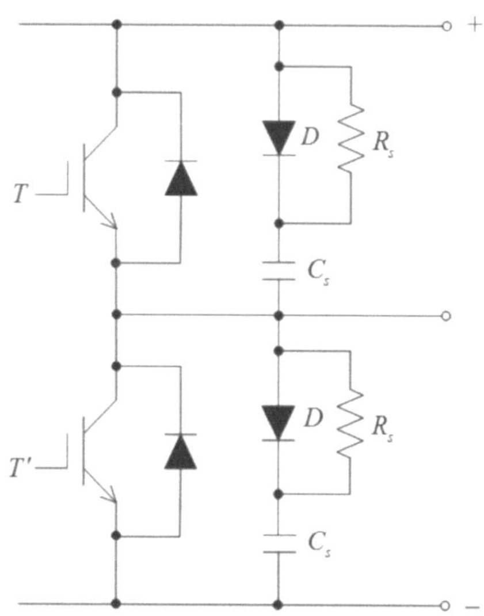

Fig. 7. The snubber circuit.

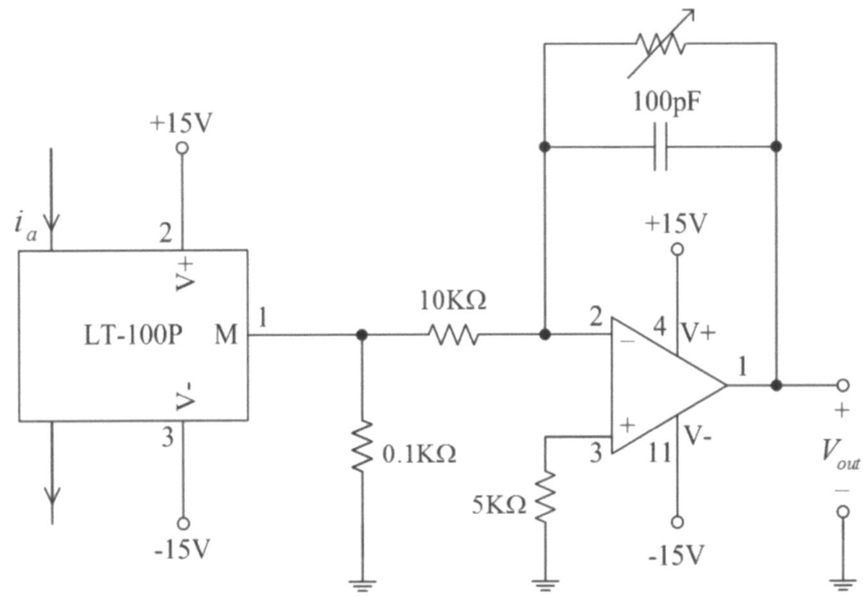

Fig. 8. The current detecting circuit.

e. The voltage detecting circuit

The voltage detecting circuit is used to sense the stator voltage of the synchronous reluctance motor, which is an important item for computing the estimated flux of the motor. A voltage isolation amplifier, AD210, is selected to isolate the input side and output side. In the chaper, R_0 and R_1 are used to attenuate the input voltage to be 0.05 v_{ab} . As a result, the input of AD 210 is limited under $\pm 10V$.

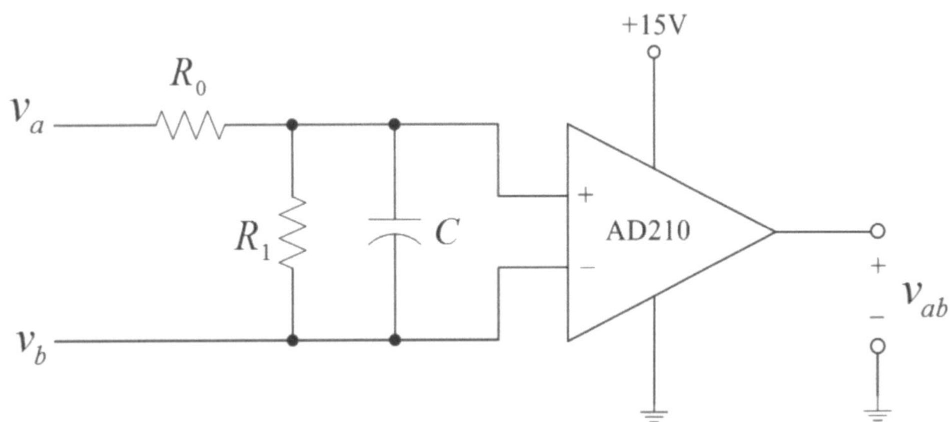

Fig. 9. The voltage detecting circuit.

f. The A/D conversion circuit

The measured voltages and currents from Hall current sensor and AD210 are analog signals. In order to be read by a digital signal processor, the A/D conversion is required. In this chapter, the 12 bit A/D converter with a 3 μs conversion time is used. The A/D converter is typed AD578. The detailed circuit is shown in Fig. 10. There are two sets: one for voltage conversion, and the other for current conversion.

When the analog signal is ready, the digital signal processor outputs a triggering signal to the A/D converter. Then, each AD578 converter starts to convert the analog signal into a digital signal. When the conversion process finishes, an EOC signal is sent from the AD578 to latch the 74LS373. Next, the digital signal processor reads the data. In this chapter, a timer with a fixed clock is used to start the conversion of the AD578 and then the digital signal processor can read the data. By using the method, we can simplify the software program of the digital signal processor.

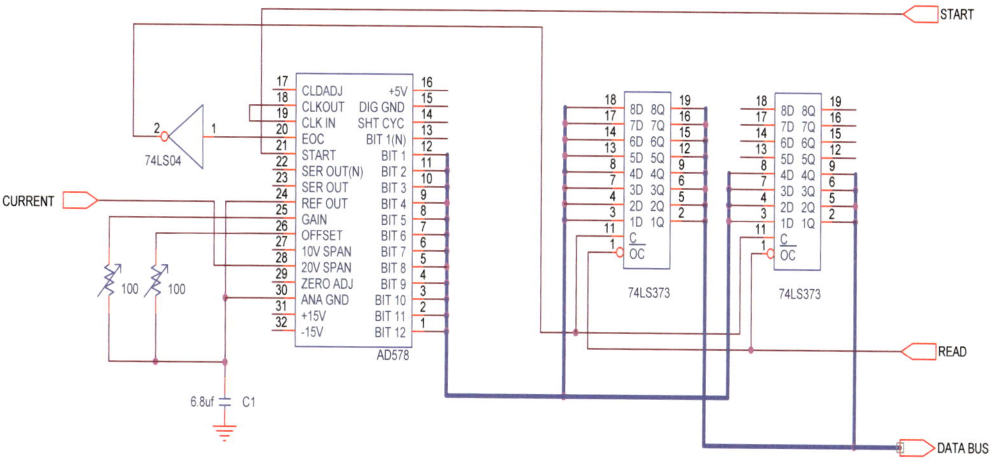

Fig. 10. A/D converter circuit.

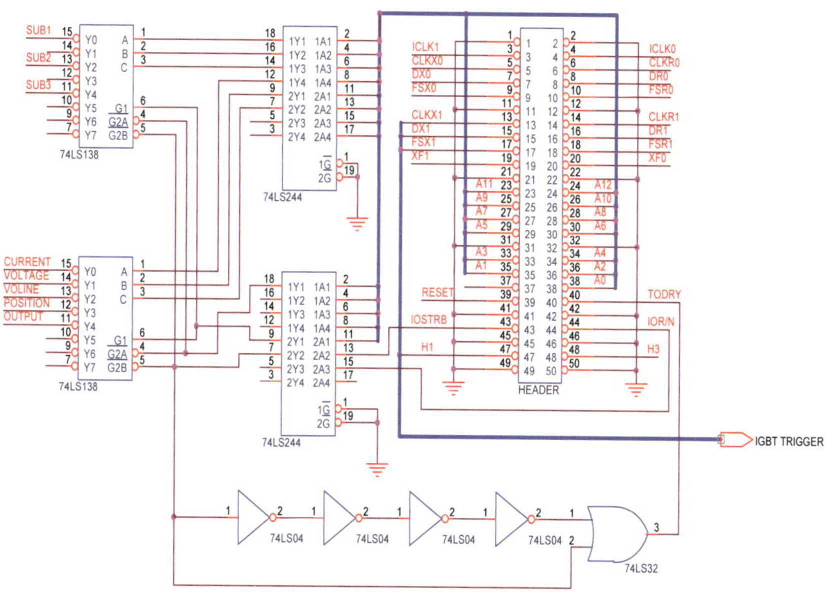

Fig. 11. The interfacing circuit of the DSP.

g. The interfacing circuit of the digital signal processor
In the chapter, the digital signal processor, type TMS320-C30, is manufactured by Texas Instruments. The digital signal processor is a floating-point operating processor. The application board, developed by Texas Instruments, is used as the major module. In addition, the expansion bus in the application board is used to interface to the hardware circuit. The voltage, current, speed, and rotor position of the drive system are obtained by using the expansion bus. As a result, the address decoding technique can be used to provide different address for data transfer. In addition, the triggering signals of the IGBTs are sent by the following pins: CLKX1, DX1, and FSX1. The details are shown in Fig. 11.

A. Software Development

a. The Main Program
Fig. 12 shows the flowchart of the initialization of the main program. First, the DSP enables the interrupt service routine. Then, the DSP initializes the peripheral devices. Next, the DSP sets up parameters of the controller, inverter, A/D converter, and counter. After that, the DSP enables the counter, and clear the register. Finally, the DSP checks if the main program is ended. If it is ended, the main program stops; if it is not, the main program goes back to the initializing peripheral devices and carries out the following processes mentioned.

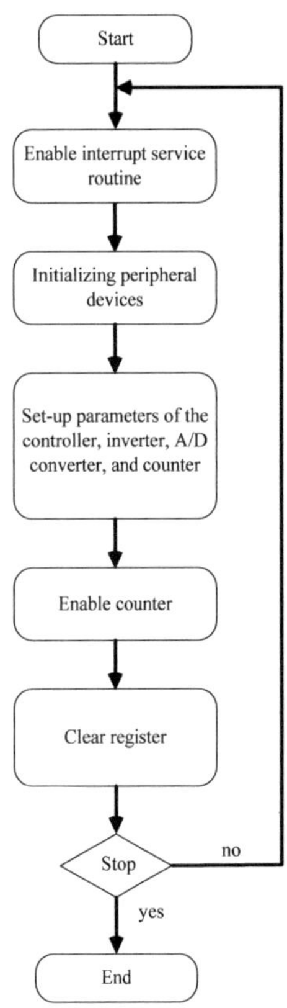

Fig. 12. The flowchart of the initialization of the main program.

b. The interrupt service routines

The interrupt service routines include: the backstepping adaptive controller, the reference model adaptive controller, and the switching method of the inverter. The detailed flowcharts are shown in Fig. 13, Fig. 14, and Fig. 15.

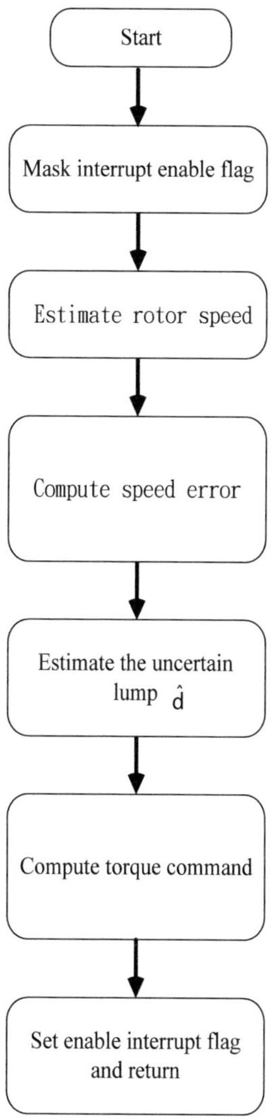

Fig. 13. The subroutine of the backstepping adaptive controller.

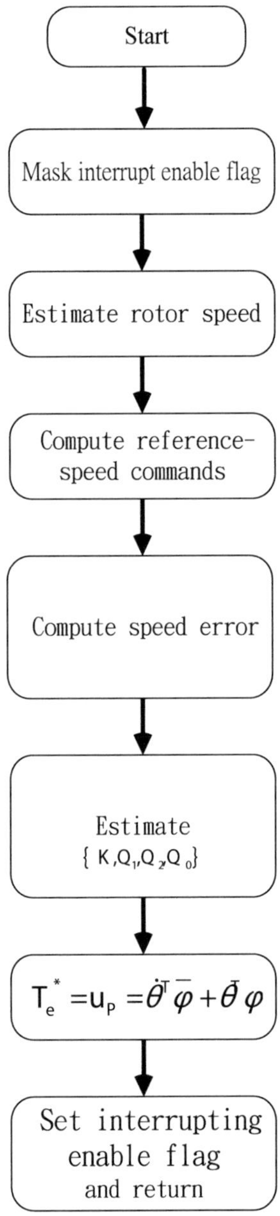

Fig. 14. The subroutine of the reference model adaptive controller.

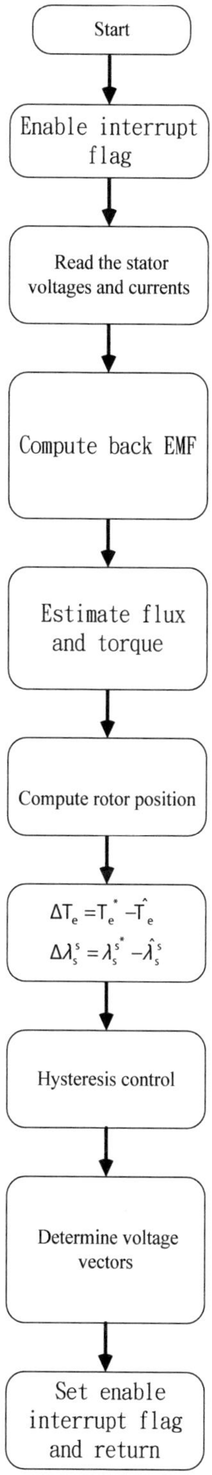

Fig. 15. The subroutine of the switching method of the inverter.

5. Experimental results

Several experimental results are shown here. The input dc voltage of the inverter is 150V. The switching frequency of the inverter is 20 kHz. In addition, the sampling interval of the minor loop is 50 μs, and the sampling interval of the speed loop is 1 ms. The parameters of the PI controller are K_P =0.006 and K_I =0.001. The parameters of the adaptive backstepping controller are M=3 and γ =0.8. The parameters of the model referencing controller are Γ = [-0.0002 -0.004 -0.004 -0.0006]. Fig. 16(a)(b) show the measured steady-state waveforms. Fig. 16(a) is the measured a-phase current and Fig. 16(b) is the measured line-line voltage, v_{ab}. Fig. 17(a) is the simulated fluxes at 1000 r/min. Fig. 17(b) is the simulated flux trajectory at 1000 r/min. Fig. 17(c) is the measured fluxes at 1000 r/min. Fig. 17(d) is the measured flux trajectory at 1000 r/min. As you can observe, the trajectories are both near circles in both simulation and measurement. Fig. 18(a) shows the comparison of the measured estimating rotor angle and the measured real rotor angle at 50 r/min. As we know, when the motor is operated at a lower speed, the flux becomes smaller. As a result, the motor cannot be operated well at lower speeds due to its small back emf. The estimating error, shown in Fig. 18(b) is obvious. Fig. 19(a)(b) show the measured estimating rotor angle at 1000 r/min. Fig. 19(a) shows the comparison of the measured estimating rotor angle and the measured real rotor angle at 1000 r/min. Fig. 19(b) shows the estimating error, which is around 2 degrees. As a result, the estimating error is reduced when the motor speed is increased. In addition, Fig. 19(b) is varied more smoothly than the Fig. 18(b) is. The major reason is that the back emf has a better signal/noise ratio when the motor speed increases. Fig. 20(a) shows the measured transient responses at 50 r/min. Fig. 20(b) shows the measured load disturbance responses under 2 N.m external load. The model reference control performs the best. The steady-state errors of Fig. 20(a)(b) are: 2.7 r/min for PI controller, 0.5 r/min for ABSC controller, and 0.1 r/min for MRAC controller, respectively. According to the measured results, the MRAC controller performs the best and the PI controller performs the worst in steady-state. Fig. 21(a)(b) show the measured speed responses at 1000 r/min. Fig. 21(a) is the measured transient responses. Fig. 21(b) is the load disturbance responses under 2 N.m. According to the measured results, the model-reference controller performs better than the other two controllers in both transient response and load disturbance response again. The steady-state errors of Fig. 21(a)(b) are: 7.3 r/min for PI controller, 1.9 r/min for ABSC controller, and 0.1 r/min for MRAC controller, respectively. As you can observe, the conclusions are similar to the results of Fig. 20(a)(b). Fig. 22(a) shows the measured external - $\hat{d}$ of the adaptive backstepping control. Fig. 22(b) shows the measured speed error of the adaptive backstepping control by selecting different parameters. Fig. 23(a)(b)(c)(d) show the relative measured parameters K, K, Q_1, Q_2, Q_0 of the model-reference controller. All the parameters converge to constant values. Fig. 24(a)(b)(c) show the measured speed responses of a triangular speed command. The PI controller has a larger steady-state error than the adaptive controllers have. Fig. 25(a)(b)(c) show the measured speed responses of a sinusoidal speed command. As you can observe, the model-reference controller performs the best. The model- reference controller has a smaller steady-state error and performs a better tracking ability than the other controllers.

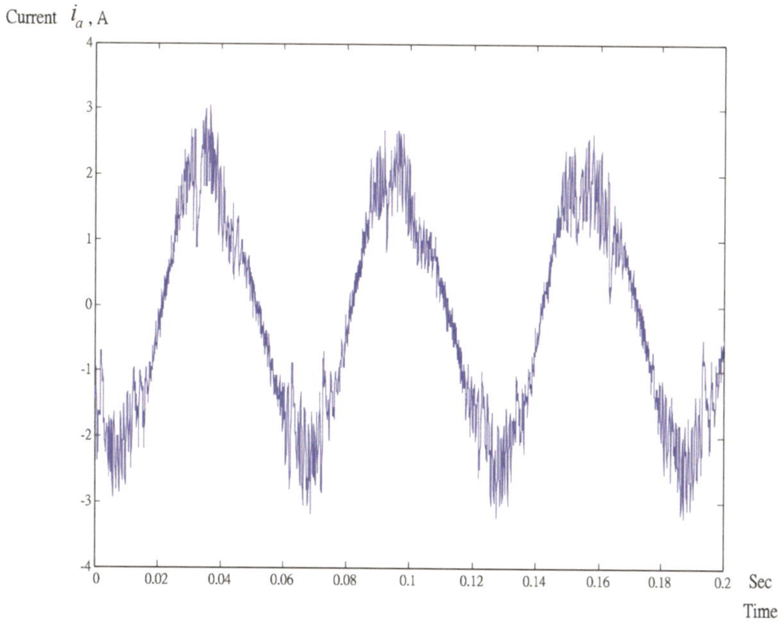

(a)

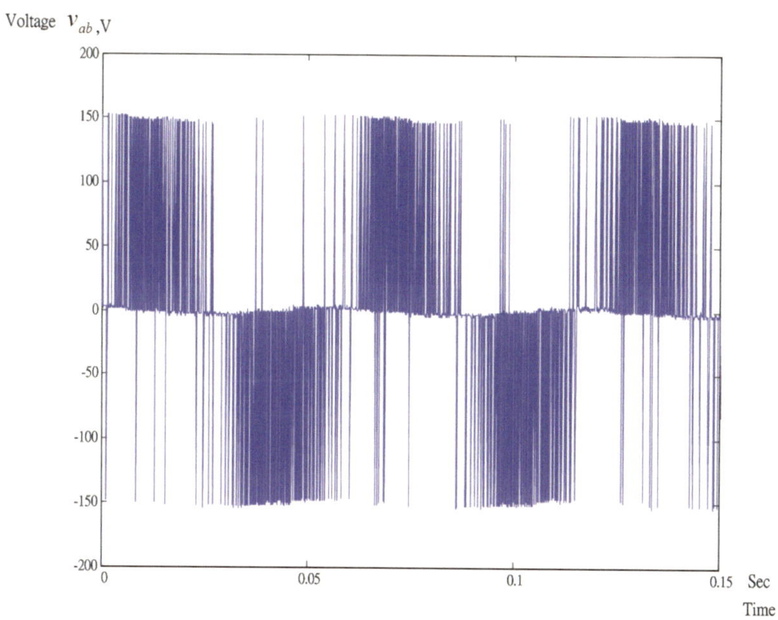

(b)

Fig. 16. The measured steady-state waveforms. (a) phase current (b) line voltage.

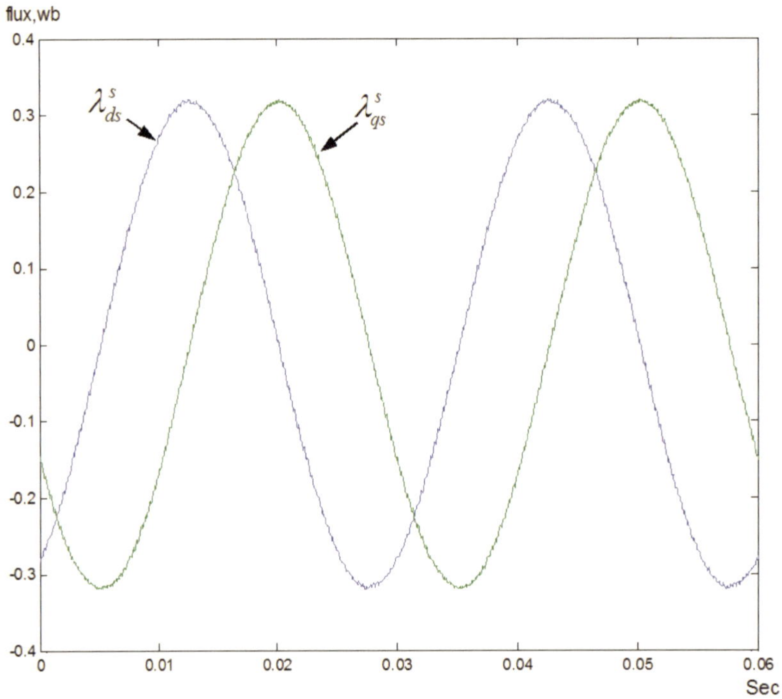
flux,wb
0.4
0.3
0.2
0.1
0
-0.1
-0.2
-0.3
-0.4
0
0.01
0.02
0.03
0.04
0.05
0.06
Sec
λ_{ds}^s
λ_{qs}^s

(a)

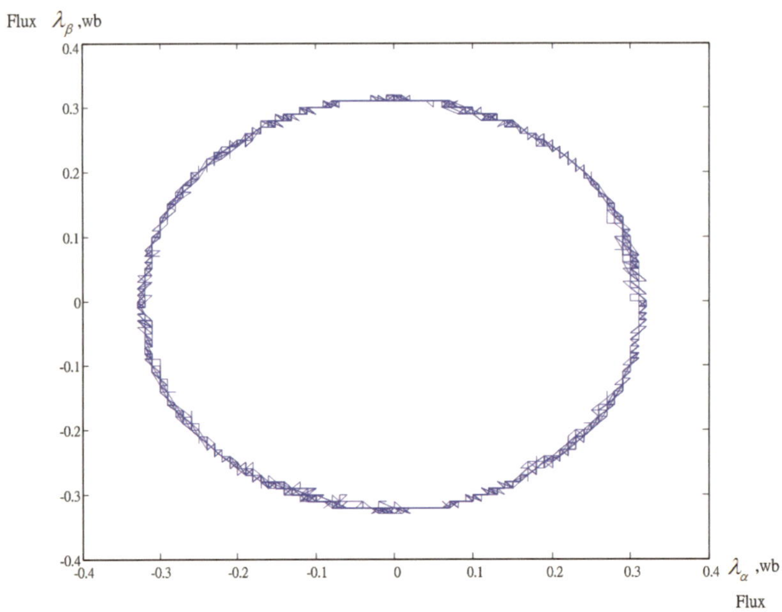
Flux λ_β,wb
0.4
0.3
0.2
0.1
0
-0.1
-0.2
-0.3
-0.4
-0.4
-0.3
-0.2
-0.1
0
0.1
0.2
0.3
0.4
λ_α,wb
Flux

(b)

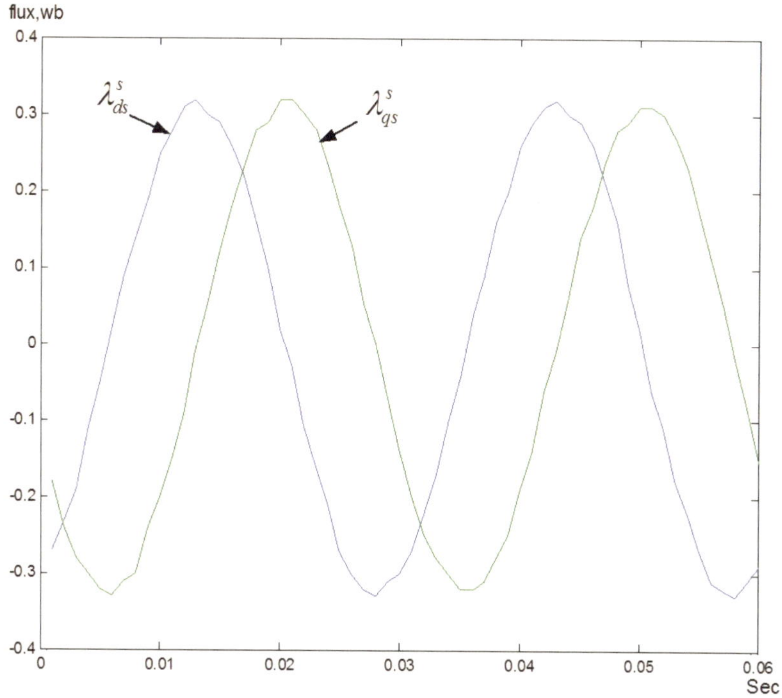

(c)

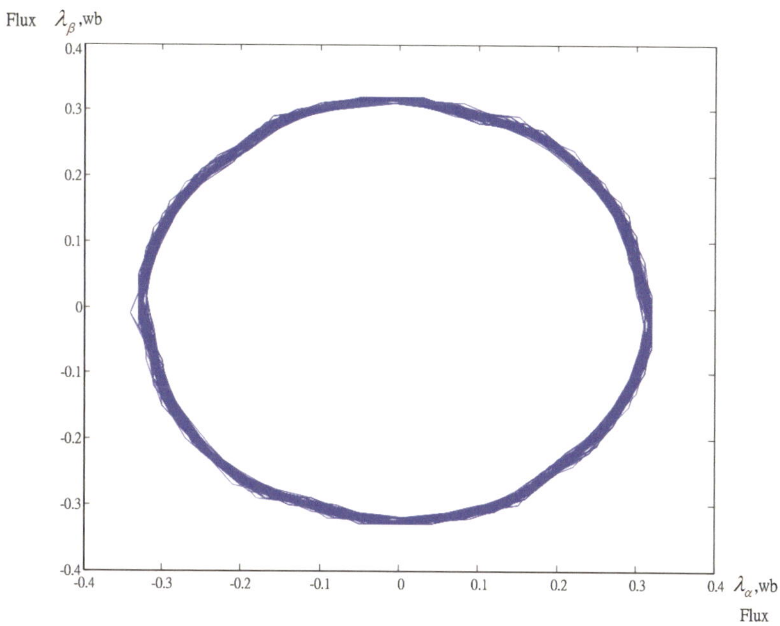

(d)

Fig. 17. The stator flux trajectories at 1000 r/min. simulated fluxes (b) simulated trajectory (c) measured fluxes (d) measured trajectory.

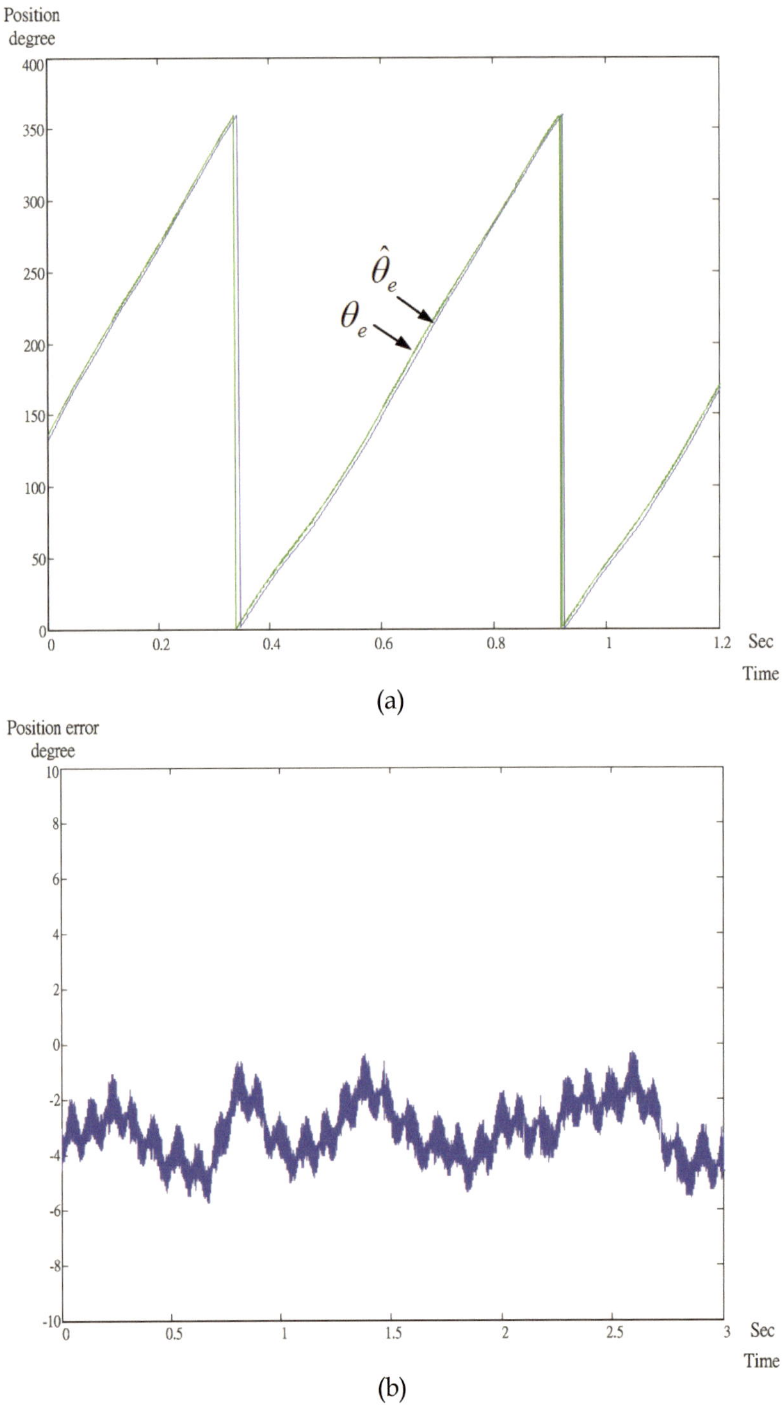

Fig. 18. The measured estimating rotor angle at 50 r/min. (a) comparison (b) estimating error.

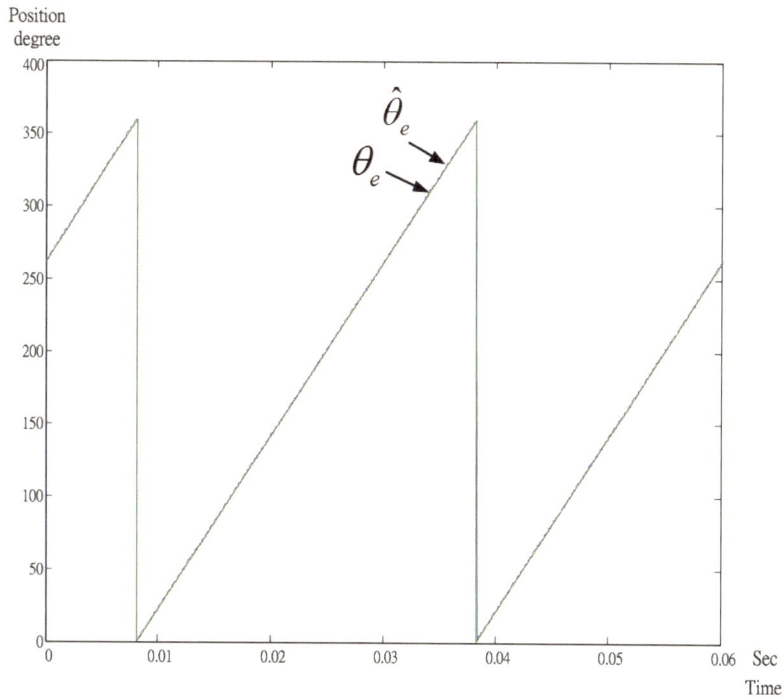

(a)

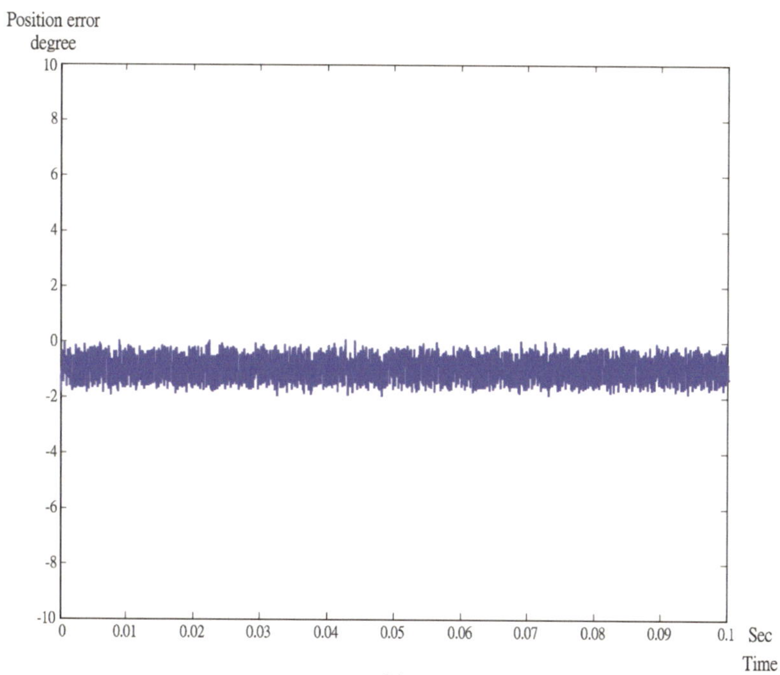

(b)

Fig. 19. The measured estimating rotor angle at 1000 r/min. (a) comparison (b) estimating error.

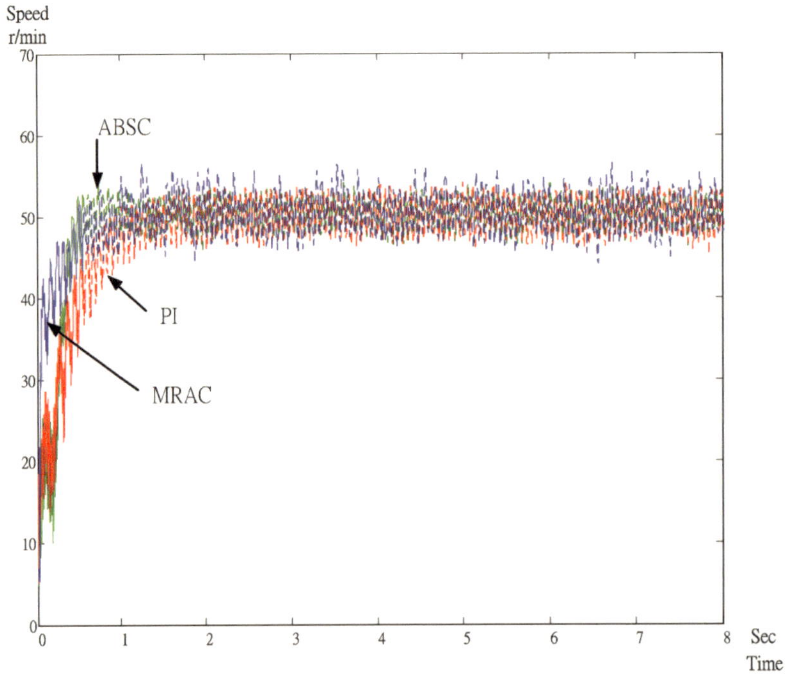

(a)

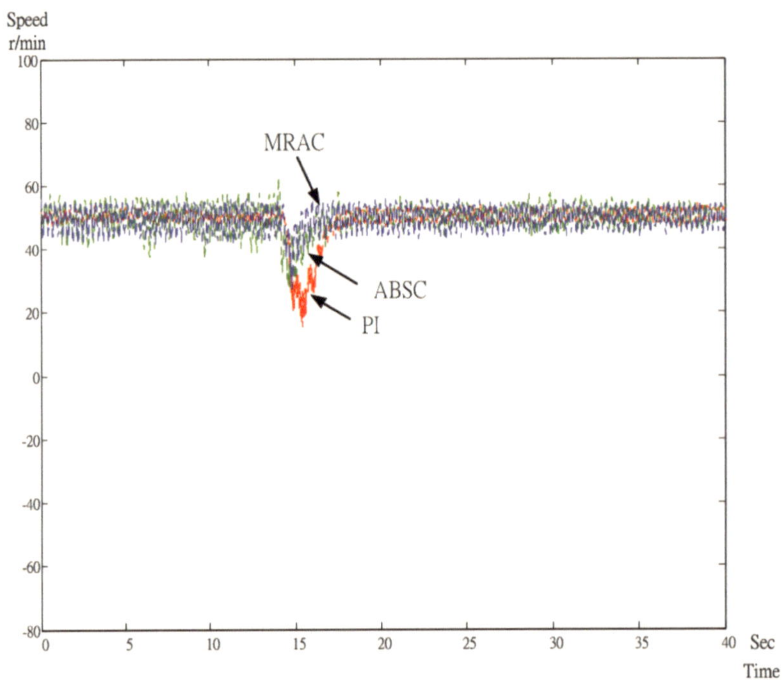

(b)

Fig. 20. The measured speed responses at 50 r/min, (a) transient responses (b) load disturbance responses

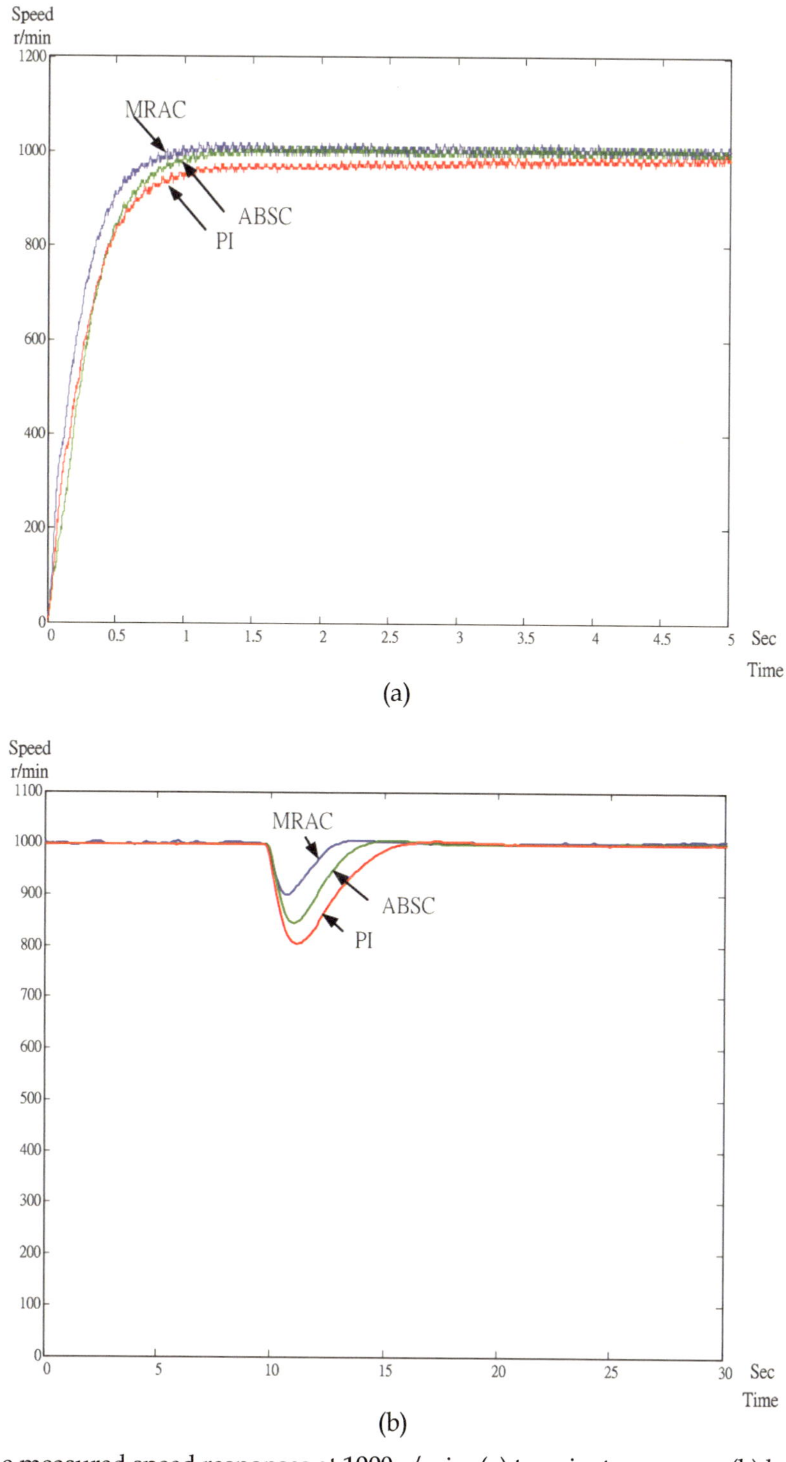

(b)

Fig. 21. The measured speed responses at 1000 r/min. (a) transient responses (b) load disturbance responses.

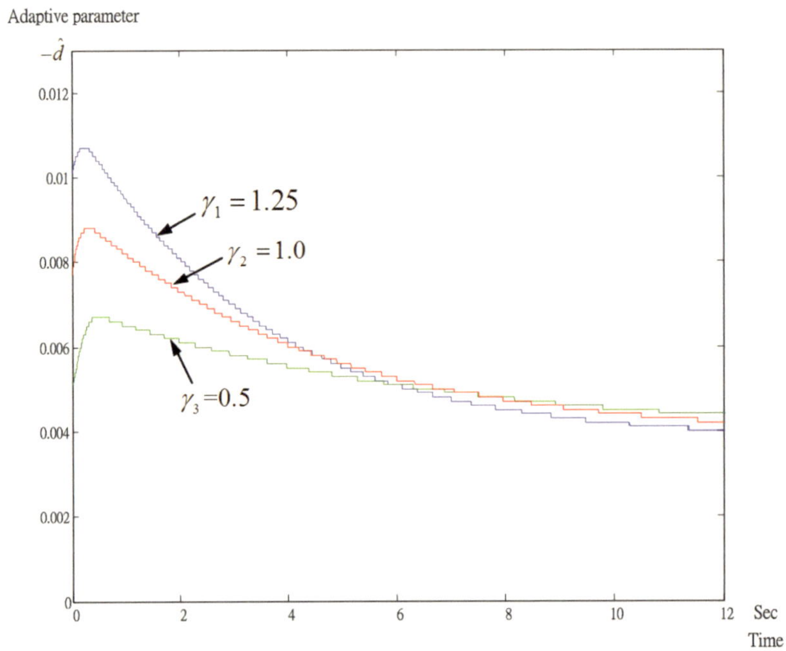

(a)

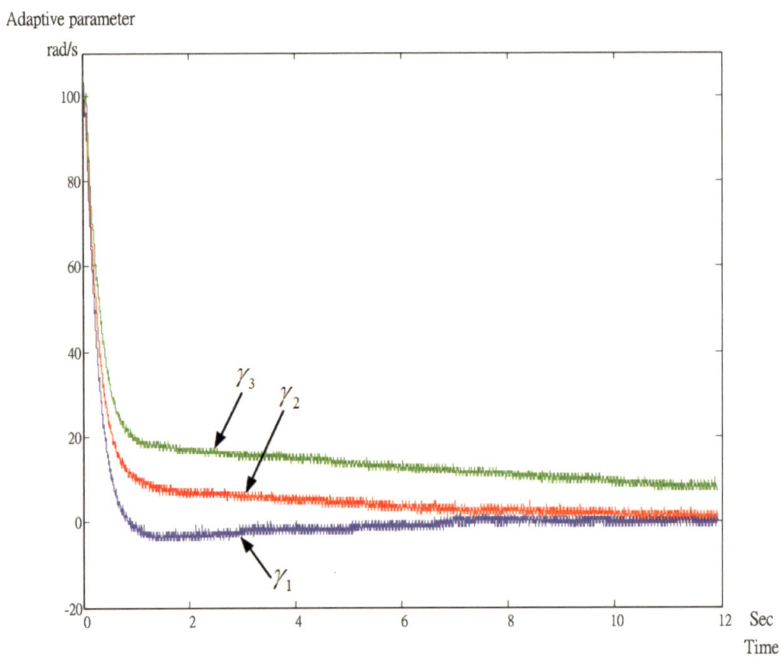

(b)

Fig. 22. The measured responses of adaptive backstepping control. (a) $-\hat{d}$ (b) speed error.

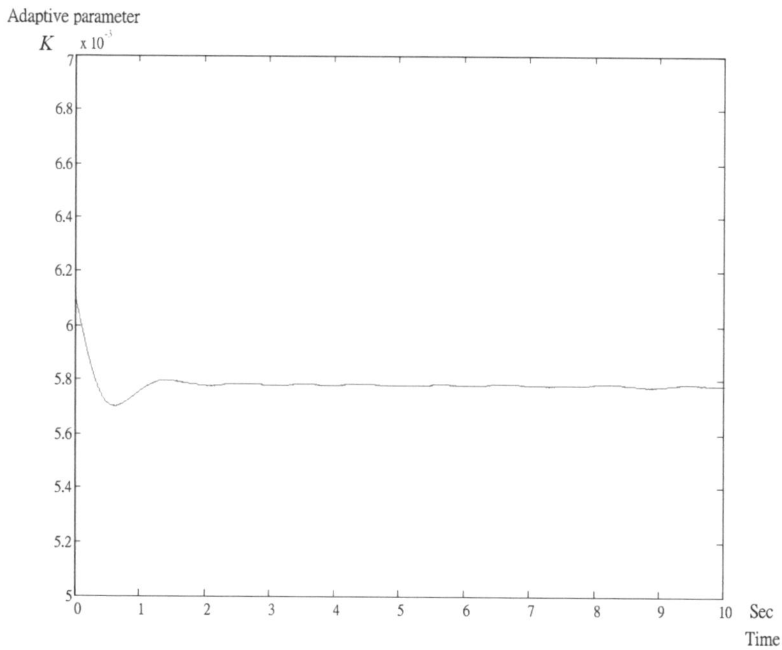
Adaptive parameter
K x 10^{-3}
7
6.8
6.6
6.4
6.2
6
5.8
5.6
5.4
5.2
5
0 1 2 3 4 5 6 7 8 9 10 Sec
Time

(a)

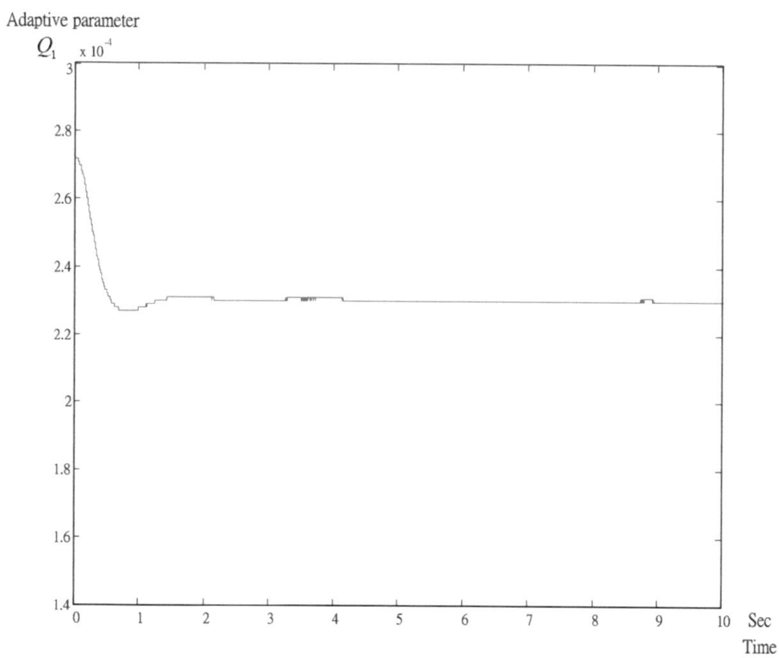
Adaptive parameter
Q_1 x 10^{-4}
3
2.8
2.6
2.4
2.2
2
1.8
1.6
1.4
0 1 2 3 4 5 6 7 8 9 10 Sec
Time

(b)

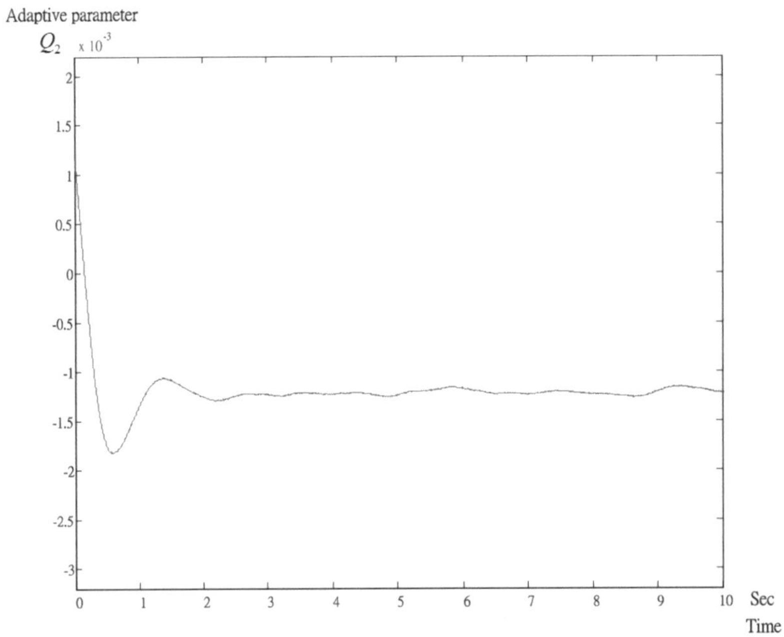

(c)

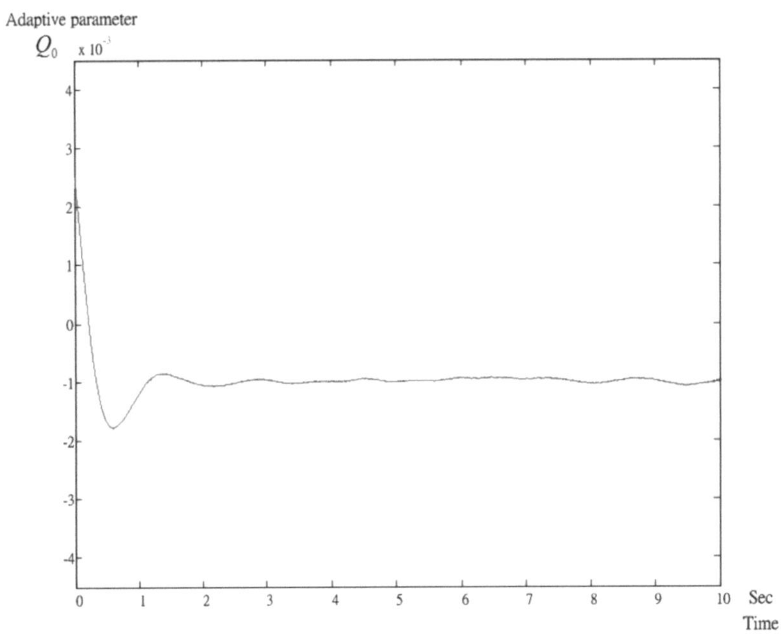

(d)

Fig. 23. The measured responses of model-reference control. (a) K (b) Q_1 (c) Q_2 (d) Q_0 .

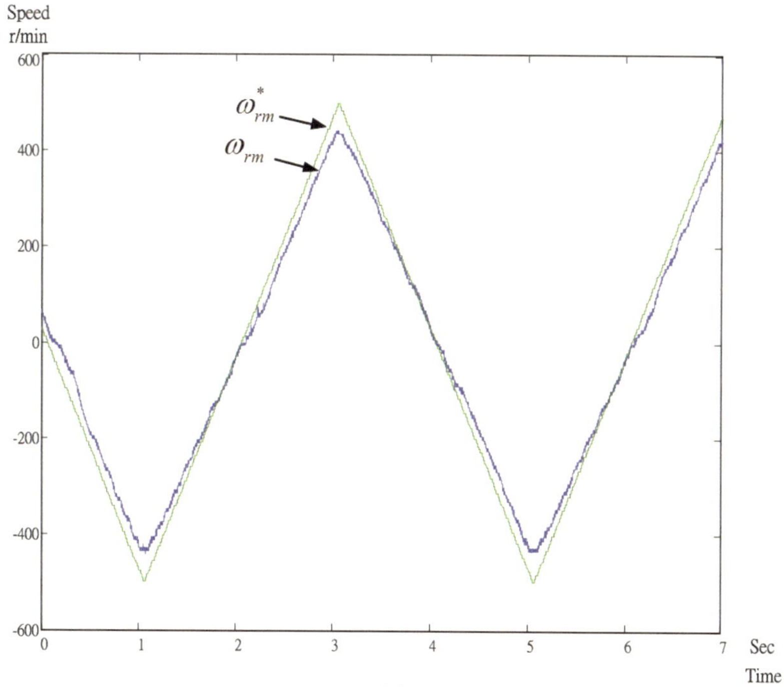
Speed
r/min
600
400
200
0
-200
-400
-600
0
1
2
3
4
5
6
7
Sec
Time

(a)

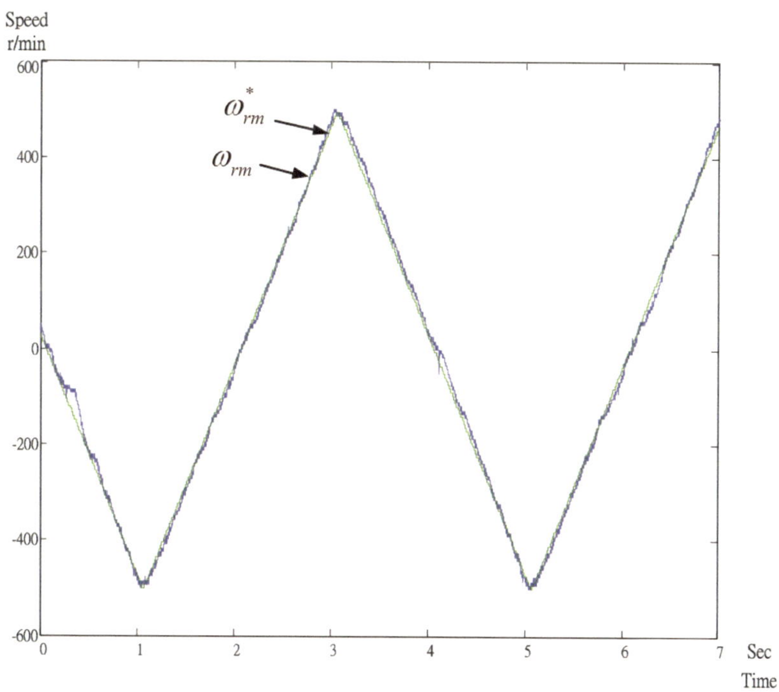
Speed
r/min
600
400
200
0
-200
-400
-600
0
1
2
3
4
5
6
7
Sec
Time

(b)

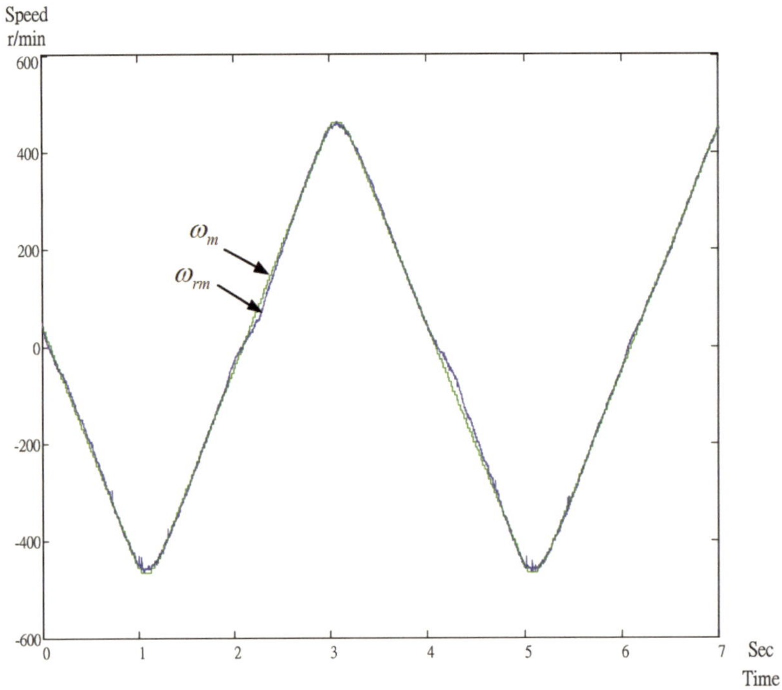

(c)

Fig. 24 The measured speed responses of a triangular speed command.
(a) PI (b) backstepping (c) model-reference.

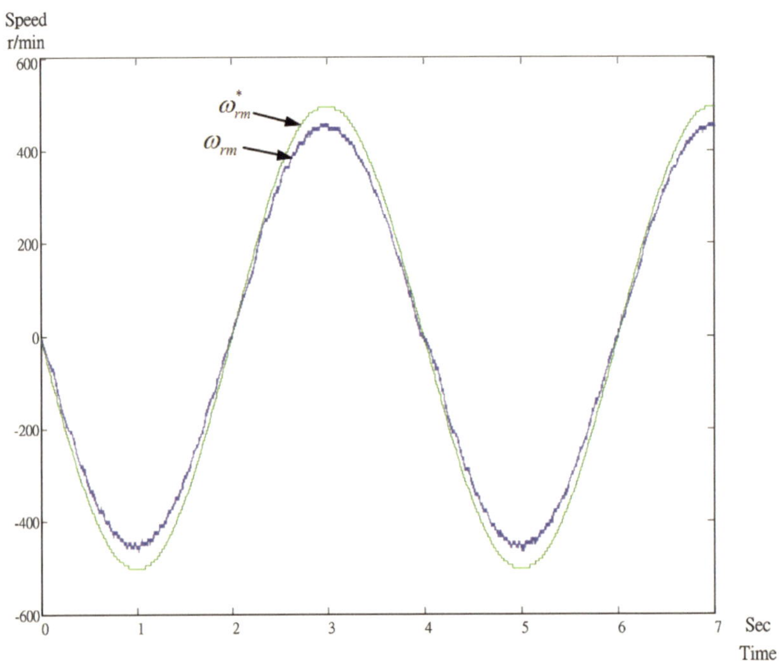

(a)

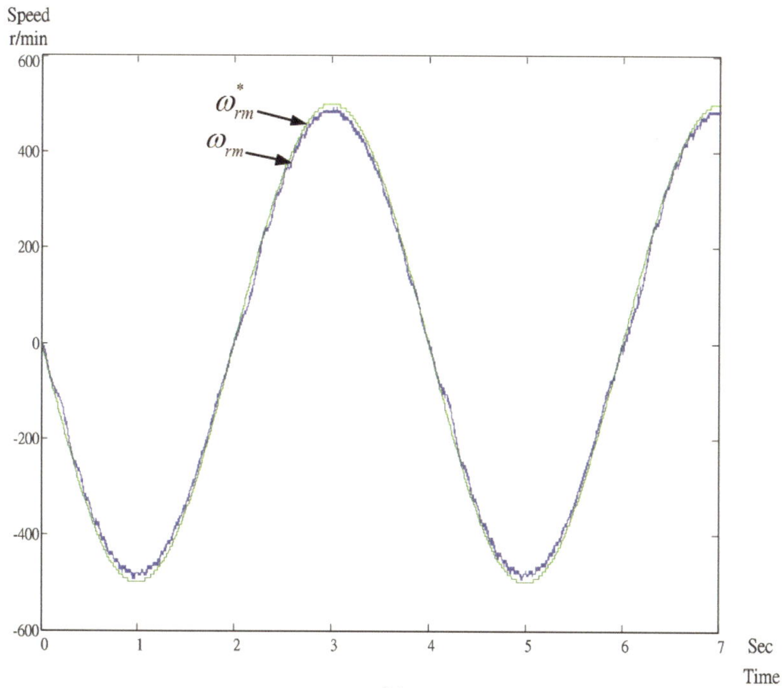

(b)

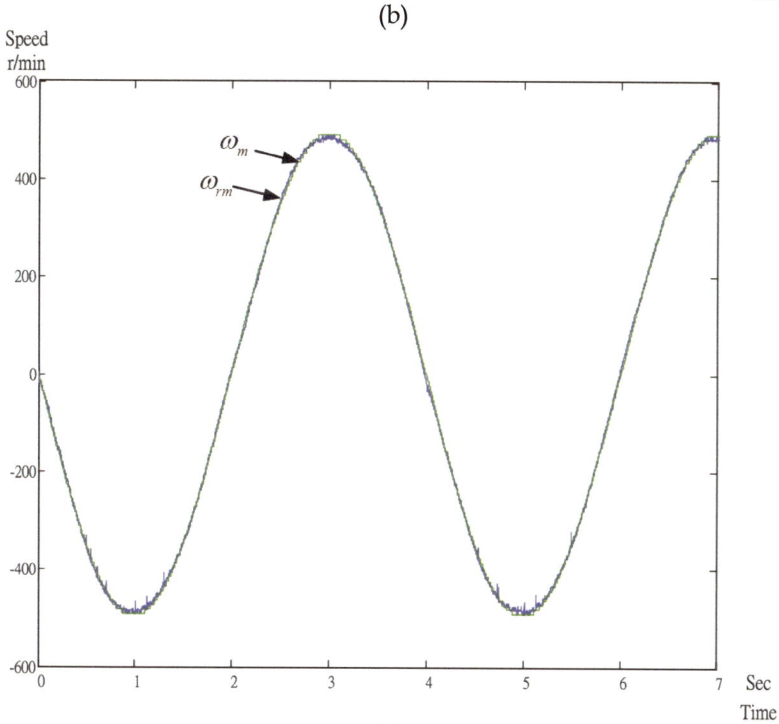

(c)

Fig. 25. The speed responses of a sinusoidal speed command. (a) PI (b) backstepping (c) model-reference.

6. Future trends

In this chapter, by using the torque control, a closed-loop sensorless speed drive system has been implemented. The proposed system can be operated from 30 r/min to 2000 r/min with satisfactory performance. Unfortunately, the proposed system cannot be operated from standstill to 30 r/min. As a result, it is necessary in the future to continuously improve the controller design, hardware design, and software design to reduce the torque pulsations and then provide better performance in low-speed operating range. In addition, it is another aim to realize a closed-loop high performance position control system by using a torque control method.

7. Conclusions

In this chapter, two different adaptive controllers have been proposed for a synchronous reluctance motor drive system. The parameters of the controllers are on-line tuned. The adaptive backstepping controller has simple control algorithm. It is more easily implemented than the model reference adaptive controller is. On the other hand, the model reference adaptive controller performs better in transient responses and steady-state characteristics. A digital signal process is used to execute the control algorithm. As a result, the hardware circuit is very simple. The implemented system shows good transient responses, load disturbance responses, and tracking ability in triangular and sinusoidal commands. This paper provides a new direction in the application of adaptive controller design for a synchronous reluctance motor drive system.

8. References

[1] Park, J. M., Kim, S., Hong, J. P., and Lee, J. H.: 'Rotor design on torque ripple reduction for a synchronous reluctance motor with concentrated winding using response surface methodology', IEEE Trans. Magnet., vol. 42, no. 10, pp. 3479-3481, 2006.

[2] G. Sturtzer, D. Flieller, and J. P. Louis, "Mathematical and experi- mental method to obtain the inverse modeling of nonsinusoidal and saturated synchronous reluctance motors," IEEE Trans. Energy Conversion, vol. 18, no. 4, pp. 494-500, Dec. 2003.

[3] Hofmann, H. F., Sanders, S. R., and Antably, A.: 'Stator-flux-oriented vector control of synchronous reluctance machines with maximized efficiency', IEEE Trans. Ind. Electron., vol. 51, no. 5, pp. 1066-1072, 2004.

[4] M. T. Lin, and T. H. Liu, "Sensorless synchronous reluctance drive with standstill starting," IEEE Aerosp. Electron. Syst. Mag., vol. 36, no. 4, pp. 1232-1241, Oct. 2000.

[5] S. Ichikawa, A. Iwata, M. Tomitat, S. Doki, and S. Okuma, "Sensorless control of synchronous reluctance motors using an on-line parameter identification method taking into account magnetic saturation," IEEE PESC '04, pp. 3311-3316, June 2004.

[6] L. Xu, X. Xu, T. A. Lipo, and D. W. Novotny, "Vector control of a synchronous reluctance motor including saturation and iron loss," IEEE Trans. Ind. Appl., vol. 27, no. 5, pp. 977-985, Sept./Oct. 1991.

[7] S. Morimoto, M. Sanada, and Y. Takeda, "High-performance current -sensorless drive for PMSM and SynRM with only low-resolution position sensor," IEEE Trans. Ind. Appl., vol. 39, no. 3, pp. 792-801, May/June 2003.

[8] C. G. Chen, T. H. Liu, M. T. Lin, and C. A. Tai, "Position control of a sensorless synchronous reluctance motor," IEEE Trans. Ind. Appl., vol. 51, no. 1, pp. 15-25, Feb. 2004.
[9] M. G. Jovanovic, R. E. Betz, and D. Platt, "Sensorless vector controller for a synchronous reluctance motor," IEEE Trans. Ind. Appl., vol. 34, no. 2, pp. 346-354, Mar./Apr. 1998.
[10] S. Ichikawa, M. Tomitat, S. Doki, and S. Okuma, "Sensorless control of synchronous reluctance motors based on an extended EMF model and initial position estimation," IEEE IECON '03, pp. 2150-2155, Nov. 2003.
[11] J. I. Ha, S. J. Kang, and S. K. Sul, "Position controlled synchronous reluctance motor without rotational transducer," IEEE Trans. Ind. Appl., vol. 35, no. 6, pp. 1393-1398, Nov./Dec. 1999.
[12] Y. Q. Xiang, and S. A. Nasar, "A fully digital control strategy for synchronous reluctance motor servo drives," IEEE Trans. Ind. Appl., vol 33, no. 3, pp. 705-713, May/June 1997.
[13] D. Telford, M. W. Dunnigan, and B. W. Williams, "A novel torque-ripple reduction strategy for direct torque control," IEEE Trans. Ind. Electron., vol. 48, no. 4, pp. 867-870, Aug. 2001.
[14] J. H. Lee, C. G. Kim, and M. J. Youn, "A dead-beat type digital controller for the direct torque control of an induction motor," IEEE Trans. Power Electron., vol. 17, no. 5, pp. 739-746, Sep. 2002.
[15] J. Beerten, J. Verveckken, and J. Driesen, "Predictive direct torque control for flux and torque ripple reduction," IEEE Trans. Ind. Electron., vol. 57, no. 1, pp. 404-412, Jan. 2010.
[16] Consoli, A., Cavallars, C., Scarcella, G., and Testa, A.: 'Sensorless torque control of synchronous motor drives," IEEE Trans. Pow. Electron., Vol. 15, no. 1, pp. 28-35, 2000.
[17] D. A. Staton, T. J. E. Miller, and S. E. Wood, "Maximising the saliency ratio of the synchronous reluctance motor," IEE Proc. Electr. Power Appl., vol. 140, no. 4, pp. 249-259, July 1993.
[18] A. Vagati, A. Canova, M. Chiampi, M. Pastorelli, and M. Repetto, "Design refinement of synchronous reluctance motors through finite-element analysis," IEEE Trans. Ind. Electron., vol. 36, no. 4, pp. 1094-1102, July/Aug. 2000.
[19] K. Uezato, T. Senjyu, and Y. Tomori, "Modeling and vector control of synchronous reluctance motors including stator iron loss," IEEE Trans. Ind. Electron., vol. 30, no. 4, pp. 971-976, July/Aug. 1994.
[20] G. Stumberger, B. Stumberger, and D. Dolinar, "Identification of linear synchronous reluctance motor parameters," IEEE Trans. Ind. Appl., vol. 40, no. 5, pp. 1317-1324, Sept./Oct. 2004.
[21] H. K. Chiang and C. H. Tseng, "Integral variable structure controller with grey prediction for synchronous reluctance motor drive," IEE Proc. Electr. Power Appl., vol. 151, no. 3, pp. 349-358, May 2004.
[22] C. H. Lin, "Adaptive recurrent fuzzy neural network control for synchronous reluctance motor servo drive," IEE Proc. Electr. Power Appl., vol. 151, no. 6, pp. 711-724, Nov. 2004.

[23] S. J. Kang, J. M. Kim, and S. K. Sul, "Position sensorless control of synchronous reluctance motor using high frequency current injection," IEEE Trans. Energy Conversion, vol. 14, no. 4, pp. 1271-1275, Dec. 1999.

[24] R. Shi, and H. A. Toliyat, "Vector control of five-phase synchronous reluctance motor with space vector pulse width modulation (SVPWM) for minimum switching losses," IEEE APEC '02, pp. 57-63, Mar. 2002.

[25] Y. Gao and K. T. Chau, "Hopf bifurcation and chaos in synchronous reluctance motor drives," IEEE Trans. Energy Conversion, vol. 19, no. 2, pp. 296-302, June 2004.

[26] N. Bianchi, S. Bolognani, D. Bon, and M. D. Pre, "Torque harmonic compensation in a synchronous reluctance motor," IEEE Trans. Energy Conversion, vol. 23, no. 2, pp. 466-473, June 2008.

[27] A. Iqbal, "Dynamic performance of a vector-controlled five-phase synchronous reluctance motor drive: an experimental investigation," IET Electr. Power Appl., vol. 2, no. 5, pp. 298-305, 2008.

[28] R. Morales-Caporal and M. Pacas, "Encoderless predictive direct torque control for synchronous reluctance machines at very low and zero speed," IEEE Trans. Ind. Electron., vol. 55, no. 12, pp. 4408-4416, Dec. 2008.

[29] J. D. Park, C. Kalev, and H. F. Hofmann, "Control of high-speed solid-rotor synchronous reluctance motor/generator for flywheel-based uninterruptible power supplies," IEEE Trans. Ind. Electron., vol. 55, no. 8, pp. 3038-3046, Aug. 2008.

[30] T. H. Liu, M. T. Lin, and Y. C. Yang, "Nonlinear control of a synchronous reluctance drive system with reduced switching frequency," IEE Electr. Power Appl., vol. 153, no. 1, pp. 47-56, Jan. 2006.

[31] S. Ichikawa, M. Tomita, S. Doki, and S. Okuma, "Sensorless control of synchronous reluctance motors based on extended EMF models considering magnetic saturation with online parameter identification," IEEE Trans. Ind. Appl., vol. 42, no. 5, pp. 1264-1274, Sep./Oct. 2006.

[32] Kristic, M., Kanellakopoulos, I., and Kokotovic, P. V. : 'Nonlinear and Adaptive Control Design, (New York: John Wiley and Sons Inc, 1995).

[33] Narendra, K. S. and Annaswamy, A. M. : Stable Adaptive Systems, (New Jersey: Prentice-Hall, 1989).

[34] Tao, G.: Adaptive Control Design and Analysis, (New Jersey: Wiley-Interscience, 2003).